Living Cosmology

Ecology and Justice

An Orbis Series on Integral Ecology

Advisory Board Members

Mary Evelyn Tucker
John A. Grim
Leonardo Boff
Sean McDonagh

The Orbis Series on Integral Ecology publishes books seeking to integrate an understanding of Earth's interconnected life systems with sustainable social, political, and economic systems that enhance the Earth community. Books in the series concentrate on ways to

- Reexamine human-Earth relations in light of contemporary cosmological and ecological science.
- Develop visions of common life marked by ecological integrity and social justice.
- Expand on the work of those exploring such fields as integral ecology, climate justice, Earth law, ecofeminism, and animal protection.
- Promote inclusive participatory strategies that enhance the struggle of Earth's poor and oppressed for ecological justice.
- Deepen appreciation for dialogue within and among religious traditions on issues of ecology and justice.
- Encourage spiritual discipline, social engagement, and the transformation of religion and society toward these ends.

Viewing the present moment as a time for fresh creativity and inspired by the encyclical *Laudato Si'*, the series seeks authors who speak to ecojustice concerns and who bring into this dialogue perspectives from the Christian communities, from the world's religions, from secular and scientific circles, or from new paradigms of thought and action.

Living Cosmology

Christian Responses to
Journey of the Universe

Edited by

Mary Evelyn Tucker and John Grim

ORBIS BOOKS
Maryknoll, New York 10545

ORBIS BOOKS
Maryknoll, New York 10545

Fathers and Brothers
MARYKNOLL

Founded in 1970, Orbis Books endeavors to publish works that enlighten the mind, nourish the spirit, and challenge the conscience. The publishing arm of the Maryknoll Fathers and Brothers, Orbis seeks to explore the global dimensions of the Christian faith and mission, to invite dialogue with diverse cultures and religious traditions, and to serve the cause of reconciliation and peace. The books published reflect the views of their authors and do not represent the official position of the Maryknoll Society. To learn more about Maryknoll and Orbis Books, please visit our website at www.maryknollsociety.org.

Copyright © 2016 by Mary Evelyn Tucker and John Grim

Published by Orbis Books, Box 302, Maryknoll, NY 10545-0302.

All rights reserved.

Selections from the Encyclical of His Holiness Pope Francis, *Laudato Si',* copyright © 2015 by Libreria Editrice Vaticana. Used with permission.

No part of this publication may be reproduced or transmitted in any form or by any means, electronic or mechanical, including photocopying, recording, or any information storage or retrieval system, without prior permission in writing from the publisher.

Queries regarding rights and permissions should be addressed to Orbis Books, P.O. Box 302, Maryknoll, NY 10545-0302.

Manufactured in the United States of America

Library of Congress Cataloging-in-Publication Data

Names: Tucker, Mary Evelyn, editor.
Title: Living cosmology : Christian responses to Journey of the universe /
    edited by Mary Evelyn Tucker and John Grim.
Description: Maryknoll : Orbis Books, 2016. | Includes bibliographical
    references and index.
Identifiers: LCCN 2015035429 | ISBN 9781626981782 (pbk.)
Subjects: LCSH: Swimme, Brian. Journey of the universe. |
    Cosmology—Religious aspects—Christianity. | Astronomy—Religious
    aspects—Christianity.
Classification: LCC QB982 .S952 2016 | DDC 261.5/5—dc23 LC record available at
http://lccn.loc.gov/2015035429

≫

To Miriam Therese MacGillis
For her remarkable work at creating
a living cosmology at Genesis Farm for thirty-five years.

And for all those Sisters and Brothers of Earth who have also been inspired
by
Thomas Berry's call
to embrace a New Story
for the Great Work
of healing the Earth Community.

Contents

Part II. Dwelling in a Cosmos

A Sacramental Universe

Cosmological Spirituality and Ecological Ritual

Part III. Participating in a Living Cosmology

Cosmology and Environmental Ethics

Part IV. Evolving Christianity within an Emergent Universe

Acknowledgments

Many people have contributed to this volume, which arose from an energizing conference at Yale University in November 2014. First, we are appreciative of the panel moderators and the participants for their excellent presentations. Next, we thank the Forum staff, including Matthew Riley, Elizabeth McAnally, Hannah Perez; and also Rachel Myslivy, for filming the panels. We are especially grateful to Tara Trapani and her husband, Paul, for their tireless efforts before, during, and after the conference. That the four hundred attendees felt so at home was due to their attention to every detail and to the healthy food! Tara has been tireless in preparing this book for publication, overseeing the formatting and editing with astute care. Once again she is indispensable in assisting the Great Work.

The Yale School of Forestry and Environmental Studies (FES) and the Yale Divinity School (YDS) are invaluable partners in this work. We acknowledge their contributions, especially through the support of Dean Peter Crane at FES and Dean Greg Sterling at YDS. The students from both schools were exceptionally helpful during the conference in welcoming visitors. Sachin Ramabhadran was generous in his able assistance throughout the conference, and Gail Briggs helped in the preparations at the Divinity School.

We would like as well to express gratitude for the support of our funders, including the Porter Fund of Berkeley Divinity School, the Germeshausen Foundation, the Engelhard Foundation, the Kalliopeia Foundation, Reverend Albert Neilson, and Marianne and Jim Welch.

It is a pleasure to be working with so many in North America in continuing the legacy of Thomas Berry, who was celebrated at this conference. We wish to honor, in particular, those Canadians who have planted the seeds of Berry's work in academia and beyond—especially Stephen Dunn and Anne Lonergan. The list of their collaborators in ecotheology and the New Story in Canada and beyond is long and distinguished. Along with many others, it includes Simon Appolloni, Mike Bell, Tom Bishop, Peter Brown, Roberto Chiotti, Jim Conlon, Paul Cusack, Anne Marie Dalton, Dawn and Steve Deme, Heather Eaton, Linda Gregg, Clare Hallward, Mark Hathaway, Mary Margaret Howard, Christopher Hrynkow, Rachel Knight-Messenger, Mary Landry, Laurent Leduc, Dorothy McDougall, Dennis O'Hara,

Edmond O'Sullivan, Jim Profit, Davileen Radigan, Michael Ross, Stephen Scharper, Henry Simmons, Don Smith, Eva Solomon, Cristina Vanin, and Maureen Wild. For efforts over many decades in seeding Thomas's work into the North American context we give thanks!

—*Mary Evelyn Tucker and John Grim*

Foreword

BRIAN THOMAS SWIMME

When did human beings become fascinated with stars? And what did they hope for as they stared at the majestic display night after night after night? Over the course of our two hundred thousand years of existence we have learned marvelous things certainly. But even with all our precious knowledge of stars and galaxies the same stunned astonishment returns whenever we contemplate their radiant presence. The three thousand stars visible to the human eye are almost manageable. But an additional 200 billion stars behind them in the Milky Way galaxy? And 400 billion more in the nearest spiral galaxy whose light travels 2 million years to reach our straining eyes? And beyond that? Can it be that there really are 1 trillion galaxies? Should we abandon this quest for a sense of the whole of things? And the exoplanets? Can there be 50 billion of them in our own galaxy? Is there life everywhere? Are there others out there just as stunned and overwhelmed and baffled by this magnificence that is so crushing and so thrilling?

For all its shortcomings and distortions, the modern scientific enterprise taken altogether has to be understood as one of the great revelatory events in human history. If the axial revelations of Israel, India, China, and Greece altered human understanding in deep and unique ways, something similar can be said about the period that began with Copernicus in 1543 and culminated with Arno Penzias and Robert Wilson in 1964. These four centuries permanently transformed our understanding of the cosmos, time, matter, energy, life, consciousness. We live in an interval similar to those earlier centuries when sensitive humans wrestled with the meanings tangled up in the Buddha, Confucius, Christ, Krishna. It seems inevitable that we, too, will need to wrestle, perhaps for centuries, with the meaning of a 14-billion-year ongoing explosive adventure.

Perhaps the deepest, certainly the most surprising, scientific discovery concerns the nature of creativity throughout the universe. This discovery has been articulated in terms coming from the complexity sciences as well as in terms coming from incarnational theologies, but I shrink from using technical phrases. I would prefer falling back on a word from our childhood.

The scientific exploration of the world has led to the discovery of a pervasive creative power that might even be called a cosmological "magic." We have discovered that molten rock is filled with a kind of magic, for molten rock has the power to become a shining ruby or a brilliant, green-flashing hummingbird. And though we cannot pretend that we understand this creative power, we do know one central fact concerning its effective operation. It draws upon the ontological power of relationship. Perhaps the simplest illustration is found at the subatomic level. A neutron, if traveling alone through the universe, will disintegrate entirely within fifteen minutes. But take that exact same neutron and bring it together with other neutrons and protons in a carbon nucleus and it will suddenly discover it has the power to endure for billions of years. Relationships open up possibilities for creativity that are absolutely unrealizable by solitary individuals.

This same cosmological creative power is at work in this book of essays, which can be considered something like an intellectual cousin to Earth's primordial molten rock. Each of the insights of these thoughtful religious thinkers shines on its own, but simultaneously each is also in search of its synergistic companions. Only when they find each other and fuse together into something new will they fully realize their deeper possibilities. If you read through these chapters, if you mull them over, if you find them taking on a life of their own in your consciousness, you also will have entered into this central creative endeavor of our moment. You will have joined in a lineage of reflection that reaches back two millennia and then reaches back another two hundred millennia to when our earliest ancestors found their destiny by wondering over the universe and the meanings it offers. They were fascinated, as are we. Perhaps only by fascinating us in this way can the universe express its unending insistence that we, too, carry this quest forward.

Introduction

MARY EVELYN TUCKER AND JOHN GRIM

Since the earliest expressions of culture, humans have sought to understand and define our place in the universe. To this end, we have developed cosmologies, which are stories that invoke our common experience of reality. These stories describe where we have come from and where we are going. The religious and philosophical traditions we have honored for millennia all bear witness to our deep desire to find orientation and meaning in what we see, experience, and feel around us. We struggle over a lifetime to know how we fit in and what our role is within a vast universe.

Over the last several centuries, however, a reductionist paradigm has taken root and, in many cases, become a dominant modern worldview. This perspective has arisen in the natural and human sciences as well as the humanities. The scientific method has contributed to this, perhaps inadvertently, and deconstruction and postmodernism have expanded it. Through the scientific method, science tends to objectivize what it describes and reduces complexity to quantifiable parts and measurable systems. While scientific objectivity is useful as a method, it becomes problematic as a worldview when it tends to view matter as inert, driven by mechanistic processes, and easily manipulated as a resource for humans.

Thus, some scientists and some deconstructionists have concluded that the universe, while following certain natural laws, is largely a random and accidental accretion of material objects, with little meaning and no larger purpose. Scientific facts are studied separately from human values. In this worldview, human life is largely dependent on mechanistic materialism, and the result is deepening alienation, disorientation, and existential angst. Moreover, this can drain a larger sense of purpose from human energies in relation to a viable ecological future.

One of the aims of *Journey of the Universe* is to counteract this view with a presentation of a dynamic, emergent, self-organizing, and sacred universe. *Journey* is a project that draws both the sciences and the humanities into fresh perspectives—investigating matter and exploring meaning as complementary and integrated modes of knowing. There are more integrated ways of looking

1

at universe and Earth processes, which is evident in systems science and in the areas of emergence, complexity, and chaos theory.[1] Similarly, in constructive postmodernism, in new materialism, and in religious naturalism, important efforts go beyond reductionism in the humanities.[2]

In modern times, until recently, scientific and religious understandings of cosmology have coexisted uneasily. There have been efforts to navigate the space between reductionism and teleology, but much of the dialogue in science and religion (in the United States especially) remains strained. When it comes to evolution this is particularly the case, with textbook controversies still a dominant feature of science teaching in public schools. But the captivating evidence of an unfolding and complexifying universe is being seeded in the human mind and heart.

This is a major shift in human consciousness. Our rational faculties and imaginative powers are coming to embrace an integrated cosmology of our 14-billion-year-old universe. Our grandparents knew very little of the size and shape of the universe. We are the first generation to inherit this new understanding and to be able to tell it as a story. This is in large measure because of the discoveries of modern science since Charles Darwin's *On the Origin of Species* in 1859. Our challenge in *Journey of the Universe* is to narrate the epic story in ways that are scientifically accurate and poetically engaging.

Relying, then, on the best of modern science and drawing on the humanities, in *Journey* we discover how we are a part of this ongoing journey and are now shaping its future form. This context can be an important one for encouraging the ecological, economic, and social transformations that are required for the continued flourishing of this remarkable evolutionary process.

Journey of the Universe *as a Living Cosmology*

Journey of the Universe narrates the 13.8-billion-year story of the universe's development, from the great flaring forth at the universe's inception, to the

---

1. See, e.g., Evan Thompson, *Mind in Life: Biology, Phenomenology, and the Sciences of Mind* (Cambridge, MA: Belknap Press of Harvard University Press, 2007); Merlin Donald, *A Mind So Rare: The Evolution of Human Consciousness* (New York: W. W. Norton, 2001); and Stuart Kauffman, *At Home in the Universe: The Search for Laws of Self-Organization and Complexity* (Oxford: Oxford University Press, 1995).

2. See, e.g., David Ray Griffin, *Panentheism and Scientific Naturalism: Rethinking Evil, Morality, Religious Experience, Religious Pluralism, and the Academic Study of Religion* (Claremont, CA: Process Century Press, 2014); Jane Bennet, *Vibrant Matter: A Political Ecology of Things* (Durham, NC: Duke University Press, 2010); and Loyal Rue, *Everybody's Story: Wising Up to the Epic of Evolution* (Albany, NY: SUNY Press, 1999).

emergence of simple molecules and atoms, to the evolution of galaxies, stars, solar systems, and planetary life of greater complexity and consciousness. This story inspires wonder as we begin to understand such complexity through science and appreciate such beauty through poetry, art, history, philosophy, and religion. The story also awakens us to the dynamic processes of evolution that are chaotic and destructive, as well as creative and life-generating.

Storytelling seems to have been with humans early on in our emergence as Homo sapiens. No doubt symbolic consciousness, language, and logical thought are closely connected with our efforts to give voice to the wonder and beauty of the world. Indeed, many indigenous peoples still speak of their place-based knowledge as forms of thinking with stories.[3] These cosmologies radiate into the axial age world religions, such as the stories in Genesis. These narratives are so much more than simple explanatory accounts; rather, they transmit values and orient communities to ways of knowing. Even scientific cosmologies, while vastly different in method, continue ancient perspectives on the unity of the universe and our human capacity for knowing that cosmic unity.

Thus, *Journey of the Universe* is a cosmology, although not just in the scientific sense of the study of the universe. Rather, it is a *living cosmology* in the sense of being an integrating story that explains from where both humans and other life forms have come. We now have the capacity to tell a comprehensive story, drawing on astronomy and physics to explain the emergence of galaxies and stars, geology and chemistry to understand the formation of Earth, biology and botany to envision life's evolution, and anthropology and the humanities to trace the rise of humans. *Journey* draws on all these disciplines to narrate a story of universe, Earth, and human evolution that is widely accessible.

Journey recognizes that evolution is governed by natural laws, discerned by scientific methods and empirical observation. However, the self-organizing dynamics of evolutionary processes are part of the remarkable creativity of evolution, which humans are also discovering. While humans are gifted with the creativity of symbolic consciousness, we know that different kinds of self-organizing creativity abound in the universe and Earth—the formation of galaxies and stars, the movement of tectonic plates, the chemistry of cells, the biological complexity of photosynthesis, the migrating patterns of birds, fish, turtles, and caribou. Creativity is also closely aligned with chaos

---

3. See, e.g., Alice Legat, *Walking the Land, Feeding the Fire: Knowledge and Stewardship among Tlicho Dene*, First Peoples, Indigenous Directions (Tucson: University of Arizona Press, 2012); and Eduardo Kohn, *How Forests Think: Toward an Anthropology beyond the Human* (Berkeley: University of California Press, 2013).

and destruction, as the universe unfolds with challenges that overwhelm our capacities to frame or explain fully. Such a cosmological perspective regarding creativity is both ancient and modern—embedded in certain aspects of world religions and philosophies and revealed anew in the scientific story of the universe.

The Cosmological Context: Evolution and Extinction

The magnitude of this Universe Story is beginning to dawn on humans as we awaken to a new realization of the vastness and complexity of this unfolding process. Christians have struggled with this realization in various dialogues trying to reconcile science and religion. But now there are further challenges.

At the same time that this story is becoming available to the human community, we are becoming conscious of the multidimensional environmental crisis and of the rapid destruction of species and habitat that is taking place around the planet. Just as we are realizing the vast expanse of time that distinguishes the evolution of the universe over some 13.8 billion years, we are recognizing how late our arrival is in this stupendous process. Just as we are becoming conscious that Earth took more than 4 billion years to bring forth this abundance of life, it is dawning on us how quickly we are foreshortening its future flourishing.

We need, then, to step back to assimilate our present context. If scientific cosmology gives us an understanding of the origins and unfolding of the universe, religious and philosophical reflection on scientific cosmology can provide a sense of our place and larger purpose in the universe. And if we are so radically affecting the story by extinguishing other life forms and destroying our own nest, what does this imply about our ethical sensibilities or our sense of the sacred? As science is revealing to us the particular intricacy of the web of life, we realize we are unraveling it, although unwittingly in part. Until recently we have not been fully conscious of the deleterious consequences of our drive toward economic progress and rapid industrialization. By exploding from 2 billion to more than 7 billion people in one hundred years we are undermining the life systems of the planet. We have lost a sense of limits. And what does this mean for Christians?

As we begin to glimpse how deeply embedded we are in complex ecosystems and dependent on other life forms, we see that we are destroying the very basis of our continuity as a species. As biology demonstrates a fuller picture of the unfolding of diverse species in evolution and the distinctive niche of species in ecosystems, we are questioning our own niche in the evolutionary process. As the size and scale of the environmental crisis are more widely grasped, we are seeing our own connection to this destruction. We humans have become a planetary presence that is not always benign. This is the ultimate challenge for Christians, to enter into the implications of an awareness of living within a sacred universe even as the Earth's ecosystems are deteriorating.

Living Cosmology Conference

To respond to this challenge, a special conference took place at Yale in early November 2014. This was held to honor the fifth anniversary of Thomas Berry's death and the hundredth anniversary of his birth.[4] Over four hundred people gathered for the conference, which was organized by the Forum on Religion and Ecology at Yale and also sponsored by the Yale School of Forestry and Environmental Studies and Yale Divinity School. It consisted of panels by scholars and activists, along with religious leaders and laity. On Saturday evening, the Thomas Berry Award was presented to James Gustave Speth, former dean of the School of Forestry and Environmental Studies and cofounder of National Resources Defense Council and World Resources Institute. A celebratory jazz liturgy, organized by Reverend Andrew Barnett and Reverend Stephanie Johnson, concluded the conference on Sunday afternoon. It was punctuated by a sequence of brief and inspiring homilies by Thomas Troeger, professor of homiletics at Yale Divinity School. Paul Winter played for the occasion.

The conference was titled "Living Cosmology: Christian Responses to *Journey of the Universe.*" This book is a collection of the papers presented at Yale.[5] Over three days many stimulating ideas converged with reflections for transformative change. The conference was intended to create a space for generative dialogue on the implications for Christians of the *Journey of the Universe* as one telling of our evolutionary epic. There will be many more narrations of this epic in different forms and by a variety of people. This particular telling has three parts: a film that was broadcast for three years on PBS, a book published by Yale University Press, and a DVD series of twenty conversations with scientists and environmentalists.[6] The environmentalists being

---

4. The Yale conference was held November 7–9, 2014. Berry's death was June 1, 2009, and his birth was November 9, 1914.

5. The oral summaries of the papers and panel discussions can be seen online at http://www.journeyoftheuniverse.org/living-cosmology-videos/.

6. See Appendix I for a fuller history of the *Journey of the Universe* project.

interviewed in the conversations illustrate how their work on the ground is inspired by this epic Universe Story.

The term "living cosmology," then, refers to the notion that this is not just a beautiful story of evolution. Rather, it is a "functional cosmology," as Thomas Berry would say, one that has implications for how we live our lives and contribute to our planetary future. As the *Journey Conservations* and the conference papers illustrate, these contributions are already taking place in areas of education, economics, energy, housing, and food, as well as in relation to race and identity issues.

Implications of Evolution and Ecology for Christians

Since the publication of Charles Darwin's *On the Origin of Species* in 1859, Christian theologians along with church leaders and laity have been struggling to absorb the implications of the science of evolution. Moreover, they are trying as well to respond to the complex ecological and social challenges of the present. The invitation, then, of the Living Cosmology conference was to bring those large questions of evolution and ecology into focus through the *Journey of the Universe* trilogy. This is of immense import for all religious traditions, but at this gathering we focused on Christianity.[7]

What was our hope for this conference at Yale? A fresh understanding of how cosmology is a way of living in the world, not just a scientific investigation of the universe. Through this lens we hope for an awakening to one of the most radical ideas humans have ever had to absorb, namely, cosmogenesis. What is it to realize that the universe is not static, but has unfolded over 13.8 billion years? What is it to absorb even a glimpse of the evolution of our planet over 4.6 billion years? What does it mean to recognize that the first cell emerged after 1 billion years and a nucleated cell took another billion years? That multicellular life evolved into such diversity and complexity and beauty surely gives us a new context for rethinking who we are—a symbol-making animal amid a vast diversity of life with powers of creativity and destruction. Does not all of this raise questions as to who we are as humans and who we may become as Christians in a world that is unraveling before

---

7. In June 2013, for one week, we concentrated on responses to *Journey of the Universe* from the various world religions. This took place at the Chautauqua Institution in upstate New York. The talks and discussions are available at http://www.journeyoftheuniverse.org/conference-at-chautauqua/.

us—with climate change, ravaged ecosystems, and broken communities on a massive level?

Conference Responses

As *Journey of the Universe* suggests, everything can be rediscovered when abiding reverence and awe reassert themselves in the face of immense mystery. This is what the Ecumenical Patriarch Bartholomew announces in his moving prologue in this volume, "And God Saw That Everything Was Good." His appreciation for the beauty of nature arises from the heart of Orthodox theology and liturgy, which celebrates the sacramental dimension of creation. A broader theology of the divine is thus possible, as John Chryssavgis observes from the Orthodox tradition and Catherine Keller notes from the perspective of process theology.

An engaged Christology emerges when fresh understanding of incarnation and immanence appears, as Kathleen Deignan and Christina Vanin observe. John's Gospel opens with, "In the beginning was the Word," and now we see how this Logos, this inner ordering principle, extends through the whole evolutionary process. It calls us to pay attention to the world here and now, as Ilia Delio writes, as well as to the promise of the future, as John Haught eloquently reminds us.

The Holy Spirit, too, is discovered anew. Thomas Berry described this as seeing the Earth as "a communion of subjects not a collection of objects." Heather Eaton, Anne Marie Dalton, Matthew Riley, and Peter Ellard explore Berry's perspective on this discovery. We begin to see how, for Berry, creation emerges as a luminous revelation to be cherished, begun with the breath of divine mystery, imprinted with creative patterns of immanence, and filled with the spirit of interrelationality. The implications of this for sacrament and ritual are significant, as Bede Bidlack and Steve Blackmer demonstrate.

In addition, ethics are expanded to include environmental ethics and ecojustice ethics, as Willis Jenkins and Fred Simmons explain, but also to an emerging cosmological ethics and Earth ethics, as Larry Rasmussen persuasively asserts. Social justice ethics joins creation justice ethics, wherein human-Earth relations are redefined and renewed in the context of a living cosmology. Carl Anthony and Paloma Pavel reveal this with their discussion of race; Mary Hunt opens up our understanding of cosmology and feminism; and Whitney Bauman highlights this with insights regarding radical pluralism.

A living cosmology thus means that we reinhabit the sacred dimensions of land, food, and water, as Chris Loughlin, James Jenkins, and Nancy Wright remind us. We are grounded in the Earth community quite literally. Moreover, we will depend on a new Earth jurisprudence for sustaining this community of life, as Brian Brown, Patricia Siemen, and Paul Waldau passionately argue.

The papers from this conference, then, are trying to break open the categories of theology and ethics, sacramental reality and spiritual practice, ritual and law, food and water, race and gender, to embrace this expansive idea of cosmogenesis, namely that we are part of a vast evolving universe and Earth system. Efficacious participation in cosmogenesis may assist the flourishing of the Earth community. That is the invitation of these articles: as we move from epic story to ecological restoration and ecojustice, each person has a unique contribution to offer.

We conclude with perspectives from various Christian communities to see how a dialogue with *Journey of the Universe* might enhance their theologies and ethics. While not exhaustive of all Christian communities, they include valuable insights on Catholicism from Dennis O'Hara, on Lutheranism from Barbara Rossing, on Methodism from Beth Norcross, on Presbyterianism from Russell Powell, on Quakerism from Laurel Kearns, and on Mormonism from George Handley. Each of these papers suggest ways for further reflection, dialogue, and implementation.

Challenges of Our Present Moment: Worldviews and Ethics

The challenges for Christian communities around the world are clearly immense. As we see our present interconnected global problems of widespread environmental degradation, climate change, crippling poverty, social inequities, rampant racism, human trafficking, and unrestrained militarism, we know that the obstacles to the flourishing of life's ecosystems and the well-being of our social systems are considerable.

Yet in the midst of these formidable problems we are being called to the next stage of evolutionary history. How are Christians responding to these problems and navigating their way? This requires a rethinking of tradition in modernity, namely, diverse responses that bring forward the rich resources of Christianity (and all the world religions) into a complex modern context. Many efforts are being made in this direction of retrieval, reexamination, and reconstruction.[8]

---

8. This includes the work of the Forum on Religion and Ecology at Yale University with its two decades of conferences and books. See fore.yale.edu.

Most particularly, the papal encyclical on the environment is a unique example of a call toward creating a diverse planetary civilization that has ecology, justice, and peace at the center. The pope is using the term "integral ecology" to emphasize that people and planet both need to be treated with respect. This involves an invitation to all peoples: (1) to be protectors of the environment, (2) to care for creation as a virtue in its own right, (3) to form a new global solidarity in our search for the common good, and (4) to cultivate a special sense of concern for the poor as they are disproportionately affected by climate change. To accomplish this we can turn toward an enlarged cosmological context of a sacred universe.

In that spirit, this book suggests that within Christianity we will need a broadening of our theologies and ethics into a living cosmology. This is what Pierre Teilhard de Chardin published in his major work, *The Human Phenomenon*, in 1959. Thomas Berry and Brian Thomas Swimme developed this further in their pathbreaking narration of *The Universe Story* in 1992. Thus we open this book by reflecting on the contributions of both Berry and Teilhard in shaping the *Journey of the Universe* perspective.

As Teilhard and Berry recognized, and as *Journey* depicts, the evolutionary life impulse transforms us from viewing ourselves as isolated individuals and competing nation-states to realizing our collective presence as a species with a common origin story and shared destiny. Christians, along with all humans, have the capacity now to realize our intrinsic unity with life in the midst of enormous biological and cultural diversity. And, most especially, we have the opportunity to see this unity as arising from the dynamics of the evolutionary process itself. We now have a scientific story of the evolution of the universe and Earth that reveals our profound connectedness to this process. We are still discovering the larger meaning of the story with the help of the humanities. That is our collective effort to enter into the implications of a living cosmology for our times, while still attentive to cultural and religious pluralism.

Our sense of the whole is becoming visible in fresh ways as we feel ourselves embraced by the evolutionary powers emerging over deep time into forms of ever-greater complexity and consciousness, as Teilhard observed. We are realizing, too, that evolution moves forward with transitions, such as from inorganic matter to organic life and from single-celled organisms to plants and animals that sweep through the evolutionary unfolding of the universe, the Earth, and the human. Just as transitions often come at times of crisis and involve tremendous cost, they result in new forms of creativity. The central reality of our times is that we are in such a creative transitional moment.

Christianity has unique contributions to make in this transition moment as it enters into an ever-richer dialogue with evolutionary biology and ecology and formulates a sacramental theology, a robust ecojustice, an inclusive environmental ethics, and a comprehensive Earth jurisprudence. That is what this book explores from diverse Christian perspectives—what Thomas Berry called the Great Work of our times.

As the *Journey Conversations* illustrate, hundreds of thousands of people around the planet are also participating in this transformative work for the environment, energy, agriculture, economics, education, the arts, sustainable cities, improved racial relations, and gender equity. The comprehensive perspective of a sacred universe depicted in *Journey of the Universe* may inspire Christians to continue contributing to this Great Work.

Prologue

I

Selections from *Laudato Si'*
On Care for Our Common Home

POPE FRANCIS

On June 18, 2015, Pope Francis issued an encyclical focused on the environment, the first in the history of the Catholic Church. At the heart of this encyclical is a profound sense of the interconnection of humans with nature and humans with each other. Undergirding this connection one can see a cosmological vision of universal communion of humans, Earth, and cosmos. For example, Francis notes that God has written a precious book "whose letters are the multitude of created things present in the universe."[1]

In addition, he draws on Thomas Aquinas, who saw that "the universe as a whole, in all its manifold relationships, shows forth the inexhaustible riches of God."[2] All of this points to the divine manifested in nature and in the cosmos as a whole. To illustrate this, the encyclical cites the hymn of St. Francis. In experiencing this unity, humans have a renewed responsibility to care for Earth and for the poor.

The interdependence, then, of people and the planet is described in the phrase "integral ecology." Francis sees environmental, economic, and social ecology as intimately interrelated. Without healthy ecosystems we cannot create just economic and social systems. Thus, combating poverty and protecting nature are two dimensions of the same challenge. Importantly, the encyclical recognizes the intrinsic value of ecosystems. While affirming their regenerative character, he notes the limits of "sustainable use" of nature. Francis also underscores the importance of local cultural voices as dynamic and participatory in protecting bioregions. This cultural ecology is found within the world's wisdom traditions and artistic heritage.

---

1. Francis I, *Laudato Si'*, June 18, 2015, chapter 2, section IV, paragraph 85. Quoting John Paul II, *Catechesis*, January 30, 2002, 6.

2. Francis I, *Laudato Si'*, June 18, 2015, chapter 2, section IV, paragraph 86.

The Gospel of Creation [Chapter Two]

IV. The Message of Each Creature in the Harmony of Creation

84. Our insistence that each human being is an image of God should not make us overlook the fact that each creature has its own purpose. None is superfluous. The entire material universe speaks of God's love, his boundless affection for us. Soil, water, mountains: everything is, as it were, a caress of God. The history of our friendship with God is always linked to particular places, which take on an intensely personal meaning; we all remember places, and revisiting those memories does us much good. Anyone who has grown up in the hills or used to sit by the spring to drink, or played outdoors in the neighborhood square; going back to these places is a chance to recover something of their true selves.

85. God has written a precious book "whose letters are the multitude of created things present in the universe."[54] The Canadian bishops rightly pointed out that no creature is excluded from this manifestation of God: "From panoramic vistas to the tiniest living form, nature is a constant source of wonder and awe. It is also a continuing revelation of the divine."[55] The bishops of Japan, for their part, made a thought-provoking observation: "To sense each creature singing the hymn of its existence is to live joyfully in God's love and hope."[56] This contemplation of creation allows us to discover in each thing a teaching that God wishes to hand on to us, since "for the believer, to contemplate creation is to hear a message, to listen to a paradoxical and silent voice."[57] We can say that, "alongside revelation properly so-called, contained in sacred Scripture, there is a divine manifestation in the blaze of the sun and the fall of night."[58] Paying attention to this manifestation, we learn to see ourselves in relation to all other creatures: "I express myself in expressing the world; in my effort to decipher the sacredness of the world, I explore my own."[59]

---

[54] John Paul II, *Catechesis* (30 January 2002),6: *Insegnamenti* 25/1 (2002), 140.

[55] Canadian Conference of Catholic Bishops, Social Affairs Commission, Pastoral Letter *You Love All that Exists... All Things are Yours, God, Lover of Life"* (4 October 2003), 1.

[56] Catholic Bishops' Conference of Japan, *Reverence for Life. A Message for the Twenty-First Century* (1 January 2000), 89.

[57] John Paul II, *Catechesis* (26 January 2000), 5: *Insegnamenti* 23/1 (2000), 123.

[58] Id., *Catechesis* (2 August 2000), 3: *Insegnamenti* 23/2 (2000), 112.

[59] Paul Ricoeur, *Philosophie de la Volonté, t. II: Finitude et Culpabilité*, Paris, 2009, 216.

86. The universe as a whole, in all its manifold relationships, shows forth the inexhaustible riches of God. Saint Thomas Aquinas wisely noted that multiplicity and variety "come from the intention of the first agent" who willed that "what was wanting to one in the representation of the divine goodness might be supplied by another,"[60] inasmuch as God's goodness "could not be represented fittingly by any one creature."[61] Hence we need to grasp the variety of things in their multiple relationships.[62] We understand better the importance and meaning of each creature if we contemplate it within the entirety of God's plan. As the Catechism teaches: "God wills the interdependence of creatures. The sun and the moon, the cedar and the little flower, the eagle and the sparrow: the spectacle of their countless diversities and inequalities tells us that no creature is self-sufficient. Creatures exist only in dependence on each other, to complete each other, in the service of each other."[63]

87. When we can see God reflected in all that exists, our hearts are moved to praise the Lord for all his creatures and to worship him in union with them. This sentiment finds magnificent expression in the hymn of Saint Francis of Assisi:

> Praised be you, my Lord, with all your creatures, especially Sir Brother Sun, who is the day and through whom you give us light. And he is beautiful and radiant with great splendor; and bears a likeness of you, Most High.
>
> Praised be you, my Lord, through Sister Moon and the stars, in heaven you formed them clear and precious and beautiful. Praised be you, my Lord, through Brother Wind, and through the air, cloudy and serene, and every kind of weather through whom you give sustenance to your creatures.
>
> Praised be you, my Lord, through Sister Water, who is very useful and humble and precious and chaste. Praised be you, my Lord, through Brother Fire, through whom you light the night, and he is beautiful and playful and robust and strong.[64]

88. The bishops of Brazil have pointed out that nature as a whole not only manifests God but is also a locus of his presence. The Spirit of life dwells

---

[60] *Summa Theologiae*, I, q. 47, art. 1.
[61] Ibid.
[62] Cf. ibid., art. 2, ad 1; art. 3.
[63] *Catechism of the Catholic Church*, 340.
[64] *Canticle of the Creatures*, in *Francis of Assisi: Early Documents*, New York-London-Manila, 1999, 113-114.

in every living creature and calls us to enter into relationship with him.[65] Discovering this presence leads us to cultivate the "ecological virtues."[66] This is not to forget that there is an infinite distance between God and the things of this world, which do not possess his fullness. Otherwise, we would not be doing the creatures themselves any good either, for we would be failing to acknowledge their right and proper place. We would end up unduly demanding of them something that they, in their smallness, cannot give us.

V. A Universal Communion

89. The created things of this world are not free of ownership: "For they are yours, O Lord, who love the living" (Wis 11:26). This is the basis of our conviction that, as part of the universe, called into being by one Father, all of us are linked by unseen bonds and together form a kind of universal family, a sublime communion which fills us with a sacred, affectionate and humble respect. Here I would reiterate that "God has joined us so closely to the world around us that we can feel the desertification of the soil almost as a physical ailment, and the extinction of a species as a painful disfigurement."[67]

90. This is not to put all living beings on the same level nor to deprive human beings of their unique worth and the tremendous responsibility it entails. Nor does it imply a divinization of the earth that would prevent us from working on it and protecting it in its fragility. Such notions would end up creating new imbalances which would deflect us from the reality which challenges us.[68] At times we see an obsession with denying any pre-eminence to the human person; more zeal is shown in protecting other species than in defending the dignity which all human beings share in equal measure. Certainly, we should be concerned lest other living beings be treated irresponsibly. But we should be particularly indignant at the enormous inequalities in our midst, whereby we continue to tolerate some considering themselves more worthy than others. We fail to see that some are mired in desperate

---

[65] Cf. National Conference of the Bishops of Brazil, *A Igreja e a Questão Ecológica*, 1992, 53-54.

[66] Ibid., 61.

[67] Apostolic Exhortation *Evangelii Gaudium* (24 November 2013), 215: AAS 105 (2013), 1109.

[68] Cf. Benedict XVI, Encyclical Letter *Caritas in Veritate* (29 June 2009), 14: AAS 101 (2009), 650.

and degrading poverty, with no way out, while others have not the faintest idea of what to do with their possessions, vainly showing off their supposed superiority and leaving behind them so much waste which, if it were the case everywhere, would destroy the planet. In practice, we continue to tolerate that some consider themselves more human than others, as if they had been born with greater rights.

91. A sense of deep communion with the rest of nature cannot be real if our hearts lack tenderness, compassion and concern for our fellow human beings. It is clearly inconsistent to combat trafficking in endangered species while remaining completely indifferent to human trafficking, unconcerned about the poor, or undertaking to destroy another human being deemed unwanted. This compromises the very meaning of our struggle for the sake of the environment. It is no coincidence that, in the canticle in which Saint Francis praises God for his creatures, he goes on to say: "Praised be you my Lord, through those who give pardon for your love." Everything is connected. Concern for the environment thus needs to be joined to a sincere love for our fellow human beings and an unwavering commitment to resolving the problems of society.

92. Moreover, when our hearts are authentically open to universal communion, this sense of fraternity excludes nothing and no one. It follows that our indifference or cruelty towards fellow creatures of this world sooner or later affects the treatment we mete out to other human beings. We have only one heart, and the same wretchedness that leads us to mistreat an animal will not be long in showing itself in our relationships with other people. Every act of cruelty towards any creature is "contrary to human dignity."[69] We can hardly consider ourselves to be fully loving if we disregard any aspect of reality: "Peace, justice and the preservation of creation are three absolutely interconnected themes, which cannot be separated and treated individually without once again falling into reductionism."[70] Everything is related, and we human beings are united as brothers and sisters on a wonderful pilgrimage, woven together by the love God has for each of his creatures and which also unites us in fond affection with brother sun, sister moon, brother river and mother earth.

---

[69] *Catechism of the Catholic Church*, 2418.

[70] Conference of Dominican Bishops, Pastoral Letter *Sobre la relación del hombre con la naturaleza* (21 January 1987).

Integral Ecology [Chapter Four]

137. Since everything is closely interrelated, and today's problems call for a vision capable of taking into account every aspect of the global crisis, I suggest that we now consider some elements of an integral ecology, one which clearly respects its human and social dimensions.

I. Environmental, Economic and Social Ecology

138. Ecology studies the relationship between living organisms and the environment in which they develop. This necessarily entails reflection and debate about the conditions required for the life and survival of society, and the honesty needed to question certain models of development, production and consumption. It cannot be emphasized enough how everything is interconnected. Time and space are not independent of one another, and not even atoms or subatomic particles can be considered in isolation. Just as the different aspects of the planet—physical, chemical and biological—are interrelated, so too living species are part of a network that we will never fully explore and understand. A good part of our genetic code is shared by many living beings. It follows that the fragmentation of knowledge and the isolation of bits of information can actually become a form of ignorance, unless they are integrated into a broader vision of reality.

139. When we speak of the "environment," what we really mean is a relationship existing between nature and the society which lives in it. Nature cannot be regarded as something separate from ourselves or as a mere setting in which we live. We are part of nature, included in it and thus in constant interaction with it. Recognizing the reasons why a given area is polluted requires a study of the workings of society, its economy, its behavior patterns, and the ways it grasps reality. Given the scale of change, it is no longer possible to find a specific, discrete answer for each part of the problem. It is essential to seek comprehensive solutions that consider the interactions within natural systems themselves and with social systems. We are faced not with two separate crises, one environmental and the other social, but rather with one complex crisis that is both social and environmental. Strategies for a solution demand an integrated approach to combating poverty, restoring dignity to the excluded, and at the same time protecting nature.

140. Due to the number and variety of factors to be taken into account when determining the environmental impact of a concrete undertaking, it is essen-

tial to give researchers their due role, to facilitate their interaction, and to ensure broad academic freedom. Ongoing research should also give us a better understanding of how different creatures relate to one another in making up the larger units which today we term "ecosystems." We take these systems into account not only to determine how best to use them, but also because they have an intrinsic value independent of their usefulness. Each organism, as a creature of God, is good and admirable in itself; the same is true of the harmonious ensemble of organisms existing in a defined space and functioning as a system. Although we are often not aware of it, we depend on these larger systems for our own existence. We need only recall how ecosystems interact in dispersing carbon dioxide, purifying water, controlling illnesses and epidemics, forming soil, breaking down waste, and in many other ways that we overlook or simply do not know about. Once they become conscious of this, many people realize that we live and act on the basis of a reality that has previously been given to us, which precedes our existence and our abilities. So, when we speak of "sustainable use," consideration must always be given to each ecosystem's regenerative ability in its different areas and aspects.

II. Cultural Ecology

143. Together with the patrimony of nature, there is also an historic, artistic and cultural patrimony that is likewise under threat. This patrimony is a part of the shared identity of each place and a foundation upon which to build a habitable city. It is not a matter of tearing down and building new cities, supposedly more respectful of the environment yet not always more attractive to live in. Rather, there is a need to incorporate the history, culture and architecture of each place, thus preserving its original identity. Ecology, then, also involves protecting the cultural treasures of humanity in the broadest sense. More specifically, it calls for greater attention to local cultures when studying environmental problems, favoring a dialogue between scientific-technical language and the language of the people. Culture is more than what we have inherited from the past; it is also, and above all, a living, dynamic and participatory present reality, which cannot be excluded as we rethink the relationship between human beings and the environment.

II

And God Saw That Everything Was Good

The Creation Story and Orthodox Theology

ECUMENICAL PATRIARCH BARTHOLOMEW

Ecumenical Patriarch Bartholomew is a leader in the movement of Christian responses to the environmental crisis. For more than two decades he has been teaching, writing, and speaking on the environment. In 1995 he began convening a series of symposia on water issues. These were called Religion, Science, and the Environment and included symposia in Europe, South America, the Arctic, and the United States. They brought together ministers from government and civil servants from the United Nations, along with environmentalists, journalists, and religious leaders. These were remarkable gatherings, generating cross-disciplinary cooperation and networking.

The Patriarch himself brings a unique spiritual vision to these symposia, inspired by the rich cosmological tradition of Orthodox Christianity. He draws on a range of theological sources from Maximus the Confessor and the Church Fathers to contemporary theologians such as John of Pergamon. This leads him to see nature as a book that needs to be read and creation as a symphony that needs to be heard.

In this selection, we see such images broadened to include the idea of story: "Each plant, each animal, and each micro-organism tells a story. . . ." Everything points to a deeper "dialogue of communication and mystery of communion." He observes, "We do not need this perspective in order to believe in God or to prove His existence. We need it to breathe; we need it for us simply to be." This is a fitting prologue to Living Cosmology, *and we are honored to have the Patriarch's comprehensive ecological and spiritual vision opening the book.*

Reading the Book of Nature

In the late third century, St. Anthony of Egypt (251–356), the "father of monasticism," described nature as a book that teaches us about the beauty of God's creation: "My book is the nature of creation; therein, I read the works

18

of God." The extraordinary spiritual collection titled the *Philokalia* records St. Anthony as saying, "Creation declares in a loud voice its Maker and master." This is how Orthodox theology and spirituality perceive the natural environment. There is, as St. Maximus the Confessor (580–662) would claim in the seventh century, a liturgical or sacramental dimension to creation. The whole world, as he observed, is a "cosmic liturgy." For, as St. Maximus observes, "Creation is a sacred book, whose letters and syllables are the universal aspects of creation; just as Scripture is a beautiful world, which is constituted of heaven and earth and all that lies in between." What, then, is the Orthodox theological and liturgical vision of the world?

As a young child, as I accompanied the priest of our local village to services in remote chapels on my native island of Imvros in Turkey, the connection of the beautiful mountainside to the splendor of liturgy was evident. This is because the natural environment provides a broader, panoramic vision of the world. In general, nature's beauty leads to a more open view of the life and created world, somewhat resembling a wide-angle focus from a camera, which ultimately prevents us human beings from using or abusing its natural resources in a selfish, narrow-minded way.

In order, however, to reach this point of maturity and dignity toward the natural environment, we must take the time to listen to the voice of creation. And in order to do this, we must first be silent. The virtue of silence is perhaps the most valuable human quality underlined in the *Philokalia*. Indeed, silence is a fundamental element, which is critical in developing a balanced environmental ethos as an alternative to the ways that we currently relate to the Earth and deplete its natural resources. *The Sayings of the Desert Fathers* relate of Abba Chaeremon that, in the fourth century, he deliberately constructed his cell "forty miles from the church and ten miles from the water" so that he might struggle a little to do his daily chores.[1] In Turkey today, the Princes Island of Heybeliada (or Halki) still forbids automobile traffic.

So if we are silent, we will learn to appreciate how "the heavens declare the glory of God, and the firmament proclaims the creation of His hands."[2] The ancient Liturgy of St. James, celebrated only twice a year in Orthodox churches, affirms the same conviction:

The heavens declare the glory of heaven; the earth proclaims the sovereignty of God; the sea heralds the authority of the Lord; and every

---

1. *Sayings of the Desert Fathers*, trans. Benedicta Ward (New York: MacMillan, 1980), 205.
2. Psalm 19:1.

material and spiritual creature preaches the magnificence of God at all times.

When God spoke to Moses in the burning bush, communication occurred through a silent voice, as St. Gregory of Nyssa (340n394) informs us in his mystical classic, *The Life of Moses*. St. Gregory believes that we can discern God's presence simply by gazing at and listening to creation. Therefore, nature is a book opened wide for all to read and to learn. Each plant, each animal, and each microorganism tells a story, unfolds a mystery, relates an extraordinary harmony and balance, which are interdependent and complementary. Everything points to the same encounter and mystery.

The same dialogue of communication and mystery of communion is detected in the galaxies, where the countless stars betray the same mystical beauty and mathematical interconnectedness. We do not need this perspective in order to believe in God or to prove His existence. We need it to breathe; we need it for us simply to be. The coexistence and correlation between the boundlessly infinite and the most insignificantly finite things in our world articulate a concelebration of joy and love.

It is unfortunate that we lead our life without even noticing the environmental concert that is playing out before our very eyes and ears. In this orchestra, each minute detail plays a critical role, and every trivial aspect participates in an essential way. No single member—human or other—can be removed without the entire symphony being affected. No single tree or animal can be removed without the entire picture being profoundly distorted, if not destroyed. When will we begin to learn and teach the alphabet of this divine language, so mysteriously concealed in nature?

The Days of Creation: Humans, Plants, and Animals

The brief yet powerful statement found in Genesis corresponds to the majesty of creation as understood in Orthodox theology, liturgy, and spirituality:

> Then God said: "Let the earth bring forth vegetation: plants yielding seed, and fruit trees of every kind on earth that bear fruit with the seed in it." And it was so. . . . And God saw that it was good. And there was evening and there was morning, the third day.[3]

---

3. Genesis 1:11.

We all know the healing and nourishing essence of plants; we all appreciate their manifold creative and cosmetic usefulness.

> Consider the lilies, how they grow: they neither toil nor spin; yet I tell you, even Solomon in all his glory was not clothed like one of these.[4]

Even the humblest and lowliest manifestations of God's created world comprise the most fundamental elements of life and the most precious aspects of natural beauty.

Nevertheless, by overgrazing or deforestation, we tend to disturb the balance of the plant world. Whether by excessive irrigation or urban construction, we interrupt the magnificent epic of the natural world. Our selfish ways have led us to ignore plants, or else to undervalue their importance. Our understanding of plants is sparse and selective. Our outlook is greed-oriented and profit-centered.

However, plants are the center and source of life. Plants permit us to breathe and to dream. Plants provide the basis of spiritual and cultural life. A world without plants is a world without a sense of beauty. Indeed, a world without plants and vegetation is inconceivable and unimaginable. It would be the contradiction of life itself, tantamount to death. There is no such thing as a world where unsustainable development continues without critical reflection and self-control; there is no such thing as a planet that thoughtlessly and blindly proceeds along the present route of global warming. There is only wasteland and destruction. To adopt any other excuse or pretext is to deny the reality of land, water, and air pollution.

Plants are also the wisest of teachers and the best of models. For they turn toward light. They yearn for water. They cherish clean air. Their roots dig deep, while their reach is high. They are satisfied and sustained with so little. They transform and multiply everything that they draw from nature, including some things that appear wasteful or useless. They adapt spontaneously and produce abundantly—whether for the nourishment or admiration of others. They enjoy a microcosm of their own while equally contributing to the macrocosm around them.

On the final days of creation, God is said to have made the variety of animals, as well as created man and woman in the divine image and likeness.[5] What most people seem to overlook is that the sixth day of creation is not

---

4. Luke 12:27.
5. Genesis 1:26.

entirely dedicated to the forming of Adam out of the Earth. That sixth day was in fact shared with the creation of numerous "living creatures of every kind; cattle and creeping things and wild animals of the earth of every kind."[6] This close connection between humanity and the rest of creation, from the very moment of genesis, is surely an important and powerful reminder of the intimate relationship that we share as human beings with the animal kingdom. While there is undoubtedly something unique about human creation in the divine image, there is more that unites us than separates us, not only as human beings but also with the created universe. It is a lesson we have learned in recent decades, but it is a lesson that we learned the hard way.

The saints of the early Eastern Church taught this same lesson long ago. They knew that a person with a pure heart was able to sense the connection with the rest of creation, especially the animal world. This is surely a reality that finds parallels in both Eastern and Western Christianity: one may recall Seraphim of Sarov (1759–1833) feeding the bear in the forest of the north, or Francis of Assisi (1181–1226) addressing the elements of the universe. The connection is not merely emotional; it is profoundly spiritual in its motive and content. It gives a sense of continuity and community with all of creation, while providing an expression of identity and compassion with it—a recognition that, as St. Paul put it, all things were created in Christ and in Christ all things hold together.[7] This is why Abba Isaac of Nineveh can write from the desert of Syria,

> What is a merciful heart? It is a heart, which is burning with love for the whole of creation: for human beings, for birds, for beasts, for demons— for all of God's creatures. When such persons recall or regard these creatures, their eyes are filled with tears.[8]

Thus, love for God, love for human beings, and love for animals cannot be separated sharply. There may a hierarchy of priority, but it is not a sharp distinction of comparison. The truth is that we are all one family—human beings and the living world alike—and all of us look to God the Creator: "These all look to you to give them . . . When you open your hand, they are filled with good things. When you hide your face, they are dismayed. When you take away their breath, they die and return to their dust."[9]

---

6. Genesis 1:24.
7. Colossians 1:15–17.
8. Abba Isaac of Nineveh, *Ascetic Treatises*, 48.
9. Psalm 104:28–29.

Social, Political, and Economic Implications

Orthodox theology takes all of this a step further and recognizes the natural creation as inseparable from the identity and destiny of humanity, because every human action leaves a lasting imprint on the body of the Earth. Human attitudes and behavior toward creation directly impact on and reflect human attitudes and behavior toward other people. Ecology is inevitably related in both its etymology and meaning to economy; our global economy is simply outgrowing the capacity of our planet to support it. At stake is not just our ability to live in a sustainable way but our very survival. Scientists estimate that those most hurt by global warming in years to come will be those who can least afford it. Therefore, the ecological problem of pollution is invariably connected to the social problem of poverty, and so all ecological activity is ultimately measured and properly judged by its impact and effect upon other people, and especially the poor.[10]

How, then, does respect for the natural environment translate into contemporary attitudes and action? The issue of environmental pollution and degradation cannot be isolated for the purpose of understanding or resolution. The environment is the home that surrounds the human species and constitutes the human habitat. Therefore, the environment cannot be appreciated or assessed alone, without a direct connection to the unique creature that it surrounds, namely humanity. Concern for the environment implies also concern for human problems of poverty, thirst, and hunger. This connection is detailed in a stark manner in the parable of the Last Judgment, where the Lord says, "I was hungry and you gave me food; I was thirsty and you gave me something to drink."[11]

Concern, then, for ecological issues is directly related to concern for issues of social justice, and particularly of world hunger. A church that neglects to pray for the natural environment is a church that refuses to offer food and drink to a suffering humanity. At the same time, a society that ignores the mandate to care for all human beings is a society that mistreats the very creation of God, including the natural environment. It is tantamount to blasphemy.

The terms "ecology" and "economy" share the same etymological root. Their common prefix "eco" derives from the Greek word *oikos*, which signifies "home" or "dwelling." It is unfortunate and selfish, however, that we

---

10. See Matthew 25.
11. Matthew 25:35.

have restricted the application of this word to ourselves, as if we are the only inhabitants of this world. The fact is that no economic system—no matter how technologically or socially advanced—can survive the collapse of the environmental systems that support it. This planet is indeed our home; yet it is also the home of everyone, as it is the home of every animal creature, as well as of every form of life created by God. It is a sign of arrogance to presume that we human beings alone inhabit this world. Indeed, by the same token, it is also a sign of arrogance to imagine that only the present generation inhabits this earth.

Therefore, as one of the more serious ethical, social, and political problems, poverty is directly and deeply connected to the ecological crisis. A poor farmer in Asia, in Africa, or in North America will daily face the reality of poverty. For these persons, the misuse of technology or the eradication of trees is not merely harmful to the environment or destructive of nature; rather, it practically and profoundly affects the very survival of their families. Terminology such as "ecology," "deforestation," or "overfishing" is entirely absent from their daily conversation or concern. The "developed" world cannot demand from the "developing" poor an intellectual understanding with regard to the protection of the few earthly paradises that remain, especially in light of the fact that less than 10 percent of the world's population consumes over 90 percent of the Earth's natural resources. However, with proper education, the "developing" world would be far more willing than the "developed" world to cooperate for the protection of creation.

Closely related to the problem of poverty is the problem of unemployment, which plagues societies throughout the world. It is abundantly clear that neither the moral counsel of religious leaders nor fragmented measures by socioeconomic strategists or political policymakers could be sufficient to curb this growing tragedy. The problem of unemployment compels us to reexamine the priorities of affluent societies in the West, and especially the unrestricted advance of development, which is considered only in positive economic terms. We appear to be trapped in the tyrannical cycle created by a need for constant productivity rises and increases in the supply of consumer goods. However, placing these two "necessities" on an equal footing imposes on society a relentless need for unending perfection and growth, while restricting power over production to fewer and fewer. Concurrently, real or imaginary consumer needs constantly increase and rapidly expand. Thus, the economy assumes a life of its own, a vicious cycle that becomes independent of human need or human concern. What is needed is a radical change in politics and economics, one that underlines the unique and primary value of the human person, thereby placing a human face on the concepts of employment and productivity.

Conclusion: A New Worldview

We have repeatedly stated that the crisis that we are facing in our world is not primarily ecological. It is a crisis concerning the way we envisage or imagine the world. We are treating our planet in an inhuman, godless manner precisely because we fail to see it as a gift inherited from above; it is our obligation to receive, respect, and in turn hand on this gift to future generations. Therefore, before we can effectively deal with problems of our environment, we must change the way we perceive the world. Otherwise, we are simply dealing with symptoms, not with their causes. We require a new worldview if we are to desire "a new earth."[12]

So let us acquire a spirit of gratitude and frugality, bearing in mind that everything in the natural world, whether great or small, has its importance *within the universe and for the life of the world*; nothing whatsoever is useless or contemptible. Let us regard ourselves as responsible before God for every living creature and for the whole of natural creation; let us treat everything with proper love and utmost care. Only in this way shall we secure a physical environment where life for the coming generations of humankind will be healthy and happy. Otherwise, the unquenchable greed of our generation will constitute a mortal sin resulting in destruction and death. This greed in turn will lead to the deprivation of our children's generation, in spite of our desire and claim to bequeath to them a better future. Ultimately, it is for our children that we must perceive our every action in the world as having a direct effect upon the future of the environment.

These sentiments were jointly communicated with Pope Francis during our apostolic pilgrimage to Jerusalem in May 2014:

> It is our profound conviction that the future of the human family depends also on how we safeguard—both prudently and compassionately, with justice and fairness—the gift of creation that our Creator has entrusted to us. Therefore, we acknowledge in repentance the wrongful mistreatment of our planet, which is tantamount to sin before the eyes of God. We reaffirm our responsibility and obligation to foster a sense of humility and moderation so that all may feel the need to respect creation and to safeguard it with care. Together, we pledge our commit-

---

12. Revelation 21:1.

ment to raising awareness about the stewardship of creation; we appeal to all people of goodwill to consider ways of living less wastefully and more frugally, manifesting less greed and more generosity for the protection of God's world and the benefit of His people.[13]

The natural environment—the forest, the water, the land—belongs not only to the present generation but also to future generations. We must frankly admit that humankind is entitled to something better than what we see around us. We and, much more, our children and future generations are entitled to a better and brighter world, a world free from degradation, violence, and bloodshed, a world of generosity and love. It is selfless and sacrificial love for our children that will show us the path that we must follow into the future.

---

13. *Common Declaration of Pope Francis and the Ecumenical Patriarch Bartholomew I*, signed May 25, 2014. Full text can be found at https://w2.vatican.va/content/francesco/en/speeches/2014/may/documents/papa-francesco_20140525_terra-santa-dichiarazione-congiunta.html.

PART I

Worldviews Shaping Journey of the Universe

❧

Thomas Berry and the New Story

✺

1

Christianity and Journey of the Universe

Heather Eaton

The deepest crises experienced by any society are those moments of change when the story becomes inadequate for meeting the survival demands of a present situation.[1]

Journey of the Universe offers a vital and compelling entry point into how we can understand the universe, Earth dynamics, and our evolution within the Earth's intimate and intricate processes, and how we might go forward as an Earth community. The importance of the project for religious traditions and spiritual sensibilities cannot be stressed enough. It is the first time in human history that we have this kind of knowledge about the immense and intimate realities in which we are embedded, and from which we can glean guidance for the short- and long-term future of the Earth community. The project of *Journey of the Universe* includes the film and accompanying book, as well as a twenty-part educational series of conversations with scientists, environmentalists, historians, and educators who explore the significance of the unfolding story of the universe and Earth and the role of the human in responding to our present challenges.[2] The contribution this project can make is far-reaching.

---

1. Thomas Berry, *Dream of the Earth* (Berkeley, CA: Counterpoint, 2015), xi.
2. The website http://www.journeyoftheuniverse.org/ provides excellent resources to appreciate the full range of this project.

Thomas Berry's insights and influences are carried forward in the *Journey of the Universe* project. This essay begins with a synopsis of what Berry meant by the need for a "new" story, which must begin with the birth of the universe. The subsequent sections explore ways in which Christianity can engage with the larger project of *Journey of the Universe*.

Thomas Berry

Thomas Berry's work is immense in scope. It requires pondering to grasp the full implications of his vision. His intellectual inquiry had breadth and depth, with an astonishing capacity to integrate knowledge. Berry's contribution is a key influence to the orientation of and perspective within the *Journey of the Universe* project.

Berry perceived that the greatest need, in light of the ecological crisis, was to foster a sense of a new story of the universe that provides a religious and cultural orientation. The new knowledge of the universe is the context for understanding the diversity and unity of religions, as well as what role religions can play in the current era. If religions could situate themselves within this Universe Story, this would become a macrophase period of development of most religious traditions. According to Berry, the traditions would be understood as dimensions of each other, of human symbolic consciousness, of Earth processes, and ultimately the universe reflecting back upon itself. It will require a process of a transformation of their deep spiritual insights. Berry returns to the necessity of the larger context, the emergent universe, as the primary reality within which the religious traditions can situate themselves, telling their religious story in the particular cultural context and narrating from the macrophase cosmological story.

The New Story

In 1978 Berry published "The New Story."[3] This essay incorporated a skeletal version of what took him decades to understand and weave together. It is quite short, had no footnotes or references, and seems straightforward. If one is aware of Berry's erudite history, then one begins to see the layers of learning and meaning that are not initially evident. "The New Story" has a lyric style. It is evocative, meaning it is appealing to mythic receptors: our dreams and

---

3. Thomas Berry, "The New Story," *Teilhard Studies* No. 1 (Winter 1978). Also published in Berry, *Dream of the Earth*.

desires, and our need for meaning and ultimacy. This essay indicates what became typical of Berry's style. It has breadth, depth, elegance, and clarity. It is incisive and profound.

Berry understood many dimensions and meanings to the necessity of "stories." He studied the need for a story about how the world came to be, how such stories function to give us an account of the world and where we fit. These stories orient us, shape our sentiments, and offer emotional, aesthetic, and spiritual fulfillment. They consecrate suffering, integrate knowledge, and provide a moral compass. In short, they allow life to function in a meaningful manner. At first blush this seems reasonable, even simple. Berry understood that the current narratives, worldviews, and cultural visions are not equipped to deal with the ecological crisis in all its dimensions. Berry describes the pathos and dysfunction connected to the current cultural orientation and story.

Berry considered that religion and science are both at an impasse—individually and together. Their sectarian horizons of interpretation are too constricted to see the inner impasse. Berry saw the need for an authentic interaction between religion and science, something the *Journey of the Universe* project invites and provides.

Berry was attentive to modes of knowing and depths of perception, realizing that the dream drives the action. He repeatedly demonstrated that cultures need orientation and functional stories; otherwise, communities are unable to derive visions or values. No community can live without a unifying story that is comprehensive and commensurate with contemporary knowledge. When stories and communities are fragmented, social and ecological problems are not resolved. This is the problem of a dysfunctional story, for Berry, because this impasse is embedded into all facts of Euro-Western culture. We cannot perceive an adequate orientation toward the planetary demands of the present. Religion and science need to collaborate to perceive the psychic-spiritual dimensions intimately interwoven in the physical-material. In order to respond to current challenges, we need to appreciate the magnitude and magnificence of existence. For Berry, the remedy is the New Story.

In "The New Story," Berry outlines his proposal of an integral vision, from the new context of seeing the emerging creative processes of the universe to the transmission of human values. He infers that this story gives a structure of knowledge, for all knowledge systems, and is a sacred story, for all religious stories. The traditional context of "story" is the best form to understand this moment in human history, as well as orient us for the future. The content of the story is the universe as an evolutionary and creative process, and is what can provide an overall orientation for a viable future. If we can discover our role in the larger evolutionary processes, there is hope. This is the impetus of the *Journey of the Universe* project. It is a unique and creative continuation of Thomas Berry's contribution.

Christianity and the *Journey of the Universe*

Any interpretation of Christianity is ambiguous at best, given the multiple historical and existing traditions, the different conceptions of the relationship between religion and culture, and the diverse emphases on the Bible, dogma, tradition, context, and contemporary concerns. For those who study and interpret Christianity, there are distinct starting points, hermeneutics, and ethical commitments. Thus, my comments are broad-spectrum, being applicable to some, but not all, contexts, denotations, and expressions of Christianity. The reflections are divided into two main sections. The first is about affinities and the second is about challenges and insights.

Part One: Affinities

Religion and Science Interchange

Journey of the Universe is rooted in science and offers essential explanations of what is now known about the universe, the Earth, and the biosphere. The tenor of the film and book helps us to see the relevance of this new knowledge for all cultures of the globe, especially in the context of worldwide ecological stresses.

Religions orient us to the larger dimensions of our lives. They offer a wide horizon or macro vision of reality as we know it. Prior to the scientific revolution, Christianity and science were allied, mutually influencing a quest to comprehend both the breadth and depth of existence. A cosmological horizon was integrated into Christian themes and provided a blueprint from which to derive an orientation to life's exigencies. The universe was relevant to religious reflections. For many Christian thinkers, including Hildegard of Bingen and Thomas Aquinas, the Earth and the entire cosmos were alive with the presence of the divine.

In general terms, from the fourteenth to the eighteenth centuries, the European worldview changed dramatically. Science, as a new mode of knowing, was rising. Galileo confirmed that the Earth is not the center of the universe. The Earth, once alive, was now seen as mechanistic, passive, and void of spirit. This cosmological horizon, as well as a reflective interchange with the natural sciences, was dropped from Christian consciousness.

The quest for understanding through science expanded. The tools were refined. Christianity and science no longer collaborated, and the former became ill equipped to address the advances of science. Christianity, overall, avoided contending with major scientific discoveries, such as evolution. Few realize how far Christian preoccupations have strayed from cosmological and Earth sciences, and the loss that that has entailed. The foremost Christian preoccupations narrowed to morality, and the dominant religiosity became human-centered, disembodied, and otherworldly. Further, it has led to a focus on sin, salvation, Christology, and biblical inerrancy, accompanied often by a rigid dogmatism.

Journey of the Universe reinvigorates a conversation among Christianity, cosmology, and evolution. The results are twofold. *Journey* tells us what the best science reveals of the universe, the Earth, the development and expansion of life, and of ourselves within these dynamic and ingenious processes. The film and book enlarge the horizon of understanding. It is a new moment of understanding to which Christianity—in fact, all of humanity—is invited to participate. From here, the tasks of religion and the Christian traditions can begin anew the search for a valid interpretation of, and orientation within, the limits of the phenomenal realm.

Second, *Journey* reveals that the universe, the Earth, and the biosphere are utterly astonishing and have a previously unknown macro and micro complexity. The macro level involves the configuration of galaxies and solar systems, the nature of gravity and dark matter, and the formation of planets and the Earth dynamics from a whole systems theory, including plate tectonics and volcanic activity and global ocean currents. The micro level includes the mysterious behavior of atomic particles, quarks, and an astonishing array of modes of attraction that range from the molecular level to the atmospheric. The film's discussion of Earth's atmospheric adaptation is a good example of micro and macro processes operating in some form of synchronicity. These interdependent processes allow us to interpret life-emergent capacities.

Journey of the Universe is illuminating and evocative. It opens our eyes to dimensions of reality not previously known. The film is revelatory. It affirms the numinous creativity that envelops us. The images and insights evoke an irrepressible sense of wonder and awe. Wonder is an authentic response to the elegant universe of which we are citizens. Wonder is also the most intimate heart of religious experience.

Herein lies an intriguing affinity between *Journey* and Christianity, and between scientific insights and religious sensibilities. *Journey* orients us toward a profound recovery of wonder and awe. As are most religions, Christianity is aware that "wonder" offers insights and guidance, and is a decisive part of human intelligence. Wonder can galvanize human energy, imagination, wisdom, insight, and spiritual receptivity.

A Creative Universe and Creation Stories

All animals require sophisticated navigation tools to survive. Canines, for example, make olfactory maps to navigate and mark territories and improve survival strategies. Humans, consistently and cross-culturally, interpret and codify our experiences within a time frame, and often in narratives. Humans represent the multiple aspects of life—personal, communal, historical—with elaborate symbolic configurations and stories. This mythmaking aspect of human functioning, which includes religious processes, is the most potent, creative, and fertile realm of human activity. Through this capacity, humans not only make meaning but also develop narratives saturated with symbols, values, ethics, and priorities within which communities navigate all dimensions of existence. Life has meaning, contours, and parameters inside of our stories. Outside the stories, there is no context in which human life can function in meaningful ways. Religions customarily provide these interpretative patterns to existence in a narrative format.

Further, we create stories or accounts of how the world came to be and how we fit into the grand scheme of things. These creation stories guide and shape our personal and collective life purposes, actions, and interactions, and offer ethical orientations. Most religions express some version of a creation or origin story: a manner of grappling with the larger horizons of time and space, and accounting for "all this reality." Creation stories can include prescripts about how to live: appropriate interactions between humans and other animals, acceptable relations among humans, and how to greet death. Some have a focus on morality or the consequences of improper living. Such stories carry a salvific purpose. There are many variations. At their base, the quest to discern our origins and purpose persists.

Christianity has both a doctrine of creation and creation accounts. The classical doctrine of creation, *creatio ex nihilo*, is a philosophical discussion about the nature, powers, and freedom of God (omnipotent, omniscient, omnipresent, omnibenevolent). It is not about "creation." The Bible offers two creation narratives in Genesis: the first about the process of creation and the second about the place of humans in the scheme of creation. The interpretations vary extensively, as do the consequences. Christianity presents one out of countless origin stories.

These creation stories do not signify science. Within the spectrum of Christianity today, some assume these biblical narratives represent science. Some believe that anthropocentric fall-redemption theologies are absolute and ultimate. People who hold these views are suspicious of evolution. They are resistant to learning what science teaches of the universe, Earth, and the biosphere, as well as our development from within these cosmic expansions. Those with examined notions of the nature of religion and scriptural texts readily embrace the new insights from cosmological and evolutionary sciences and allow these to influence their understanding of "creation." Overall, the significance of creation and Christian creation stories are currently in dispute.

Journey of the Universe is also an origin story, about the quest to understand the scope of reality through scientific inquiry. Twentieth-century science has learned more about the nature of the universe and Earth processes than all previous generations combined. For example, instead of considering the universe and the Earth to be mechanistic, static, and passive, the stories are best understood as organic, dynamic, and active. "Reality" is more of a verb than a noun—a journey more than a place. Processes are interconnected in unimaginable ways. The atomic, molecular, galactic, planetary, and biotic activities of emergent complexity are unmistakable, even if not fully apparent. Furthermore, the parts and the whole are reciprocally interacting. The line between life and inert matter is inexact, as are many of the demarcations we accentuate between humans and other animals, and between plants and animals. The *Journey* film and book stress that from the origins of the universe to the present, these dynamics and processes, including ourselves, are symbiotic, interconnected, interrelated, and interembedded in each other. This is a key insight.

Journey of the Universe tells "a story," chronologically, of the origins and development of the universe. We, too, are in the story. This is the creation and origin story that best represents what scientists know of these realities. Again, there is an affinity in that both Christianity and *Journey of the Universe* represent forms of creation stories. The film and book invite Christians to expand and align their understanding of "creation" with a scientific appreciation. This includes a challenge of how to best understand the meaning and purpose of such "stories."

Role and Responsibility of Humanity

A third affinity between Christianity and *Journey* is the significance of humanity within the Earth community. Christianity affirms that humanity is in the image of God, and is particularly relevant to the spiritual expression of creation. This significance can be interpreted in a narrow sense, wherein humanity is of greatest relevance, or in a broad sense, wherein humanity is one incarnation of divine presence, among others. Either way, humanity is significant. The evolutionary perspective of *Journey* reveals that the particularity of humans, among mammals, is an extended childhood, a supple imagination, and a symbolic consciousness. These allow for an adventuresome spirit, an ability to extend our awareness, a capacity for intense intrapsychic experiences, and the ability to navigate the world within rich symbolic images that inform, enrich, and deepen our involvement with life. From *Journey* we can deduce that this form of self-consciousness evolved from Earth processes. Thus, humanity, as a species, is a form of self-consciousness of the Earth.

From a different angle, the complex capacities of the human come with a commensurate responsibility. Given the anthropogenic causes of the current ecological stress, even crisis, it is evident that humanity must play a decisive role in any ecological recovery or sustainable future. *Journey* shows well that the ecological crisis is an unintended consequence of the expansion of human imagination, capacities, and impact. We thus must become conscious that the Earth is our origin, our home, and where we belong. We must therefore become responsible citizens within this community of life.

The stream within Christianity that addresses the ecological crisis is explored below. In brief, it offers myriad metaphors of stewardship, Earth-keeping, Earth ethics, ecojustice, animal rights, and ecofeminism to remind people of their responsibilities to the biosphere. Ecological Christianities can be anthropocentric, Earth-centric, or cosmologically informed. They can speak of a sacred Gaia or be within a traditional fall-redemption theological paradigm. Any form of ecologically sensitive Christianity, in spite of major theological differences, stresses that humanity has an obligation to sustain the well-being of the planet and reduce our ecological footprint.

Thus, both *Journey* and the Christian tradition recognize that humanity has a particular status within the biosphere and must assume some responsibility for the ecological crisis and its potential recovery.

Part Two: Challenges and Insights

The *Journey of the Universe* project has brought forth new knowledge. We are only beginning to apprehend the evidence and have not yet discerned the significance or realized the implications. The challenges are immense. It requires awakening to a profoundly new understanding of the fundamental dynamics of reality. It also indicates that to perceive what is essential and vital about humanity, we must appreciate the cosmological context and our multilayered integration with the cosmic processes as well as the Earth's self-organizing and life-giving systems. A second theme of *Journey* is that to comprehend the magnitude of the ecological crisis and to respond with competence, the cosmic, Earth, human, and ecological contexts need to be integrated and understood as parts of a cohesive whole.

Christianity and Ecological Issues

Christianity has been grappling with the ecological crisis for several decades in multiple ways, referred to as "ecotheology." It became clear early on that certain aspects of Christianity are implicated in ecological ruin. There is a correlation between Christian orientations toward dominion and ecological exploitation and degradation. Christian-influenced cultures are entrenched in excess economic modes of production, extreme consumption, and climate emissions. In tandem, Christian churches have done little to restrain deforestation, mining, species extinction, water contamination, and so on. As a result, ecotheology has become a comprehensive reform of Christianity, as well as a new expression.

Christian ecotheology is a multipronged endeavor. The reform part examines aspects of Christianity that neglect contemporary sciences and ecological decline in favor of shrinking horizons and narrow intellectual sources. Ecotheologians associate the normative interpretations of doctrines, biblical texts, revelations, and beliefs with a fundamentalism ubiquitous across Christian theologies. This narrows the religious horizon, perception, and preoccupations. The persistent promise of a better world elsewhere is objectionable, especially when joined with ideas that salvation is either an escape from this Earth or limited to morality. These together have made Christianity defensive, incapacitated, and insular in the face of the massive social, ecological, and global upheavals.

A second part of ecotheology is rejuvenating the tradition. It is focused on retrieving ecological wisdom, reinterpreting teachings and doctrines, and reconstructing spiritual sensibilities. Extraordinary and extensive work is being done all over the world on identifying, expanding, and engaging symbolic, scriptural, and ethical dimensions. Such efforts focus on reinterpreting the tradition and addressing ecological issues and their connection with economic, social, or political concerns. Christians join with others working on ecojustice, ecofeminism, environmental racism, biofuels, GMOs, and food insecurity. The goals are to revive Christianity to enable engagement with the ecological crisis and to strengthen its cultural influence. Such efforts are expanding and involve scholars, religious leaders, and adherents in main-stream Christianity.

A third ecotheology prong comes from those who are grounded in the same insights as are found in *Journey of the Universe*. They are more interested in creation than classical notions of sin and salvation. The work of Thomas Berry, a cultural historian and Catholic priest, is an important inspiration to both projects. Here is the closest alignment between *Journey* and Christianity. Much can be said about Berry and his predecessors, such as Pierre Teilhard de Chardin, Thomas Aquinas, and others who share a spiritual and scientific quest to bring forth a comprehensive synthesis of what we can perceive from cosmology to human existence. Behind this synthesis is the perception that there is a fundamental and intrinsic unity to reality, and there is coherence to the whole.

Nothing about these notions of unity and coherence is facile. They are spiritual intuitions that all religions share, although with distinct explanations. These are also key insights in the scientific discoveries presented in *Journey of the Universe*.

From *Journey* it is obvious that such a fundamental unity is claimed only with a demonstration of countless differentiations and precisions, yet the processes of the universe undeniably gave birth to galaxies, planets, this Earth, and life, here at least. It is equally incontrovertible that a synergetic web of life exists: interconnecting biosystems that enhance the emergence of increasingly complex life that then interrelates in more intense and astonishing ways. Thus, the discernment of unity and coherence—of differentiation yet intimate interaction—is foundational to the message of *Journey of the Universe* and of insightful and perceptive Christian leaders.

Scientific and Spiritual Knowledge

A cluster of Christian thinkers and theologians have never shied away from the natural sciences. This is not only a "dialogue with science." Science and theology are not two knowledge systems, as is often expressed. For these people, errors of reasoning about the natural world lead directly to errors concerning the divine world. To understand anything about larger contexts and processes requires a comprehensive integration of science and theology. Science and theology address one phenomenon because of this fundamental unity with reality. The vastness of reality is best described as an interactive weave of psychic, spiritual, and material phenomenon.

For Jesuit priest and paleontologist Pierre Teilhard de Chardin, and now countless more ecotheologians, evolution is not a speculative theory among other theories of creation, or an interesting exchange, or to be debated with the Bible as a parallel mode of revelation. Evolution is the primary context out of which everything, including theology, emerges, and to which it needs to refer. For Teilhard, "Evolution is a general condition to which all theories, all hypotheses, all systems must bow and which they must satisfy hencefor-ward if they are to be thinkable and true. Evolution is a light illuminating all facts."[4]

This is also a key insight within *Journey*, an insight that is only beginning to enter human consciousness in either science or Christianity. Humanity, the Earth, and the universe are part of one reality, immersed in the same pro-cess, the same creative energies, the same evolutionary dynamics and orienta-tion. The notions of deep time, process, emergent complexity, Earth dynam-ics, and interrelatedness are only beginning to have an impact in theological circles. They offer a new sensibility to Christianity, opening the possibility of deciphering a sacred or holy dimension within the dynamics and structures of the universe and the Earth. It provides a context out of which to reflect theologically. Such a view also has theological implications. It means that the immersion into the deep creative powers of the universe is the most direct contact a human can have with the divine.

This is the greatest challenge between the insights of *Journey* and Chris-tianity. Much of Christianity is far from this comprehension. Christian churches often focus on rituals. Mostly, theology is preoccupied with doc-trine and biblical interpretation. Other forms of Christian engagement are about recruiting converts or tackling social issues. It takes time for these sci-entifically informed Christian views to be understood, as they cannot be con-tained in the usual preoccupations. It is a different foundation and scaffold for theology, and for Christianity.

Influential academic theologians are engaging with this cosmic, evolu-tionary view, often inspired by Thomas Berry. This group includes liberation theologians such as Charles Birch, Leonardo Boff, and Ivone Gebara; system-atic theologians such as John Haught, Elizabeth Johnson, Sallie McFague, and Ilia Delio; process thinkers such as Jay McDaniel and John Cobb; ethi-cists such as Barbara Holmes and Larry Rasmussen; biblical scholars such as Norman Habel; feminists such as Ursula King, Rosemary Radford Ruether, and myself; and many more. These works may differ from Berry's in some aspects, but they have interacted with central ideas or insights.

---

4. Pierre Teilhard de Chardin, *The Phenomenon of Man*, trans. Sarah Weber (New York: Harp-er, 1959), 219. Now available as Pierre Teilhard de Chardin, *The Human Phenomenon*, trans. Sarah Appleton-Weber (East Sussex, UK: Sussex Academic Press, 1999).

That Christianity needs to learn from *Journey of the Universe* is also evident in the efforts of popular ecotheology authors, including Diarmuid Ó'Murchú, Cletus Wessel, and Miriam McGillis, to name a few.

Symbolic Consciousness, Religions, and Christianity

Another insight from the *Journey of the Universe* project that challenges Christianity is in how to understand the nature of religion itself. *Journey* tracks the emergence of consciousness and self and symbolic consciousness. The dynamics between such consciousness and symbolic expression involves self-amplifying loops that enhance and intensify human experiences. Consciousness births symbols that amplify consciousness and then bring forth new symbols. These symbols shape our grasp of reality, our depth of experience, and our actions. Religions are borne of the amplifying energy of symbolic consciousness.

Further, human presence is so dominant now that the Earth is being shaped by human consciousness and our symbolic images, and it is not constructive. Christianity has played a role here, with its lack of attention to a relationship between the Earth and religion, science and spirit. *Journey* reveals that religious symbolism is a powerful form of symbolic engagement with the world and thus has vital importance. Our images of God, of Christ, and of the relationships between a divine presence and the Earth have consequences. Furthermore, to reconsider religion in an ecological context requires an in-depth appreciation of symbolic consciousness and the consequences of symbolic activity. For that, one needs to study the genesis and expansion of religions. Here again, Christianity is challenged to become informed about the history of religions and the breadth and diversity of religions. Overall, Christianity has assumed sovereignty and remained insulated from a basic awareness. Few engage in any renovating interreligious exchanges. Most Christians do not have an adequate theory of religion or a suitable intellectual grasp on the experiential bases, symbolic and mythic nature, and cultural scope of religion(s).

Conclusion

A final comment pertains to an implicit challenge that the *Journey of the Universe* project poses to religions. We can, and should, be dazzled by the ingenuity of the cosmic and Earth processes. We can perceive dimensions of reality never previously appreciated. We can understand ourselves as part of this immense, epic journey. Wonder and awe are appropriate religious responses. We can further deduce that all disciplines need to refer to these new insights for greater self-understanding, and that humanity is invited into a new orientation: a New Story. This changes our consciousness forever.

In addition, we are entering a most difficult ecological era where much is at stake. The demand for transformation of cultural practices, ideas, symbols, and meaning is massive. It becomes clear that we require spiritual and ethical change. Religions need to be transformed and transformative and assist in the work of knowing ourselves to be part of, embedded in, and participants in this great journey.

While the best of religious ideals are rarely embodied, religions can be movements of liberation, moral cohesiveness, and life-affirming orientations. The *Journey of the Universe* project is a concise and profound appeal to engage with the breadth, depth, and intricacies of numerous aspects of reality. It carries the challenge of a dual awakening to a devastating ecological crisis and to an alluring, yet still unfathomable, universe. It is a call to Christianity to awaken to both.

2

The New Story and Journey of the Universe *as* Habitus

The Power of Ecopoetics

ANNE MARIE DALTON

It was an early afternoon in May when I first wandered down the incline, crossed the creek and looked over the scene. The field was covered with white lilies rising above the thick grass. A magic moment, this experience gave me something that seems to explain my life at a . . . profound level. It was not only the lilies. It was the singing of the crickets and the woodlands in the distance and the clouds in a clear sky. . . . This early experience has become normative for me throughout the entire range of my thinking.[1]

Do poetry, narrative, visual images, and other forms of art evoke, provoke, inspire, or motivate human action? More precisely, are we more likely to act in ecologically responsible ways if we experience the world through artistic expression? Do the tellers of Earth stories, the makers of films about the cosmos, the artists for whom ecological devastation is their subject matter, the poets who decry the present consumerist culture, the photographers who celebrate in image the majesty of creation do so in vain or do they make a difference? At least theoretic evidence shows that they are likely to effect change. Certainly cultural ecologists such as Thomas Berry were convinced that they are a critical component of change, that they are necessary for the reinhabiting of the Earth in more ecologically sustainable ways.

This chapter briefly examines a number of contemporary conversations around the notion of *habitus,* the field of ecopoetics, and the investigation of mirror neurons, all of which together support Berry's view that stories,

---

1. Thomas Berry, *The Great Work: Our Way into the Future* (New York: Bell Tower, 2000), 12.

poems, and images inform human action and can also bring about change. I argue that Berry's proposals for a new story of the universe, the articulation of that story in Brian Swimme and Berry's *The Universe Story*, and the film *Journey of the Universe* contribute to *habitus* in the manner in which ecopoetics does, in general.[2] Recent work on mirror neurons indicates a fluidity in *habitus* and supports further the role of ecopoetics in effecting change in human practice.

What is *habitus*? The word "habitus" received its first systematic treatment in the work of Thomas Aquinas. *Habitus* referred to the "state of possessing or having" (root *habere*) an acquired, trained "disposition" to engage in certain modes of activity when encountering particular objects or situations. For Aquinas, *habitus* grounded the propensity to act morally or to be a virtuous person.[3] Aquinas based his account of *habitus* on Aristotle's treatment of *hexis*, also carrying the meaning of having or possessing. In both cases, *habitus* and *hexis* include the sense of being in a state of possessing something while that state is not understood as passive. Rather, Aristotle and Aquinas both use the words to denote more than a disposition with the meaning we attribute to that word today. It includes both a connection to the nature of a person and the active way in which humans choose to express their character through emotion and practices. Hence, *habitus* carries with it the idea that one's *habitus* can be affected by the external environment, through socialization, but also in the active engagement of that environment. The difference in the sense in which *habitus* was originally understood compared to the modern, post-Kantian sense of virtue or habit (as it is commonly translated) is that modern usage connotes a repetitive, passive (with respect to reflection), and an almost complete reliance on training as the source of habits or virtues. The traditional sense carries both a connection to nature (in one's natural capacities and dispositions) and a vitality related to the effects of socialization and of free will. Thus, for Aristotle, temperance, fortitude, justice, and prudence are *hexeis*, which are not equivalent to the modern rendering of "habits." Likewise, for Aquinas, such virtues are *habitus*.[4]

2. Thomas Berry, "The New Story" in *Dream of the Earth* (Berkeley, CA: Counterpoint, 2015); Brian Swimme and Thomas Berry, *The Universe Story* (San Francisco: HarperSanFrancisco, 1992); Brian Swimme and Mary Evelyn Tucker, *Journey of the Universe*, DVD directed by David Kennard and Patsy Northcutt (2011; New York: Shelter Island, 2013); and Brian Swimme and Mary Evelyn Tucker, *Journey of the Universe* (New Haven, CT: Yale University Press, 2011).

3. Omar Lizardo, "*Habitus*," Notre Dame University, 2012, http://www3.nd.edu/~olizardo/papers/habitus-entry.pdf.

4. Cf. Joseph Torchia, "The Limits of Virtuous Action: Aquinas on the Subjective Conditions of Moral Virtue," *Angelicum* 79 (2002): 855–77; Clare Carlisle, "The Ques-

The term *habitus* itself has been used in modern times by social scientists to refer primarily to the process of adaptation. The work of Pierre Bourdieu has expanded the meaning of *habitus* to include not only the arbitrary conditions of the world in which one experiences reality, adapts to it, and so acts in accordance, but also to the nonarbitrary conditions into which one is born. In fact, any contemporary discussion of the term *habitus* includes some response to Bourdieu's account. The frequent critique of Bourdieu's account of *habitus* relates to his view that the *habitus* is virtually unchangeable in the adult. Born into a certain milieu, socialized by family and immediate community, one lives that *habitus* for life. Even if one manages to change the milieu (economic class, for example), one then becomes subject to the *habitus* of a different milieu in which one successively attunes one's own *habitus* to the new one, or doesn't. It might simply just not fit; hence the person fails in the new milieu. In critiquing Bourdieu, several scholars point out that, for him, the agency of the person is confined to the *habitus*. In other words, contingency and conscious reflection do not figure substantially in one's behavior. Thus, Dave Elder-Vass and Anthony King argue for a more inclusive account of *habitus* cognizant of the ever-changing milieus in which humans act throughout their lives.

Recent work on mirror neurons, those primal neurons that seem to be the prelingual (both in time and precedence) access to learning, has further contributed to the consideration of how a *habitus* is created and expressed in human practice.[5] Mirror neurons work in encounters. The simplest example is the way infants imitate those around them and give back their version of the smile, the frown, and so on. Research uncovers the flexibility, agility, and changeability of the neurons themselves. They can self-organize in the process of encounter and interaction as we engage with the world. So there is a spiral process of influence and expression; we are active in the reorganization of structures that we both inherit and receive from socialization. The physical structures themselves, in this case the mirror neurons, are fluid and responsive. They operate in the creation and re-creation of *habitus*. The *habitus* is, however, substantially formed by the interaction of our physical being and our experience. It can be addressed and substantially constructed by sophisticated techniques, such as those that form us into the capitalist,

---

tion of Habit in Theology and Philosophy: From Hexis to Plasticity," *Body and Society* 19, nos. 2–3 (2013): 30–57.

5. This account of mirror neurons relies on Romand Coles, "The Neuropolitical Habitus of Resonant Receptive Democracy," *Ethics in Global Politics* 4, no. 4 (2011), http://www.ethicsandglobalpolitics.net/index.php/egp/article/view/14447.

consumerist, political ideologues we habitually become. The account of *habitus* that emerges both from the philosophical arguments and neurobiological research restores to the concept the more vital and encompassing sense of usage by Aristotle and Aquinas. Romand Coles has used this revitalized sense of *habitus* by mirror neuron research to argue for a cultivation of new democratic practices across and beyond dominant ideologies.[6] In this same sense, Berry's proposal is an incursion into the habitual life of modern humanity and the creation of a new *habitus* to generate an ecological morality and practice.

Berry's proposal calls for an ecopoetic response to the ecological crisis.[7] *The Universe Story* related by Brian Swimme and Berry as well as the film *Journey of the Universe*, written by Mary Evelyn Tucker and Brian Swimme, are attempts to put that new story into language and practice.

Ecopoetics is a crossover of sorts between ecology-themed poetry and literary criticism.[8] It belongs to the postmodern focus on textuality—the significance and life of a text on its own. Ecopoetics asks the question, What is it that is addressed and created by the text itself? However, the strictly postmodern literary critic focuses on the text alone as real, based on the assumption that all reality is radically constructed. Ecopoetics, on the other hand, has the proviso that "nature is simultaneously real and constructed," in the words of Bruno Latour in his book *We Have Never Been Modern*.[9] Scott Knickerbocker comments on Emily Dickinson's poem "A Bird came down the Walk." The bird is domesticated, he claims, by Dickinson's language and familiar metaphors in the beginning of the poem, but repels our understanding in the latter part of the poem, a sense captured by the metaphors of distancing.[10] We relate to the otherness of nature in this dialectic of familiarity or intimacy and the mysterious otherness. There is something that escapes our domestication, our construction. David Abram expresses this sense, similarly, in *The Spell of the Sensuous*, when he writes, "It is the inanimate earth

---

6. Ibid.

7. This is not a claim that Berry recognizes only ecopoetic response, but that his major contribution, the proposal of the New Story, is an ecopoetic response.

8. Introductions to ecopoetics can be found in several volumes, including those by Jed Rasula and Scott Knickerbocker referenced below.

9. Bruno Latour, *We Have Never Been Modern*, trans. Katherine Porter (Cambridge, MA: Harvard University Press, 1993), 7, quoted in Scott Knickerbocker, *Ecopoetics: The Language of Nature, the Nature of Language* (Amherst: University of Massachusetts Press, 2012), 9.

10. Knickerbocker, *Ecopoetics*, 10–13.

that speaks: Human speech is but a part of that vaster discourse."[11] Likewise, Gary Snyder concurs on the role of language as a participator in a larger natural phenomenon when he refers to his poetry as "a practice of the wild."[12]

So ecopoetics claims more than merely making nice language or talking about nature. It claims to represent something (nature) that is both constructed by, as well as constructing, humans. Nature has a being "other than" that which we create. Yet we are nature also. Nature is, as Patrick Murphy has contended, not really "other" but "ANother" to us.[13] Ecopoetics challenges the notion that the map precedes the territory as the postmodern critic and public commentator Jean Baudrillard described our relationship to the world.[14]

Furthermore, ecopoetics has an ethical edge, not in an overt preaching sense, but in the way in which language is understood to operate and is then intentionally employed. It asks from an environmental perspective: Where does language fit in? Can language weave us into nature? Can it open our hearts to the "other" that is the natural world? Or to use Berry's way of describing an appropriate relationship to the other: "The deepest values and most profound commitments of one seem to be challenged by the other. At this moment everything depends on mutual confidence and hospitality."[15] He was referring to the meeting of cultures, but think for a moment of the natural world that became the sublime other in his later writings. "The ultimate goal of any renewal process," he writes in *Evening Thoughts*, "must be to establish a mutually enhancing mode of human-Earth relations."[16]

In its ethical dimensions, ecopoetics examines the function of language as a vehicle of transformation. As Knickerbocker explains, environmental

---

11. David Abram, *The Spell of the Sensuous: Perception and Language in a More-than-Human World* (New York: Random House, 1996), 179, quoted in Knickerbocker, *Ecopoetics*, 2.

12. Gary Synder, *A Practice of the Wild* (Berkeley, CA: Counterpoint, 1990), quoted in Jed Rasula, *This Compost: Ecological Imperatives in American Poquetry* (Athens: University of Georgia Press, 2002), 7.

13. Patrick Murphy, "Anotherness and Inhabitation in Recent Multicultural American Literature," in Richard Kerridge and Neil Sammells, eds., *Writing the Environment: Ecocriticism and Literature* (London: Zed Books, 1998), 40–51.

14. Richard Kerridge, "Small Rooms and the Ecosystem: Environmentalism and De-Lillo's *White Noise*," in Kerridge and Sammells, *Writing the Environment*, 183. Reference is to Jean Baudrillard, "Simulacra and Simulations," in *Selected Writings*, ed. M. Poster (Cambridge, UK: Polity Press, 1988), 166.

15. Berry, "Education in a Multicultural World," in *Approaches to the Oriental Classics*, ed. Theodore deBary (New York: Columbia University Press, 1959), 22.

16. Berry, *Evening Thoughts: Reflecting on Earth as Sacred Community*, ed. Mary Evelyn Tucker (Berkeley, CA: Counterpoint, 2015), 22.

poets see themselves as curbing the excesses of poststructuralism (hermetically sealed textuality) that ignores the connection to the social, cultural, and physical environment. Rasula agrees, citing Neil Evernden's words, "Environmentalism without poetry is merely regional planning."[17] Ecopoetic language or figuration (meaning poetry, metaphors, stories, images, and so on) calls on the power of the imagination to re-create our manner of being in the world. Hence, it creates *habitus.* In its response, ecopoetic language reacts to established modes of living in a techno-scientific world in which the understanding of the humanities is confined to practices of leisure. In a manner akin to feminism, ecopoetic language challenges the dominant ideology of dividing the world according to dualistic categories. Poetics is not separate from science or politics any more than nature can be separated from human life, or your backyard can be immune to pollution.[18] Martin Heidegger compared poetry to the act of dwelling because he claimed poetics creates presence, a form of being, not a representation such as a map.[19]

To understand adequately Berry's proposal to change human consciousness with regard to the natural world, one has to look to his own performance in the use of language. To say that Berry's work has a resonance with contemporary work in ecopoetics is not to claim that he is writing poems (although he did write a few) or to say that all of his writing fits this designation.[20] Poetics, in general, and ecopoetics in particular deal with a wide notion of poetic language. The quotation with which I began this paper is a sample of poetic writing. There is a certain cadence to the text and an overflow of language, the use of language unnecessary to the literal meaning; there is a sense of intimacy and mystery at once. It plays with imagination and invites one in. Berry does not write merely, "The inspiration for my life work arose from the destruction of the natural world by the rise of industry." Rather, he wandered (one pauses to dwell on this word) into the field, noticing the creek as he went, saw not flowers or plants in general but lilies arising above the thick grass, heard not just insects but crickets; it was not merely a clear day—there was a blue sky with drifting white clouds.

The manner in which we engage this text is different from how we engage a simple abstracted statement. Most of Berry's texts contain such passages.

17. Neil Evernden, "Beyond Ecology: Self, Place and the Pathetic Fallacy," *North American Review* (December 1978): 103. Quoted in Rasula, *This Compost*, 7.

18. Kerridge and Sammells, *Writing the Environment*, 7.

19. Ibid. Reference is to Martin Heidigger, *Poetry, Language, Thought,* trans. Albert Hofstadter (New York: Harper Collins, 1971), 141–60.

20. See Berry's essay "Returning to Our Native Place," in *Dream of the Earth*, 1–5, for an example of an extended ecopoetic articulation.

Just to cite a few at random: We find in "Wonderworld as Wasteworld," "Scientists themselves are *awakening* to the wonder and the mystery of the universe, even to its numinous qualities."[21] Earlier, in 1972, in an essay titled "Traditional Religion in the Modern World," we find a dramatic contrast. The optimistic role modern humans understand themselves playing in the healing of the world, for example, is set against the tragedy of modern human despair. "This is mirrored profoundly," he writes, "in the theatre and literature of the absurd where we find the revelation of the human as a despicable reality, ignoble even in his highest aspirations, a disoriented, deteriorated being."[22]

Berry's program for ecological responsibility is weighted heavily on the power of the story. All cultures had stories, he observes, which answered the basic questions of life—where we came from, what we are doing here, and where we are going. What is important in life? But what is it about a story that can create or change a worldview? The New Story is based on the scientific evolutionary account, but as he significantly points out, it requires a certain way of telling. The scientific account needs a retelling in a new language, one that captures the imagination and promotes an intimacy with the natural world—science told in poetics. So Swimme and Berry write, "The power at the beginning of time drew forth a universe unfurling vast regions of space within itself." Later they reflect, "The power shaping life is a wild energy; an inner urgency to pursue a particular path in life."[23] Clearly the intent is not merely to explain the science.

The New Story, the Universe Story, and *Journey of the Universe* are intended to give a functional meaning to lifestyle change, systems change, religious change, and personal change in addressing the ecological crisis. We contemplate *Journey of the Universe* to find the contextual meaning to energize our efforts, save the trees in our cities, challenge the waste of natural resources, march in support of sustainable systems, and so on. We also, however, contemplate *Journey of the Universe* in all the ways that are expressed as something akin to a sacred ritual or an exercise in contemplation, a conversion or transformation of consciousness. In turn this change of consciousness ought to incise, cut its way into, how we live and what we fight for. That is why the exercises in ecopoetics participate in *habitus*.

In a Foreword to Dorothy McLean and Kathleen Thornod Carr's *To Honor the Earth*, Berry writes,

---

21. Berry, "Wonderworld as Wasteworld: The Earth in Deficit," *Cross Currents* 35 (Winter 1985–1986): 421.

22. Berry, "Traditional Religion in the Modern World," *Cross Currents* 22 (Spring 1972): 136.

23. Swimme and Berry, *The Universe Story*, 32, 136.

Even though we foster ecological and environmental movements throughout the planet, even though we seek to save the rainforests and to renew the regions we have devastated, none of this will ultimately succeed unless it expresses a true intimacy with this larger Earth community. Such intimacy requires an awareness of the unique aspects of each region of the Earth. It requires a consciousness of the many varied species, and of the individuals within each species, as these speak to us from the inner depths of their reality. . . . Every single life form has its own personality, its own voice, its own spirit reality. Each communicates its unique mystery that we never quite comprehend.[24]

Recall Knickerbocker's comment above on Emily Dickenson's poem "A Bird came down the Walk." The bird is both domesticated by our language or made intimate in Berry's terms and yet never totally comprehended. In another instance of ecopoetics, the film *Journey of the Universe* moves from images of the magnificent and vast cosmos to images of individual creatures and to intimate images of markets and folk dancers. In the end, we are all stardust. The film also sets the poignancy of ecological destruction against the grandeur of cosmic evolution and the large scale of time involved in the process. The writers use the highly intellectual pursuit of knowledge about the emergence of the cosmos but translate it by word and moving image to a form intended for a wide public audience. Amid dazzling images and scientific facts about the universe, it raises significant issues designed to evoke changes of minds and hearts. Thus, the book that accompanies the film states, "Perhaps human consciousness has a much greater significance than earlier philosophers could imagine. Could it be that our deeper destiny is to bring forth a new coherence within the planet as a whole, as the human community learns to align itself with the underlying dynamics of earth's life?"[25] Such questions arise frequently in this text and are suggested in the film. They speak to the heart of Berry's mission to re-create human presence on the Earth and in the whole cosmos. If the research to date is correct, and it is convincing, ecopoetic language contributes to changes in our very physical being, our mirror neurons at least, and to a milieu with which our established *habitus* engages, reflects, and is transformed.

As was the case with Aquinas's account of *habitus*, virtuous living is for all. The practices and milieu in which *habitus* is created intend, but do not determine, outcomes in human practice. Human reflectivity engages *habi-*

---

24. Thomas Berry, "Foreword," in Dorothy McLean and Kathleen Thornod Carr, *To Honor the Earth: Reflections on Living in Harmony with Nature* (Forres, UK: Findhorn Press, 1990), ix.

25. Swimme and Tucker, *Journey of the Universe*, 66.

tus freely, yet the creation of *habitus* makes possible—and, in the light of all research to date, more likely—the intended human response. Berry referred to the creation of what this paper has argued is a new *habitus* for ecological responsibility as the Great Work of our time. Relying on an extensive intellectual exploration of the great religious cultures of East and West, Berry concluded that the Great Work of our time was to tell a new story in a new language—a poetic account of science, the story of the universe. There is no other way to capture the mysterious and numinous evolving process of the universe but in poetics, no better way to evoke a change in human consciousness intentional on reinhabiting the Earth in more responsible ways than to participate in the journey of the universe through ecopoetics. In concluding a poem about children, "It Takes a Universe," Berry wrote,

> So now we write our own
> verses, bringing
> the child and the universe
> into their mutual fulfillment.
> While the stars ring out in the heavens.[26]

In such brief form, Berry articulates here an ecopoetic rendition of his proposal for the future, tells the story in every way possible, and creates the *habitus*. In the tradition of Thomas Aquinas, he places his bets that such a new consciousness, an active state of engagement with the whole universe, will issue into virtues, habits, and intentional action for an ecologically viable future.

---

26. Thomas Berry, "It Takes a Universe" (unpublished poem, November 1996).

3

Becoming Planetary in a Planetary Crisis

Reflections on Thomas Berry
and the Metaphor of Journey

MATTHEW T. RILEY

"This book is an invitation to a journey into grandeur," write Brian Swimme and Mary Evelyn Tucker on the opening page of the *Journey of the Universe* book.[1] Humans have, they observe, a "desire to journey and to experience the depths of things."[2] But what does it mean to "journey"? Where is humanity going? Thomas Berry observed that humanity finds itself to be uprooted at this moment in the story of the universe. We are alienated and at odds with the cosmos, our home. We are *wanderers*, bodies seemingly set adrift in the vastness of the cosmos as we travel without reference to our origin or knowledge of our end. In some sense we are planetary beings—like planets, the lost wanderers of the celestial realm observed by the ancient Greeks (literally πλανήτης/*planētēs*, or "wanderers" in the original Greek language).[3]

But perhaps we are not lost. Wandering, in this sense, can be understood as purposeful wandering—as journeying. In this chapter I look into Berry's writings to explore the participatory nature of wandering and the great celebratory experience of "the depths of things"[4] that journeying can bring one toward.

---

1. Brian Thomas Swimme and Mary Evelyn Tucker, *Journey of the Universe* (New Haven, CT: Yale University Press, 2011), 1.

2. Ibid., 112.

3. I first discussed the relationship between being "planetary" and "wandering" in a paper given at a regional meeting of the American Academy of Religion in 2010: Matthew T. Riley, "A Change of Planetude: Attaining a Sense of Place in the Universe Story, Quantum Physics, and Process Philosophy," New England-Maritimes and Mid-Atlantic American Academy of Religion, Regional Meeting, New Brunswick, March 11, 2010.

4. Swimme and Tucker, *Journey*, 112.

Berry's own life, one might observe, was marked by journeys. Early in his life, when he was just a young boy, Berry's family moved to the edge of town in Greensboro, North Carolina. Then, at just eleven years old, Berry remembered how he "wandered down the incline, crossed the creek, and looked out over the scene."[5] Before him lay a field of white lilies. Beyond it was a woodland and clouds floating in a blue sky. He describes his experience in this field of lilies as a "magical moment," a moment when he was transformed and his entire life was moved to reorientation. In this act of wandering, in this small boyhood journey, Berry's "own understanding of the Great Work began."[6] Also in this place his personal journey came full circle. Some seventy years later, Berry returned to his childhood home in Greensboro. There he reflected on his life and continued with his writing.

Three Ways of Understanding "Journey": Itinerary, Travel, and Symbol

Looking beyond Berry's personal life, it is illuminating to examine Berry's use of the term "journey" in his writings. In this discussion I borrow from religious studies scholar Thomas Tweed's theory of religion to propose three ways in which Berry's use of the term "journey" can be understood: journey as *itinerary*, journey as *travel* itself, and journey as *symbol*.[7]

The first way in which Berry used the metaphor of journeying is to describe an *itinerary*, or a suggested route. In this sense, the term "journey" is used to describe a plan for travel, a proposed set of normative guidelines for action, or a plan for personal and spiritual growth. Journeys as itineraries therefore can be geographically and metaphorically directional. For instance, Berry frequently used the metaphor of the exodus to describe how humanity might transform its collective worldviews in order to journey figuratively out of the terminal Cenozoic period and into new modes of human-Earth flourishing in the Ecozoic.

Journeys, as we see, are therefore more than just descriptions of, and plans for, physical journey. One can also conceive of them in terms of guidelines for moral actions and as proposed sets of ethical principles. Berry's telling of the New Story is perhaps the most salient example of this. It is a story about

---

5. Thomas Berry, "The Meadow across the Creek," in *The Great Work: Our Way into the Future* (New York: Bell Tower, 1999), 12.

6. Ibid.

7. I draw inspiration, and borrow terminology, from Thomas Tweed's *Crossing and Dwelling: A Theory of Religion* (Cambridge, MA: Harvard University Press, 2006).

where humanity has been and where it is going. It is, in a sense, an itinerary for a great transformative journey across all scales: the physical, the spiritual, and also at the emotional, psychological, and intellectual levels. To borrow language from John Grim and Mary Evelyn Tucker's book *Ecology and Religion*, the New Story provides a mode for orienting, grounding, nurturing, and transforming.[8]

Journeys, as evidenced in Berry's writings, can be further understood in a second way: as embodied *travels*, or as movement through time and space. In regard to the journey of the universe itself, he spoke of the transformation of the cosmos on a grand scale:

> We are speaking here of the more comprehensive journey of the emergent universe, a unique and irreversible journey of galactic systems, of Earth formation, of living forms, of human community—a journey passing through a sequence of unpredictable discontinuities.[9]

But one must exercise caution when speaking of journey only in terms of temporal development. Berry noted that "for the Western world the basic symbolism is more temporal than spatial"[10] and that the evolutionary journey of the universe has largely been understood in the West within a "time developmental context."[11] But, he warned, emphasizing the temporal over the spatial has led to alienation in the Western world. Berry contrasted these temporal understandings of journeying with the spatial understanding found in Chinese philosophy and religion. Here, journey can be understood not just as a journey through space, as a traveling, but also as grounding and orienting in a physical location set meaningfully in spatial relation with the cosmos itself.

Journeys can also be understood in a third way as *symbols*, or symbolic representations. This way of understanding the metaphor of journey in Berry's writings can denote narratives of journeys and ritual enactment of travels and transformations. These can have an emotive or a spiritual character to them. Most notably he drew upon the heroic journey archetype[12] and the symbolic

---

8. John Grim and Mary Evelyn Tucker, *Ecology and Religion* (Washington, DC: Island Press, 2014).

9. Thomas Berry, "Individualism and Holism in Chinese Tradition: The Religious Cultural Context," in *Confucian Spirituality*, vol. 1, ed. Tu Weiming and Mary Evelyn Tucker (New York: Crossroad, 2003), 52.

10. Ibid.

11. Thomas Berry, "Reinventing the Human," in *Great Work*, 162.

12. Ibid., 160.

journey of the shaman as a means of discussing ways of knowing, of connecting to something larger.[13]

From the example of the symbolic journey, Berry established the need for, and efficacy of, the New Story. In his words,

> The story represents a transition in human awareness from the universe as cosmos to the universe as cosmogenesis. It represents a shift in the spiritual path from a mandala-like journey to the center of an abiding world to the great irreversible journey of the universe itself as the primary sacred journey. This journey of the universe is the journey of each individual being in the universe. So this story of the great journey is an exciting revelatory story that gives us our macrophase identity—the larger dimension of meaning that we need.[14]

The New Story, therefore, is journeying in all three regards: it is an *itinerary*, or a plan for ethical and communal transformation. As a journey from the great flaring forth to the emergence of the Ecozoic, it is a journey that is actual *travel* through reality in both its spatial and temporal aspects. And finally as a *symbolic* journey, it is a spiritual transformation and a reconnecting of the human with the more-than-human. Journeying, however, is also something that is greater than the sum of these parts. The journey described in the New Story can be, in other words, all of these things and more.

Conclusion: Our Journey into the Future

I end this chapter as I began: by turning to the *Journey of the Universe* book. There, Swimme and Tucker observe that in the quest to become a planetary presence, humanity has learned to live *off* the Earth, but we haven't yet learned to live *on* the Earth. Intimacy has been abandoned for progress. "We are finding ourselves in the midst of a vast transition," they write. "How are we to respond? . . . With what shall we navigate?"[15] If humanity is to enter into the Ecozoic, I suggest in response to this inquiry, then perhaps we must do more than wander—we must learn to *journey*.

---

13. See, for example, Mircea Eliade, *Myths, Dreams, and Mysteries: The Encounter between Contemporary Faiths and Archaic Realities* (New York: Harper, 1960).

14. Thomas Berry, "Appendix: Reinventing the Human at the Species Level," in *The Christian Future and The Fate of Earth*, ed. Mary Evelyn Tucker and John Grim (Maryknoll, NY: Orbis Books, 2009), 121.

15. Swimme and Tucker, *Journey*, 111.

In conclusion, I draw upon Berry's poignant words that are reprinted in *Evening Thoughts*. In that talk, he invited his audience to "recall the ancient law of hospitality, whereby the *wanderer* was welcomed."[16] Just as Ulysses— even after the greatest of struggles—was welcomed home from his long travels, Berry suggested that we too will be welcomed home by the universe in our own journeys. In his words,

> As a final reflection, I would suggest that we see these early years of the twenty-first century as the period when we discover the great community of Earth, a comprehensive community of all the living and nonliving components of the planet. We are just discovering that the human project is itself a component of the Earth project, that our intimacy with Earth is our way to intimacy with each other. Such are the foundations of our journey into the future.[17]

In the final remarks of the *Journey of the Universe* film and book, Swimme and Tucker suggest that a sense of wonder, an opening of the self to the universe, will guide us into the future. But, to be planetary beings, to fully participate in the larger journey and communities of which we are a part, also requires more than awe and wonder. Humanity must also become *wanderers*.

---

16. Thomas Berry, "Evening Thoughts," in *Evening Thoughts: Reflecting on Earth as Sacred Community*, ed. Mary Evelyn Tucker (Berkeley, CA: Counterpoint, 2015), 140 (emphasis added).
17. Ibid., 141.

4

Thomas Berry and Journey of the Universe

Recognizing the Great Self in the Great Work

PETER ELLARD

The turning point in Disney's 1994 animated film *The Lion King* is when the wayward lion cub, Simba, in a moment of crisis, looks up into the billowing clouds of the night sky that morph into the face of his father and we hear the bellowing voice of James Earl Jones as Mustafa pronounce to his son, "You have forgotten who you are" (the voice is all the more menacing since it is also the voice of Darth Vader). Simba was experiencing an identity crisis, and the story of his place in the pride of lions—and the circle of life—was needed to help him through the crisis. Humanity today is experiencing a similar crisis. We have forgotten who we are. Thomas Berry and the new story of creation as outlined in *Journey of the Universe* offer a very clear answer to this identity question. Indeed, Berry argues that understanding the answer is an essential part of the solution to solving our current and coming environmental challenges. It is an essential part of what Berry called the Great Work that will enable us to "carry out the transition from a period of human devastation of the Earth to a period when humans would be present to the planet in a mutually beneficial manner."[1]

At Siena College I teach a course called Religion and the Environment. At the onset, we review a host of reasons that scholars have presented as to why humanity is destroying many of the life systems of the planet. We look at commentaries on the role of patriarchy with Sallie McFague and Carolyn Merchant. We read Lynn White Jr. and his critics. We read about the Green Pope and Green Patriarch—and their critics. We read Leopold and Schweitzer. We look at the rise of capitalism and consumerism. And then I introduce Thomas Berry and *Journey of the Universe* with the notion that

---

1. Thomas Berry, *The Great Work: Our Way into the Future* (New York: Bell Tower, 1999), 3.

the problem is at heart a cosmological one. The thesis for that section of the course is that the ecological crisis is a result of a cosmological identity crisis.

In short, the argument is that we destroy the planet because we don't know who we are. For Berry, until we solve this cosmological crisis and fully embrace the New Story we cannot hope to solve the ecological crisis. The ecological crisis is vast. But the cosmological crisis is larger still. It will take a radical shift in thinking to understand that it is indeed "all a question of story."[2] Still, it was Berry's conviction that by careful analysis and study of the story, and by mindful experience of the journey, we can rise to meet our current challenge. When we view the story of the universe, the journey of the universe, we are engaged in self-reflection and self-discovery. In doing so, we are required to reimage what it means to be human.

Humans exist on many levels. We exist on the level that you might imagine me now writing this essay or as you see yourself reading it. We also exist—that is, we have an identity—on the cellular level. Microbiologists tell us that there are around 37.2 trillion cells in my body.[3] Actually, that is not quite right. Let me rephrase that. Part of my identity is that I exist *as* 37.2 trillion cells. That's one way of addressing who I am. Smaller than that, we can say that we also exist on the molecular level. On this atomic level we exist as 7×10^{27} number of molecules. These molecules, made from twenty-six different atoms, are written out as 7,000,000,000,000,000,000,000,000,000.[4] We also exist smaller than that on the quantum level, where we function as waves and particles, where we tunnel, jump, entangle, and exist in superposition in ways that bear little resemblance to the reality we perceive with our eyes.[5] These levels of our existence—the quantum, the molecular, the cellular, and the ordinary "human" level we see through the reflection in the mirror—we might refer to as various aspects of our micro-level selves. Thomas Berry and *Journey of the Universe* remind us that we also exist on several macro levels, including the level of what Berry calls the Great Self.

---

2. Thomas Berry, "The New Story," *Teilhard Studies* 1 (1978): 1. Also published in *Dream of the Earth* (Berkeley, CA: Counterpoint, 2015).

3. Eva Bianconi et al., "An Estimation of the Number of Cells in the Human Body," *Annals of Human Biology* 40, no. 6 (November/December 2013): 463–71. The number cited does not count the number of bacteria cells "in" our bodies which is said to outnumber the "human" cells and without which we would die. See Melinda Wenner, "Humans Carry More Bacterial Cells Than Human Ones," *Scientific American* (November 30, 2007).

4. Forrest H. Nielsen, "Ultratrace Minerals," in *Modern Nutrition in Health and Disease*, ed. Maurice E. Shils (Baltimore: Williams and Wilkins, 1999), 283–303.

5. Brian Greene, *The Fabric of the Cosmos: Space, Time, and the Texture of Reality* (New York: Knopf, 2004).

This bringing together of the macro and micro is a central part of Berry's contribution. Berry was the one who called on us to acknowledge that on this macro level we exist as the Earth itself, as the galaxy itself, and as the universe itself. We are, as Berry often wrote, the universe conscious of itself. We are the universe expressed in a unique, conscious way. For Berry, all that is and all that will ever be in the phenomenal world was present at the moment of the Big Bang, the great elastication, the primordial flaring forth, and we have been transforming and evolving ever since. This includes spirit, mind, consciousness, and that which some would call a soul. This psychic/spiritual aspect of the universe was present from the beginning—as was all that makes up our creativity and our destructiveness. We are 4.5 billion years old in our Earth self and 13.8 billion years old in our universe self.

This is not theoretical or metaphorical for Berry. We are not talking in similes. This is the science revealed by the Universe Story, and according to Berry, this is also revealed religious truth. If we see this purely as an intellectual leap, we miss the central point, and our identity crisis remains. The journey of the universe is our journey. Before we existed in our current form, we existed as the body of plants and other animals, of rocks, rivers, and stars. The idea that the iron in my blood and the iron in my chair along with most of the other elements of my body were forged in the belly of a star billions of years ago is part of our collective, our communal, family heritage. Moreover, we continue to exist as the universe right now. In both some deep mystical and scientific sense, we are still stardust, a quasar, and a black hole. We are still gravity and dark matter. We are still our neighbors and our enemies. Christianity and the world's other religions have not absorbed these truths, and according to Berry, if we want to tackle our environmental crisis, we need to.

On a foundational level, Berry is pointing at the need for humanity to recognize our true place in the story as the conscious aspect of the universe. In terms of our relationship to the rest of the planet, we must come to understand that we do not live on the Earth. We are not part of the Earth. We are the Earth. We are the Earth, conscious of itself. It is our macro self and an expression of our Great Self. Berry says, "Every individual self of the Earth finds fulfillment in the Great Self of the Earth and beyond in the universe itself as the primary and ultimate expression of that mystery out of which all things come forth and then return in the full wonder of existence."[6] The

---

6. Thomas Berry, *The Sacred Universe*, ed. Mary Evelyn Tucker (New York: Columbia University Press, 2009), 168–69.

molecular, the cellular, and the "human" are derivative. The universe is our primary self. It is the primary expression of our Great Self—and we all share the same Great Self.

This is the framework from which Berry says that the Great Work of reimagining our sense of self and our relationship to the rest of the planet must flow. We are the universe woken up, evolved to consciousness and reflecting on itself. Again, this was, for Berry, both fundamental science and fundamental religious doctrine. Berry tells us that as we move forward to tell the story of the universe and combat the environmental devastation that we are unleashing on ourselves, we must awaken from our identity crisis and recognize our true place in the journey of the universe.

When we look at this from the level of the Earth, our conclusions take on deep ethical implications. As well, when we see the Earth as an expression of divine outflowing—as Berry does—it takes on significant spiritual implications. If we are the Earth conscious of itself, then when we spew carbon into the atmosphere, we are hurting ourselves and silencing a divine expression. This mindless geocide that Berry talked about is also a form of suicide and deicide. In destroying our world, we destroy ourselves, and in some way we destroy a manifestation of the divine itself—"the mystery out of which all things come forth."

This is why the success of the *Journey of the Universe* film, publications, conversations, and website are so important. They present what was for Berry the only context within which to understand reality, and the best hope for finding a solution to the crushing ecological crisis at hand. Brian Swimme and Mary Evelyn Tucker make this point succinctly in the *Journey of the Universe* book when they note that "every time we are drawn to look up into the night sky and reflect on the awesome beauty of the universe, we are actually the universe reflecting on itself. And that changes everything."[7]

It changes everything because the revelation of the New Story upends our anthropology. It upends our theology. It upends our economics, our jurisprudence, our educational philosophies, and our religious sensibilities. It changes fundamental Christian ways of interpreting reality. Berry and the story of the universe both call on us to free ourselves from the problematic and destructive thinking regarding the primacy of an isolated, independent, micro human self, and open ourselves up to the right understanding regard-

---

7. Brian Thomas Swimme and Mary Evelyn Tucker, *Journey of the Universe* (New Haven, CT: Yale University Press, 2011), 2.

ing the reality of our Earth self, our universe self, and our Great Self. Indeed, this is a central teaching of the greatest story ever told—the Universe Story.

It was from Pierre Teilhard de Chardin that Berry learned to see the scientific story of cosmic unfolding as a spiritual story. Of the Universe Story, Berry tells us, "For the first time the entire human community has, in this story, a single creation or origin myth. Although it is known by scientific observation, this story also functions as myth. In a special manner it is the overarching context for any movement toward one creative interaction of peoples, cultures, and religions."[8] The story is the overarching context for any religious or environmental movement.

This cosmogenesis in which we are still immersed contains within itself the most mysterious, the most awesome, the most powerful and most joyful revelation that humanity—that the universe—has ever known. The universe, Berry informs us, is the "primary religious reality," "the primary sacred community," "the primary revelation of the divine," and "the primary unit of redemption."[9] This primacy needs to be understood and accepted. This is also where we find that Berry is arguably more radical, more revolutionary, than Teilhard.

It was Berry who made the connection between the New Story and the devastation of the planet—between cosmology and the environment. This realization leads to the radical solution that is required of those aligned with religious traditions of the world—including those who come out of the Christian tradition: that is, we need to change everything. This is what Larry Rasmussen, updating James Baldwin, suggested in stating that we need "to do our first works over."[10] Christian ideas about creation, sin, grace, redemption, salvation, and ritual need to be reconstructed. Indeed, this reconstruction has already begun, and however these and other ideas come to be understood, it will need to be within the context of the New Story. As we move forward, we need to keep in mind that radical transformations are not easy.

I was giving a talk at a Catholic Theological Society of America conference a few years ago in Cleveland on Thomas Berry's work, and during the question session a Catholic priest from Tanzania stood up and said something to the effect that he loved Berry's work and how I had presented it. However, Berry's turn to the cosmic Christ, to cosmology itself as the primary source

---

8. Thomas Berry, *The Christian Future and the Fate of Earth,* ed. Mary Evelyn Tucker and John Grim (Maryknoll, NY: Orbis Books, 2009), 25.

9. Ibid.

10. Larry Rasmussen, "Doing Our First Works Over," *Journal of Lutheran Ethics* 9, no. 4 (April 2009): para. 1, http://downloads.elca.org/html/jle/www.elca.org/what-we-believe/social-issues/doing-our-first-works-over.aspx.htm.

of revelation, and to seeing humans as the Earth conscious of itself was so far beyond what his parishioners could handle that he feared it would never take root in them. I had to respond that the parishioners in the parish in Brooklyn, New York, where I was raised would also have a hard time with this approach. It was a very sobering exchange. Sobering also is the awareness of the current and coming devastation of climate change. This reflects the vastness of the challenge that the turn to the New Story requires.

Still we plod on. And we plod on because as the planet heats up and the poles and glaciers melt and sea levels rise, we can still prevent the worst-case scenarios from happening. We plod on because, as the planet continues to change, we will need a spirituality, and we will need a worldview. The Great Work of humans living in a "mutually enhancing manner" with the rest of the planet will need a plan rooted in the New Story of how to think, how to relate, how to commune, how to celebrate, how to continue the journey, and how to thrive on the new "Eaarth."[11] Thomas Berry can be our guide and our prophet as we move into the future. But in truth, it is the journey of the Universe Story itself that is our guide, our prophet, and our future.

---

11. Bill McKibben added the extra "a" to Earth as a way to get us to see that the planet we live on now has been so transformed by our actions that it needs to be thought of as a different planet, and thus have a different name.

Evolutionary Perspectives of Pierre Teilhard de Chardin

✺

5

Influences of Pierre Teilhard de Chardin on Journey of the Universe

Mary Evelyn Tucker

The cosmic sense must have been born as soon as humans found themselves facing the forest, the sea, and the stars. And since then we find evidence of it in all our experience of the great and unbounded: in art, in poetry, in religion. Through it we react to the world as a whole as with our eyes to the light.[1]

How might a comprehensive vision of the cosmos ground a life? How does the perspective of evolutionary time heal the anguish of our present social, political, and ecological challenges? How can a sense of evolutionary purpose bring renewed energies in the midst of existential doubt? These burning questions first led me to the thought of Pierre Teilhard de Chardin and continued with four decades of commitment to his vision. What I explore here is how he inspired several of us in creating *Journey of the Universe*.

The contributing partners to developing *Journey of the Universe* were Thomas Berry, Brian Swimme, myself, and John Grim. Teilhard was a central inspiration to each of us. As a scientist, Teilhard provided the evolutionary

---

1. Pierre Teilhard de Chardin, *Human Energy* (New York: Harcourt Brace, 1962), 82.

context in *The Human Phenomenon*[2] as well as in his other writings, such as *The Divine Milieu* and *Hymn of the Universe.* Thomas Berry, as a historian of religion, realized that to be accessible for more people, evolution needed to be told as a story. He outlined this in "The New Story," an essay first published as a Teilhard Study in 1978 and ten years later included in his book of essays, *Dream of the Earth.*

As vice president of the American Teilhard Association while Thomas was still president, I was helping to midwife this evolutionary story and also beginning to connect Teilhard with environmental issues, writing about the "Ecological Spirituality of Teilhard."[3] John was grounding the work as president of the Teilhard Association after Thomas stepped down in 1987. He and I were also writing on Teilhard's life[4] and his vision of evolution,[5] along with describing Berry's journey into this new cosmology.[6] Brian Swimme contributed the richness of the complex scientific information, working for a decade with Thomas on writing *The Universe Story*, which was released in 1992. This was the first epic narrative of evolution in book form; *Journey of the Universe* was its first telling in film nearly twenty years later (2011).

The contributions of each of these individuals were crucial to the *Journey* project. The intellectual friendship that grew through exploring and expanding the evolutionary vision of Teilhard over four decades has been life-sustaining. Several dimensions of the Teilhardian vision helped to ground and shape *Journey of the Universe.* These dimensions are also in concert with certain key ideas in East Asian Confucian thought, my primary area of scholarly research.[7] These include:

- Orienting to the universe: Cosmogenesis—emergent change (*sheng sheng*)

2. *The Phenomenon of Man*, completed in 1940, first published in English in 1959, was painstakingly retranslated by Sarah Weber and published in 1999 with the new title, *The Human Phenomenon.*

3. Mary Evelyn Tucker, "Ecological Spirituality of Teilhard," *Teilhard Studies* 13 (Spring 1985); reissued as *Teilhard Studies* 51 (Fall 2005).

4. John A. Grim and Mary Evelyn Tucker, "Teilhard de Chardin: A Short Biography," *Teilhard Studies* 11 (Spring 1984).

5. John A. Grim and Mary Evelyn Tucker, "Teilhard's Vision of Evolution," *Teilhard Studies* 50 (Spring 2005).

6. John A. Grim and Mary Evelyn Tucker, "Thomas Berry: Reflections on His Life and Thought," *Teilhard Studies* 61 (Fall 2010).

7. In creating the *Journey* film and book we consciously built in the cosmological and ecological perspectives of the world religions, including Western, Asian, and indigenous traditions. These were highlighted in the Harvard conference series on "World Religions and Ecology" (1996–98) and published in the ten volumes arising from the series (1997–2004).

- Grounding in the Earth: The within of things—matter/energy (*ch'i*)
- Nurturing within fecundity: Self-organizing dynamics—pattern/order (*li*)
- Transforming in action: Hominization—humaneness (*ren*)

Orienting to the Universe: Cosmogenesis— Emergent Change (Sheng Sheng)

Journey of the Universe draws on Teilhard's comprehensive vision of cosmogenesis, namely, that we dwell in a vast and evolving universe. The fact that we are living, not just in a cosmos, but in a cosmogenesis, an unfolding universe, is one of Teilhard's central realizations. As he understood it, the idea of evolution (or cosmogenesis) requires a major change of consciousness to which we are presently awakening. He describes this as a new dimension:

> Whereas for the last two centuries our study of science, history, and philosophy has appeared to be a matter of speculation, imagination, and hypothesis, we can now see that in fact, in countless subtle ways, the concept of Evolution has been weaving its web around us. We believed that we did not change; but now, like newborn infants whose eyes are opening to the light, we are becoming aware of a world in which neo-Time [evolution] is endowing the totality of our knowledge and beliefs with a new structure and a new dimension.[8]

Indeed, he sees this as a condition of all experience:

> For our age, to have become conscious of evolution means something very different from and much more than having discovered one further fact, however massive and important that fact may be. It means (as happens with a child when he acquires the sense of perspective) that we have come alive to a new dimension. The idea of evolution is not, as sometimes said, a mere hypothesis, but a condition of all experience.[9]

The understanding of cosmogenesis permanently changes our perception and way of being: "[Cosmogenesis] is something we must understand once

---

8. Pierre Teilhard de Chardin, *Future of Man* (New York: Harper and Brothers, 1964), 85.
9. Pierre Teilhard de Chardin, *Science and Christ* (New York: Harper and Row, 1968), 193.

and for all: for us and for our descendants, there is henceforth a final and permanent change in psychological times and dimensions."[10]

For Teilhard, what is crucial is that this evolving universe is not perceived as random or purposeless. Rather, it is organic and involves a dynamic unfolding process, which can be imaged like the tree of life, working toward a larger purpose:

> The world is abuilding. This is the basic truth which must first be understood so thoroughly that it becomes habitual and as it were a natural springboard for our thinking. At first sight, beings and their destinies might seem to us to be scattered haphazardly or at least in an arbitrary fashion over the face of the Earth; we could very easily suppose that each of us might *equally well* have been born earlier or later, at this place or that . . . as though the universe from the beginning to the end of its history formed in space-time a sort of vast flowerbed in which the flowers could be changed about at the whim of the gardener. But this idea is surely untenable. The more one reflects, with the help of all that science, philosophy, and religion can teach us . . . the more one comes to realize that the world should be likened not to a bundle of elements artificially held together but rather to some organic system animated by a broad movement of development which is proper to itself. . . . A process is at work in the universe . . . which can best be compared to the process of gestation and birth. . . . We are not like the cut flowers that make up a bouquet: we are like the leaves and buds of a great tree on which everything appears at its proper time and place as required and determined by the good of the whole.[11]

Deep time and vast space can be comprehended within the metaphor of that which is in process of gestation and birth, as this passage reveals. Awakening to the reality of cosmogenesis will take humans decades, if not centuries. We are in the beginning stages of understanding what evolution means. The Darwinian revolution that began 150 years ago is still penetrating human consciousness. *Journey* is one telling of the story, which evokes our understanding of and participation in this vast unfolding universe.

The universe is far greater in time and space than humans could have imagined, and we are birthed from this process. This is Teilhard's remarkable insight—that the increasing complexity and consciousness of the universe

10. Pierre Teilhard de Chardin, *Activation of Energy* (New York: Harcourt Brace Jovanovich, 1971), 264.

11. Pierre Teilhard de Chardin, *Hymn of the Universe* (New York: Harper and Row, 1965), 92.

gave rise to the human and awakens us to the depths of developmental time. He sees humans as a phenomenon of this universe, namely, intrinsic to the whole and not an epiphenomenon or an addendum. As *Journey* shows, we are part of this immense 14-billion-year-old evolutionary journey. Our orientation to the universe is the foundation of who we are. We are not at the margins of this process by virtue of our unique consciousness, but embedded in it, central to it, arising from it, and responsible for its future flourishing.

We are, as the Confucians suggest, and as we say in the film, "the mind and heart of Heaven and Earth." Here Heaven indicates the universe itself and Earth refers to the fecund processes of nature. In Chinese thought, humans needed to be in harmony with these dynamic processes. This gave rise to the classical text *The Book of Changes* (*I Ching*), where humans tried to ascertain their appropriate place and efficacious action amid these transformations. In the later neo-Confucian tradition, the changes in nature were seen as the generative cycles of life itself (*sheng sheng*). This outpouring of fecundity or cosmogenesis was celebrated as a source of human creativity.

Journey of the Universe explores this idea of cosmogenesis, inviting us to imagine how this vast process arose from the initial flaring forth some 14 billion years ago. It draws us into the almost miraculous rate of expansion of the early universe, illustrating that the conditions for life were so carefully enfolded into this process that one cannot help but be filled with amazement. The age of the universe, its vast scale, and its rate of expansion were only discovered within the last several decades. These discoveries are what *Journey* is bringing forth in a way that humans can comprehend and absorb, not as an abstraction or scientific fact but as a lived reality in which we live and breathe and have our being.

To see Earth as a story in which humans emerge is the intent of *Journey*. The universe is not simply a background or a platform. Rather, as a cosmogenesis it is an epic process in which we participate. Our role matters, and we are discovering ourselves anew in this epic context. Our individual and collective journeys are aligned with the journey of the universe. We are birthed into deep time and vast space. Our coming alive to these new dimensions of being human in an evolving universe is one of the great challenges of our time. We are acquiring a sixth sense, an integrated understanding of expanding space and developmental time. This consciousness changes everything—who we are and why we are here.

Grounding in the Earth:
The Within of Things—Matter/Energy (Ch'i)

Such a growing consciousness is connected to the understanding that we are part of a living universe. For Teilhard the universe had a "within," what Thomas Berry called "interiority" or "subjectivity." In *The Human Phenomenon* Teilhard writes,

> It is a fact beyond question that deep within ourselves we can discern, as though through a rent, an "interior" at the heart of things; and this glimpse is sufficient to force upon us the conviction that in one degree or another this "interior" exists and has always existed in nature. Since at one particular point in itself, the stuff of the universe has an inner face, we are forced to conclude that in its very structure—that is, in every region of space and time—it has this double aspect. . . . In all things there is a Within co-extensive with their Without.[12]

If we are orienting ourselves to this universe, we are grounding ourselves on the Earth because this is where we can experience the within of things. This within of things (or subjectivity) is similar to what religious traditions refer to as spirit within matter or scientists speak of as energy and matter. The Confucians tell us that we are part of a dynamic universe permeated with *ch'i*, a vital force–infusing matter. *Ch'i* is what awakens us to the power of nature—as it constantly reveals itself to us, in the sunrise or sunset; in the beauty of oceans, lakes, or rivers; in the luminous display of vegetative life— trees, plants, and flowers—and in the complex quality of fish, animals, and birds. *Ch'i* is cultivated in exercises such as *t'ai chi* and *chigong*. Chinese acupuncture is a method of increasing the flow of *ch'i* for long-term health.

In *Journey of the Universe* we use the term "creativity," which is analogous to the within of things. The creativity of the universe brought forth galaxies and stars and solar systems. In the living world, "sentience" in different forms permeates all life from the single-celled creatures to self-reflective humans. Ursula Goodenough suggests that even a cell has a kind of self with its ability to move, to protect, to replicate, and to discern what it takes in. We are grounded in and resonating with the creativity and sentience of matter. But creativity requires an inner nurturing as well.

---

12. Pierre Teilhard de Chardin, *The Human Phenomenon*, trans. Sarah Appleton-Weber (East Sussex, UK: Sussex Academic Press, 1999), 56. Originally published in English as Pierre Teilhard de Chardin, *The Phenomenon of Man* (New York: Harper and Brothers, 1959).

Nurturing in Fecundity:
Self-Organizing Dynamics—Pattern/Order (Li)

The creativity of the universe is not undifferentiated; it is not simply a smudge. It is pregnant with pattern. The neo-Confucians call this the *li* within *ch'i*—the inner pattern of things, which needs to be intuited, sensed, studied, and understood. The Chinese character for *li* refers to the pattern in jade; it is like the form in matter. In traditional Confucian thought, to perceive this patterning is both a spiritual act of cultivation and a mental act of perception. Sensing the self-organizing dynamics of nature is resonating with it; understanding it intellectually is the task of natural historians who catalogue plants, animals, fish, shells, rocks, and so forth.[13]

Teilhard connected the creativity of the universe that lead to greater complexity and greater consciousness over time with energy—radial energy, which centers matter and draws it forward, and tangential energy, that gravitational pull on matter to interact with other forms of matter.

Teilhard also had a sense of the creativity of matter as animated by the Logos, an inner ordering pattern. This is not pattern imposed from the outside as proponents of "intelligent design" would have it. In "The Spiritual Power of Matter," Teilhard writes, "I acclaim you as the divine *milieu*, charged with creative power, as the ocean stirred by the Spirit, as the clay molded and infused with life by the incarnate Word."[14]

In *Journey*, we call this inner ordering dimension of matter its self-organizing dynamics. These are within matter, not exterior to it—embedded with creativity in universe processes and sentience in life processes. We describe how this gave rise to the structures of the universe and then to life forms. We participate in these unfolding patterns of the universe because, in some sense, our challenge is as follows: How do we work with these self-organizing dynamics of evolution in a way that will benefit the life of the Earth community? This can be in how we construct our agricultural systems, how we build our cities, how we engineer transportation, how we foster renewable energy, and how we embrace sustainable development. As the film asks at its conclusion, "Can we align ourselves with the grain of cosmic evolution . . . and thus help to transform the face of the Earth?"

---

13. See *The Philosophy of Ch'i: The Record of Great Doubts*, trans. and with an introduction by Mary Evelyn Tucker (New York: Columbia University Press, 2007).

14. *Hymn of the Universe*, 67.

Transforming in Action: Hominization—Humaneness (Ren)

To align ourselves with evolutionary processes suggests that we need to shift from our current destructive tendencies to ways of enhancing Earth processes. How to do that remains a challenge. We have created an environmental crisis of such proportions that humans are becoming paralyzed by the size and complexity of the problems. Can we find the perspective and the energy to meet these challenges without losing hope? Teilhard's concern about activating human energy for creating the future was ever present in his thinking, especially in the face of the existentialist angst and postwar despair that gripped Europe.

Clearly Teilhard could not have anticipated Earth's present population explosion to more than 7 billion people and the accompanying environmental destruction that has occurred in the last sixty years. He had a more optimistic sense of progress and of the power of technology to assist humans in "building the Earth." His optimism needs to be tempered by a new understanding of the limits of growth and of planetary boundaries. We can move forward with an ecological economics that recognizes Earth's limits. We can draw on indicators of human happiness that are measured not by Gross Domestic Product but rather by the Gross National Happiness Indicator, such as offered by Bhutan. This suggests that the quality of a person's life is not measured in material wealth.

Yet Teilhard's love for the Earth and his hope for the future are invaluable for inspiring humans to reshape our future. He encouraged humans not to become distorted, disgusted, or divided toward the world.[15] Indeed, he maintained that his faith in the world was his dominant faith. He spoke of the need for the perspective of hominization, whereby human reflection and action complete the evolutionary process.

> Let us just consider whether we might not be able to escape from the anxiety into which the dangerous power of thought is now plunging us—simply by improving our thinking still more. And to do this let us begin by climbing up till we tower over the trees which now hide the forest from us; in other words let us forget for a moment the details of the economic crises, the political tensions, the class struggles which block out our horizon and let us climb high enough to gain an inclu-

---

15. Pierre Teilhard de Chardin, *The Divine Milieu: An Essay on the Interior Life* (New York: Harper and Brothers, 1960), 21.

sive and impartial view of the whole process of *hominization* as it has advanced during the last fifty or sixty years.[16]

For Teilhard, this perspective is indispensable for seeing how we can relate the meaning of human endeavors to evolution. He urges us to see ourselves as an invaluable part of evolution—woven into its processes, not extrinsic to its long unfolding. Without this perspective he feels it is impossible to answer the question of our life purpose, which is now a central question of the modern period.

> Is life an open road or a blind alley? This question, barely formulated a few centuries ago, is today explicitly on the lips of mankind as a whole. As a result of the brief, violent moment of crisis in which it became conscious at once of its creative power and of its critical faculties, humanity has quite legitimately become hard to move: no stimulus at the level of mere instinct or blind economic necessity will suffice for long to goad it into moving onward. Only a reason, and a valid and important reason, for loving life passionately will cause it to advance further. But where, at the experiential level, are we to find, if not a complete justification, at least the beginnings of a justification of life? Only, it would seem, in the consideration of the intrinsic value of the human phenomenon.[17]

We can move forward once we humans are able to resituate ourselves within these dynamic processes of life and to sense how our actions matter for the future of nature. This is the hope of Teilhard that is highlighted in *Journey of the Universe*. As we say at the end of *Journey*, "We belong here, we have always belonged here." Teilhard puts it this way:

> Humans have every right to be anxious about their fate so long as they feel themselves to be lost and lonely in the midst of the mass of created things. But let them once discover that their fate is bound up with the fate of nature itself, and immediately, joyously, they will begin again their forward march. For it will denote in them not a critical sense but a malady of the spirit if they were doubtful of the value and the hopes of an entire world.[18]

As we seek to create the grounds for a common future in a shared story of the universe, we realize that the basis for a vibrant Earth community rests with

---

16. *Hymn of the Universe*, 118.
17. Ibid., 108.
18. Ibid., 109.

such a vision as Teilhard articulated. While it may appear utopian, our effort to extend comprehensive compassion to humans and other species is at the heart of our destiny as the conscious "mind and heart of Heaven and Earth."

For the Confucians this expression implied realizing one's humaneness (*ren*) in relationship to others. With continued moral learning and spiritual cultivation, one could enter into resonance with the concentric circles of self and society. For the Confucians this meant that humans are not isolated individuals, but inherently interconnected within the circles of family, education, society, politics, nature, and the cosmos itself. To realize one's full humanity and thus become a humane person (*ren*) is to foster the common good for society as a whole. For Confucian scholars and teachers, this frequently implied serving in public office. Becoming ever more concerned about the well-being of the whole (humans and nature) was the aim of Confucian education, self-cultivation, and civil service. As one became less self-centered, one moved from being a "small person" to becoming a "noble person" (*junzi*) and entering into one's great self. Indeed, this implied that the deepening of comprehensive compassion (*ren*) was a gateway to realizing one's larger cosmological self. The humane person is thus able to enter into the "transforming and nourishing powers of Heaven and Earth" and be sustained by them. At the same time, practicing humaneness contributes to the well-being of these processes. For example, Confucians understood that a healthy agricultural system is required as a basis of a healthy society. Thus, personal and communal flourishing are both considered central. This aspiration for the flourishing of the whole is also key for Teilhard:

> Humanity, the spirit of the Earth, the synthesis of individuals and people, the paradoxical conciliation of the element with the whole, of the one with the many: all these are regarded as utopian fantasies, yet they are biologically necessary; and if we would see them made flesh in the world what more need we do than imagine our power to love growing and broadening till it can embrace the totality of men of the Earth?[19]

To do this, for Teilhard, is to foster the hominization of our activities, to assist the means by which the powers of the universe may be alive in us. This is Teilhard's hope and *Journey*'s as well—to see our efforts as fostering the vibrancy of life.

> The human soul . . . is inseparable, in its birth and in its growth, from the universe into which it is born. . . . It is we who, through our own activity, must industriously assemble the widely scattered elements.

---

19. Ibid., 151.

The labor of seaweed as it concentrates in its tissues the substances dispersed, in infinitesimal quantities, throughout the vast layers of the ocean; the industry of bees as they make honey from the juices scattered in so many flowers—these are but pale images of the continuous process of elaboration which all the forces of the universe undergo in us in order to become spirit.[20]

To assist this process, Teilhard puts before us the challenge of individual work, what Thomas Berry calls our "Great Work." Teilhard sees this as indispensable to completing the world, and *Journey* sees this as essential to transforming the face of the Earth:

Thus every person, in the course of their life . . . must construct—starting from the most natural zone of their own self—a work, an *opus*, into which something enters from all of the elements of the Earth. *We make our own soul* throughout all our earthly days; and at the same time we collaborate in another work, in another *opus*, which infinitely transcends, while at the same time it narrowly determines, the perspective of our individual achievement: the completion of the world.[21]

The completion of the world relies on each of our efforts. It is an ongoing journey in which we contribute to the flourishing of our sacred Earth community.

---

20. *Divine Milieu*, 29.
21. Ibid.

6

Teilhard de Chardin, Thomas Nagel, and Journey of the Universe

JOHN F. HAUGHT

New York University philosopher Thomas Nagel's latest book, *Mind and Cosmos*, is so unexpected, and so annoying to many of his fellow philosophers of mind, that we may assume he is onto something. The book's subtitle is especially provocative: *Why the Materialist Neo-Darwinian Conception of Nature Is Almost Certainly False.*[1] The publication of Nagel's book provides me with the opportunity to reflect not only on the cosmic story as set forth in *Journey of the Universe* but also on the enduring importance of Pierre Teilhard de Chardin (1881–1955).

On the occasion of our celebrating and reflecting on *Journey*, I bring up the names of Nagel and Teilhard simultaneously because, as far apart as the two thinkers may be theologically, each agrees that a purely materialist metaphysics cannot make sense of the new scientific cosmic story in general and the fact of mind in particular. Like Brian Swimme, Mary Evelyn Tucker, and Teilhard, Nagel has no problem with the standard scientific narrative of mind's gradual evolutionary emergence. Rather, his concern is whether the materialist metaphysics in which that narrative is still being packaged, especially in the academic world, can make the story of the human mind's emergence in the cosmic story intelligible. He argues convincingly that it cannot, but he provides no coherent alternative to his formerly espoused materialism. Here I want to suggest that Teilhard provides a largely implicit metaphysical worldview that *can* contextualize contemporary versions of the cosmic story such as *Journey* in a manner that avoids the materialist and atomistic reduction of the universe to a primordial and (eventually) final state of mindlessness.

What bothers Teilhard most is the materialist assumption that consciousness, which evolution has labored so long to produce, will eventually be lost

---

1. Thomas Nagel, *Mind and Cosmos: Why the Materialist Neo-Darwinian Conception of Nature Is Almost Certainly False* (New York: Oxford University Press, 2012).

altogether in a final energetic collapse of the universe. Ever since childhood Teilhard was preoccupied with the fact of perishing.[2] Were he to read *Journey*, or any other account of natural history, he would wonder, above all, what the authors think about the materialist prediction of the final perishing of life and consciousness. He would ask why the authors think the story matters at all, and if it does, whether it matters everlastingly. What if the cosmos comes to nothing in the end, and consciousness perishes along with it? Can we really love a universe whose destiny is the abyss of nonbeing? Teilhard writes,

> Man, the more he is man, can give himself only to what he loves; and ultimately he loves only what is indestructible. Multiply to your heart's content the extent and duration of progress. Promise the earth a hundred million more years of continued growth. If, at the end of that period, it is evident that the whole of consciousness must revert to zero, *without its secret essence being garnered anywhere at all*, then, I insist, we shall lay down our arms—and mankind will be on strike. The prospect of a *total death* (and that is a word to which we should devote much thought if we are to gauge its destructive effect on our souls) will, I warn you, when it has become part of our consciousness, immediately dry up in us the springs from which our efforts are drawn.[3]

It is not the objective of *Journey* to consider Teilhard's preoccupation formally, of course, and I am not bringing up his concern as a criticism of the book. Yet any narration of the story of the universe does raise, at least in the minds of some readers, the question Teilhard is putting to us: Does the story really matter if absolute perishing is the end of it all?

The materialist is quite content to respond that everything including consciousness will eventually slip into final nothingness. Teilhard is not. So he would be happy to discover that Nagel, a staunch materialist throughout his academic career, now agrees with the Jesuit geologist that the union of science and materialism is an intellectual dead-end. Like Teilhard, Nagel allows that evolutionary science is correct as far as it goes, but he claims that *evolutionary materialism* is unscientific and logically self-contradictory since it cannot account for the reality of mind. Nagel agrees with contemporary cosmology that mind is stitched seamlessly into the fundamental physical features of the

---

2. Pierre Teilhard de Chardin, *The Heart of Matter*, trans. René Hague (New York: Harcourt Brace Jovanovich, 1978), 18.

3. Pierre Teilhard de Chardin, *How I Believe,* trans. René Hague (New York: Harper and Row, 1969), 43–44.

Big Bang universe, but he denies that a purely materialist account of mind's emergence can make sense of our capacity for thought.

Like Teilhard, and unlike most other scientists and philosophers, Nagel proposes instead that before we tell the cosmic story we must first look closely at the splendid properties of mind to which the story has recently led. He means by this that any causal narrative of the Universe Story must be loaded from the start with a sufficiently rich explanatory content to account for the eventual production of thought. Such a cosmic narrative would differ decisively from the standard materialist version according to which "what is *more*" is always coming out of "what is *less*." Materialist claims about mind clearly violate the principle of sufficient reason. Whatever else it may be, after all, the cosmic journey is at the very least a mind-making enterprise, and we need to ask why. According to evolutionary materialists, however, life and mind are simply unintended arrays of the mindless material bits they take to be the ultimate ground and final destiny of the cosmic journey. In other words, to materialists the emergence of mind in natural history is incidental, nothing more than a fluke that tells us little or nothing about what the universe really is. Interpreted materialistically, mind's evolutionary emergence and its amazing properties are not enough to disturb the fundamental pointlessness of the cosmic story.[4]

However, in view of the narrative togetherness of mind and cosmos as depicted in *Journey* as well as in the writings of Teilhard (and now acknowledged in Nagel's recent book), it seems arbitrary to overlook the extraordinary properties of mind while we are engaged in the business of telling the cosmic story. Doesn't the fact of mind tell us something vital about the *whole* cosmic story, something that materialism inevitably leaves out?

Nagel suspects it does, but in acknowledging the inseparability of mind and the whole cosmic journey, he faces two problems. On the one hand, isn't materialism warranted if life and mind arose historically, without any sharp breaks, out of a fundamentally mindless cosmic process? On the other hand, how can theology be avoided if the universe is biased from the start toward the making of minds, as Nagel now suspects it is?

The fatal flaw Nagel now sees in any materialist cosmology is that it makes the universe look so fundamentally mindless from start to finish that, if true, we have no good reason to trust our own minds. The physical series of events leading to life and mind, materialists claim, has no strain of directionality or intentionality, even though Darwinian evolution has brought about intel-

---

4. See, for example, Peter W. Atkins, *The 2nd Law: Energy, Chaos, and Form* (New York: Scientific American Books, 1994), 200.

ligent, intentional human subjects. Speaking on behalf of materialist naturalism, Duke University philosopher Owen Flanagan, for example, writes
that human intelligence, an admittedly splendid evolutionary product, can
be accounted for in terms of blind physical laws and purely Darwinian processes. "Evolution," he announces, "*demonstrates* how intelligence arose from
totally insensate origins."[5] And philosopher Daniel Dennett agrees: "The
designs in nature are nothing short of brilliant, but the process of design that
generates them is utterly lacking in intelligence of its own."[6]

A main theme in Nagel's new book, however, is that if indeed our minds
have emerged from, and can be adequately explained in terms of, mindless
physical stuff processed by blind laws of nature and aimless Darwinian processes, then why should we pay any attention to these minds? So Darwin
would appreciate Nagel's point. "With me the horrid doubt always arises,"
Darwin wrote to one of his friends, "whether the convictions of man's mind,
which has been developed from the mind of the lower animals, are of any
value or at all trustworthy. Would any one trust in the convictions of a mon-
key's mind, if there are any convictions in such a mind?"[7] Nagel is now asking
his fellow materialists—think, for example, of Flanagan and Dennett—to
look carefully at Darwin's doubts: How can evolutionary materialists justify
their own intellectual self-confidence and expect us to take them seriously if
they really believe that the ultimate explanation of all mental functioning lies
in a purely mindless process emanating from aimless physical stuff rushing
toward a final abyss of nonbeing? I wonder how Swimme and Tucker, along
with readers of *Journey,* would respond to this question.

For his part, Nagel now finds materialism intellectually unacceptable. Yet
he wants nothing to do with theology either. Surely he realizes that endow-
ing nature with even the thinnest vein of teleology opens up the universe to
a theological interpretation. He suspects that the cosmic story has a kind of
inbuilt purpose, namely, to produce minds, but he has no explanation for
why the cosmos would be so endowed. Still clinging to the tattered remnants
of his lifelong affair with materialism, he is not ready to link up with theology. Again, I wonder what response Swimme and Tucker, along with readers of *Journey,* might make to Nagel's dilemma. *Journey* rightly refrains from
dealing formally with this larger question, but it seems natural that in *reflect-*

---

5. Owen Flanagan, *The Problem of the Soul: Two Visions of Mind and How to Reconcile Them*
(New York: Basic Books, 2002), 11 (emphasis added).

6. Daniel Dennett, "Intelligent Thought," in *The Third Culture,* ed. John Brockman (New
York: Touchstone Books, 2006), 87.

7. Letter to W. Graham, July 3, 1881, *The Life and Letters of Charles Darwin*, ed. Francis Darwin (New York: Basic Books, 1959), 285.

ing on the story *Journey* narrates, as we are doing in this volume, we would wonder about its theological implications.

To support its pessimistic assumptions about the universe, evolutionary materialism, as both Teilhard and Nagel would agree, turns its focus away from the most palpable and *immediately real* of all phenomena: the interior fact and performance of our own minds. Instead of asking what an examination of our subjectivity might tell us about the universe, materialism reduces the subjective dimension of our being to an epiphenomenon with no intelligible connection to the presumably more "real" mindlessness of physical stuff. In their attempt to make the cosmos appear completely objectifiable, some scientists and philosophers today even deny that subjectivity or consciousness has any real existence at all.[8] Their materialist worldview in effect leaves no room for subjects of any kind. Consequently, to fit human beings into their "objectively" mindless universe, scientific materialists must first strip themselves of the very minds they are using to objectify the world and teach us the truth! Nagel now quite understandably wants nothing to do with this mindless self-subversion of human mental existence and functioning.

Unlike Nagel, who has no alternative metaphysics to substitute for his formerly materialist worldview, Teilhard realized that the fact of interiority and consciousness requires a nonmaterialist metaphysics as the setting for any coherent telling of the cosmic story. Because he was a geologist and not a philosopher, he was never fully successful in articulating an alternative worldview. However, I believe he was leaning toward what I would call a *metaphysics of the future.* Here I want to suggest that such a worldview may also intelligibly frame *Journey of the Universe.* If Swimme and Tucker, like Teilhard, do not want their narrative to be interpreted materialistically, then they might consider an *anticipatory vision,* one that understands the cosmos as a drama whose intelligibility, or what we might call its *narrative coherence,* emerges only "up ahead," in the future.

Such a vision would be consistent with contemporary scientific cosmology in allowing no sharp breaks physically and historically between the early universe and the eventual arrival of mind. But instead of diminishing mind to virtual nothingness and final insignificance, a metaphysics of the future validates the splendor of consciousness by situating *the whole cosmic story* (in which mind is embedded) within the setting of a worldview that identifies what is really real, fully intelligible, ideally good, and maximally beautiful

---

8. For instance, Paul M. Churchland, *The Engine of Reason, the Seat of the Soul: A Philosophical Journey into the Brain* (Cambridge, MA: MIT Press, 1995).

with *what is coming* from the future, rather than basing the reality of emergent phenomena solely on what is or what has been.

Put more succinctly, the world, as Teilhard writes, "rests on the future as its sole support."[9] In other words, the fullest way to contact the world's true being, intelligibility, value, and beauty is to turn our attention toward the future, putting on the habit of hope. Hope then is not only a theological virtue but also an epistemological necessity that, far from being an illusory escape from "reality" as materialists always take it to be, is essential to the orienting of our hearts and minds—and the universe along with us—toward the domain of *fuller being* arising uncertainly on the horizon of the "up ahead." In this metaphysical setting, the reality and value of mind consist of its being an especially sensitive anticipation of the infinite being, meaning, value, and beauty arriving from the future that prophetic faith traditions refer to as "God."

Again, I am not suggesting that *Journey*—since it is not a work of philosophy or theology—needs to engage in this kind of metaphysical inquiry. However, stories are always tacitly carried along by one metaphysical vision or another, so at some point, as we reflect on a story's meaning or intelligibility—as we are doing here with the cosmic story—we need to make its implicit metaphysics explicit. In the case of *Journey*, the implied worldview must be wide enough to encompass not only the "objective" discoveries of natural science but also the anticipatory character of conscious subjectivity.

Teilhard's implicit metaphysics of the future seems to me to have the breadth that Nagel is looking for but that remains largely unavailable in the contemporary philosophical world in which he still dwells. No doubt, what I am suggesting on the basis of my reading of Teilhard will surely seem foreign to Nagel. So, to bring out the distinctiveness of a metaphysics of the future, we would have to contrast it with at least two other, more familiar worldviews, both of which fail to render intelligible the dramatic character of the mind-making natural world that *Journey* depicts. The first of these is the persistent Platonic "metaphysics of the eternal present" assumed by most prescientific Christian thought and still influential among contemporary Christian philosophers and theologians. According to this vision, which still dominates Christian spirituality, the world's being, intelligibility, value, and beauty are the result of finite being's "vertical" sacramental participation in the infinite God who is taken to be the eternal source of all finite being. The fact of mind is valued and becomes intelligible in this "fixist" worldview on the assump-

9. Pierre Teilhard de Chardin, *Activation of Energy*, trans. René Hague (New York: Harcourt Brace Jovanovich, 1971), 239.

tion that it is a special kind of finite participation in an infinite intelligence. This vision of the relationship of finite being to a timeless infinite is familiar to Christian believers as well as scientific skeptics, so I need not develop it further here. I only want to note that I undertake a critique of it in a recent book, arguing at length that an exclusivist religious devotion to an "eternal present" is incapable of providing an intelligible context for telling the new cosmic story in which mind comes into existence *gradually*, over the course of many millions of years.[10]

A second alternative to my proposed metaphysics of the future is what I like to call a "metaphysics of the past." This is the worldview taken for granted by most materialist scientists and philosophers. It strives, unsuccessfully I believe, to make the world, life, and mind intelligible by reducing them analytically to the lifeless and mindless physical units that inhabited the remote cosmic past. Trying to understand everything in a purely reductive way, however, only makes the world look less, not more, coherent.[11] The further back our analytical journey takes us into the cosmic past, in other words, the more we see things falling apart. Neither conscious subjectivity nor any other emergent realities will ever show up in a landscape reduced analytically to atomic bits.[12]

We need to turn around, therefore, after making our analytical journey into the cosmic past, and scan the horizon for what is yet to come, that is, if we are interested in looking at the *real* world.[13] Teilhard, of course, did not oppose scientific analysis as a method of inquiry but only the atomistic metaphysics that materialists have arbitrarily built on it. He realized that it is only by breaking things down into their component parts that the stages of the cosmic journey can be laid out for us chapter by chapter. *Journey* rightly embraces the same methodological point of view. Still, a purely reductive analysis cannot by definition yield a coherence that fully satisfies the human drive to understand, as Nagel's new suspicions indicate. If the cosmos is a story, analysis alone cannot tell us what the story is about. Instead we have to follow the story as it moves forward. If the universe is a drama still unfolding, after all, how can we grasp its intelligibility without watching where

---

10. John F. Haught, *Resting on the Future: Catholic Theology for an Unfinished Universe* (New York: Bloomsbury Press, 2015).

11. Pierre Teilhard de Chardin, *Human Energy*, trans. J. M. Cohen (New York: Harvest Books / Harcourt Brace Jovanovich, 1962), 172–73. For an extended critique of the materialist assumptions underlying what Teilhard calls the "analytical illusion" see my book *Is Nature Enough? Meaning and Truth in the Age of Science* (Cambridge: Cambridge University Press, 2006).

12. *Human Energy*.

13. See, for example, Pierre Teilhard de Chardin, *The Future of Man*, trans. Norman Denny (New York: Harper Colophon, 1964).

it is going?[14] "Like a river which, as you trace it back to its source, gradually diminishes till in the end it is lost altogether in the mud from which it springs, so existence becomes attenuated and finally vanishes away when we try to divide it up more and more minutely in space or—what comes to the same—to drive it further and further back in time. The grandeur of the river is revealed not at its source but at its estuary."[15]

By looking toward the cosmic delta, an anticipatory metaphysics of the future also shows itself to be confluent with an Abrahamic theology in which God, the creative source of all being, draws all things toward coherence from out of the future. God is not so much a governor, overseer, or an "eternal now" cleansed of any real contact with time. Rather God, who creates the world from out of the future, is the stimulus and not yet fully enfleshed Goal of all cosmic becoming.[16] Accordingly the patient, long-suffering hope prescribed by the prophetic traditions as essential to our going out to meet the coming of God seems congruent with a cosmological imperative to *wait* patiently for the world's intelligibility and fuller being to arise on the horizon of the not-yet.

I have not the space here to develop further what I am calling a metaphysics of the future. I merely want to suggest that the anticipatory vision I am sketching here in its barest bones on the basis of my reading of Teilhard needs to be distinguished carefully not only from the reductive and backward-looking perspective of analytical science and evolutionary materialism but also from the otherworldly metaphysics of traditional philosophy and theology. An otherworldly Platonic metaphysics may give us a sense of the imperishability of being, but it uproots us from the flow of cosmic time. A materialist metaphysics of the past acknowledges that we are immersed in the flow of time, but its focus is on what *has been* rather than on what is not-yet. Teilhard's anticipatory vision, one that he never fully articulated, recommends itself in my opinion since it locates the enduring foundation of the universe in the realm of the "up ahead." That is, it looks toward the infinite resourcefulness of the God-Omega whose reality cannot be apprehended apart from our assuming the open posture of hope. In reading *Journey*, therefore, I can make most sense of its narrative if I place it against the backdrop of a metaphysics of the future.

---

14. Pierre Teilhard de Chardin, *Christianity and Evolution*, trans. René Hague (New York: Harcourt Brace & Co., 1969), 79–86, 131–32.

15. Pierre Teilhard de Chardin, *Hymn of the Universe*, trans. Gerald Vann (New York: Harper Colophon, 1969), 77. See also Teilhard de Chardin, *Activation of Energy*, 239.

16. As implied in Teilhard de Chardin, *Activation of Energy*, 139, 239.

7

Teilhard's Deep Catholicity and Conscious Evolution

ILIA DELIO

As *Journey of the Universe* unfolds, Brian Swimme strolls along the shores of the Greek island Samos, recounting the fantastic discoveries of twentieth-century science. At one point he says there are now billions if not trillions of galaxies in the universe. He asks, "What is the creativity that brought forth a trillion galaxies?" One could almost hear a small French voice in the background, saying, "Mais oui! C'est l'amour!" Brian's question is the question, I think, that Pierre Teilhard de Chardin grappled with his entire life: What accounts for the creativity of cosmic, emergent life? For Teilhard, the answer was quite simple and profound: love. Love is a passionate force at the heart of the Big Bang universe, the fire that breathes life into matter and unifies elements, center to center. He spoke of love as a cosmological force, and by this he meant an attractive force of energy deeply intrinsic to cosmic life. Cosmological and biological evolution is marked by an increase in complexity and consciousness, and love is the energy of relatedness by which consciousness rises. Cosmos is not a given; it is what emerges (by way of evolution) through a consciousness of the whole or what the ancient Greeks called "catholicity." Thus we find in Teilhard a dynamic interplay of love, consciousness, and catholicity as a triadic complex undergirding creativity in the cosmos. Here I want to focus on love and the emergence of consciousness in the human person and how these play out in human creativity today, especially on the level of artificial intelligence. Using Teilhard's paradigm of ultrahumanity, I discuss how a consciousness of love in the noosphere can help deepen life ahead.

Catholicity and Consciousness

The word *cosmos* emerged among the Greeks and referred to the "whole" or that which makes the world a "world"; the "whole" flowed from human consciousness and attention. The pre-Socratic philosopher Heraclitus of Ephesus (c. 540–480 BCE) referred to *ta panta* (the "all things"), which another pre-Socratic, Empedocles (c. 490–430 BCE), transformed into the singular *to pan*, or "the all." In this respect the Greek word "catholicity" finds its root meaning, which is, "according to the whole." The word "catholicity" was coined by the Greeks to define a sense of the whole. In its original meaning, catholicity was related to cosmology.

Teilhard de Chardin was a deeply catholic thinker. Like the ancient Greeks, he had a sense of the whole, a grasp of divine, human, and cosmic realities intertwined in an ongoing development of life and consciousness. Around the time of Teilhard, physicist David Bohm was grappling with quantum physics and the unity of existence. He spoke of a quantum potential that underscores unbroken wholeness in nature despite quantum fluctuations. This principal of quantum wholeness undergirds an "implicate order": in our everyday lives we seem separate, but in our roots we are part of the same cosmic process. Similarly, Teilhard described a unitive principle undergirding wholeness and called this principle "Omega." Omega is the most intensely personal center that makes everything personal and centered.[1] It is both in evolution and independent of evolution, within and yet distinct from the process itself.[2] As the principle of centration that is both within and ahead, Omega emerges from the organic totality of evolution.[3] It is operative from the beginning of evolution, acting on preliving cosmic elements even though they are without individualized centers, by setting them in motion from the beginning, a single impulse of energy.[4]

Teilhard spoke of two fundamental types of energy in evolution: *tangential energy*, making the element interdependent with all elements of the same order in the universe as itself, what we might call "bonding" energy, and *radial energy*, attracting the element in the direction of an ever more complex and centered state, toward what is ahead which is "psychic" or conscious energy.[5]

---

1. Pierre Teilhard de Chardin, *Activation of Energy*, trans. René Hague (New York: Harcourt Brace Jovanovich, 1971), 112.

2. In this respect, Omega is like the strange attractor of chaos theory: within the system and yet different from the system's pattern of behavior. For a discussion on chaos theory, see Ilia Delio, *The Emergent Christ: Exploring the Meaning of Catholic in an Evolutionary Universe* (Maryknoll, NY: Orbis Books, 2011), 26–27.

3. Teilhard de Chardin, *Activation of Energy*, 114.

4. Ibid., 121.

5. Pierre Teilhard de Chardin, *The Human Phenomenon*, trans. Sarah Appleton-Weber (East Sussex, UK: Sussex Academic Press, 1999), 29–30; Harold J. Morowitz, Nicole Schmitz-Moormann, and James F. Salmon, "Looking Again at Teilhard, Tillich and Haught: Teilhard's Two Energies," *Zygon* 40, no. 3 (September 2005): 721–32.

While tangential energy follows the second law of thermodynamics, dissipating into entropy, radial energy defies this second law and increases with complexity. The increase in radial energy, the energy of consciousness, led Teilhard to suggest that consciousness is the core of evolution. It is this energy of evolving consciousness that gives evolution its qualitative direction. The centration of everything that exists is not a static whole but an emerging one, an underlying force of attraction that Teilhard described as "love-energy." While love-energy may not explicitly show itself on the level of the preliving and the nonreflective, it is present, inchoately, as the unifying principle of wholeness, as entities evolve toward greater complexity.[6] Love-energy marks the history of the universe; it is present from the Big Bang onward, though indistinguishable from molecular forces. "Even among the molecules," Teilhard wrote, "love was the building power that worked against entropy, and under its attraction the elements groped their way towards union."[7] In Teilhard's view, "If there was no internal propensity to unite, even at a rudimentary level—indeed in the molecule itself—it would be physically impossible for love to appear higher up, in a hominized form."[8] He proposed a philosophy of love to support the principal features of evolutionary convergence, complexity, and consciousness, which, he indicated, reflected something deep and profound at the heart of nature. Love draws together and unites and, in uniting, generates something new. Because of the primacy of love-energy, reality is intrinsically relational; "being" is "being with another" in a way that is open to more being and more union. Evolution is the movement toward more being and consciousness—that is, greater awareness of the whole and deeper connection to the whole. An Omega-centered evolution, according to Teilhard, can be no other than an evolution toward the fullness of love.

Cosmic Personalization

The beautiful images of *Journey of the Universe* show the immense, rich diversity of life, simple atoms and cells emerging into complex mammals and humans. This emergent creative diversity is, in Teilhard's view, the outflow of love energy. Nature does not readily disclose its secrets to us, but Teilhard looked deeply into the depths of nature. He wrote in one of his essays that it

---

6. Pierre Teilhard de Chardin, *Human Energy*, trans. J. M. Cohen (New York: Harcourt Brace Jovanovich, 1969), 119.

7. Ibid., 33.

8. Pierre Teilhard de Chardin, *The Phenomenon of Man*, trans. Bernard Wall (New York: Harper and Row, 1959), 264.

is not *something* but *Someone* who is forming in evolution. The transcendent depths of nature and the emergence of new being led him to describe the process of evolution as "deep incarnation," meaning that Jesus Christ is present from the Big Bang onward and evolution finds its meaning and direction in the Christ. What the biological sciences identify as the forces of nature, Christian faith sees as the incarnation of divine love. In Teilhard's view, evolution is the process of cosmic personalization. Evolution is not sheer blind, random chance but a process of personalization toward unity in love, what he called an "irreversible personalizing universe." This personalization of the universe, marked by an ascent of love and consciousness, is the evolution of Christ.

Created Cocreators

Teilhard did not see the human person lost or insignificant in light of evolution; rather he saw the human person as one truly unique, not a chance arrival but an integral element of the physical world. The human person is vital to evolution in that we rise from the process, but in reflecting on the process we stand apart from it. Teilhard defines reflection as "the power acquired by a consciousness to turn in upon itself, to take possession of itself *as of an object* endowed with its own particular consistence and value: no longer merely to know, but to know that one knows."[9] He quotes a phrase of Julian Huxley: the human person "is nothing else than evolution become conscious of itself." To this idea Teilhard adds, "The consciousness of each of us is evolution looking at itself and reflecting."[10] Thus the human person emerges from the evolutionary process and is integral to evolution. She/he is "the point of emergence in nature, at which this deep cosmic evolution culminates and declares itself."[11]

Teilhard saw the process of unification in and through the human person who is the growing tip of cosmic evolution. The destiny of humanity and the aim of the cosmos are intertwined in the mystery of Christ. Teilhard wrote that God and creation evolve to *pleroma* in and through human efforts to unite through love and thus to unite the multiple into unity. Humanity plays a distinct role in the Christogenic process. Christ is the physical and personal center of an expanding universe. But Christ cannot be the energy of evolution unless the incarnation is allowed to continue in us. As the apex of the created world, the spiritualized, self-reflective human person is open to the

---

9. Ibid., 165.

10. Ibid., 221.

11. Teilhard de Chardin, *Human Energy*, 23.

infinite, seeking fulfillment in the absolute mystery of God. In light of an evolutionary universe, Teilhard spoke of the human person as a "cocreator." God evolves the universe and brings it to its completion through the instrumentality of human beings. The human person is called to cooperate with God in the transformation of the universe.[12] Thus, it matters what a human person does and how a person lives in relation to God, for only through our actions can she or he encounter God.[13]

"*Trans*" *Human or* "*Ultra*" *Human?*

Teilhard was acutely aware of internal forces that could thwart the direction of evolution toward the Omega Point. He was concerned about the use of resources, limited food supplies, and whether an expanding population would be able to live amiably and in peace with each other under conditions that he no longer described as "convergence" but "external compression."[14] He questioned whether the human race, having experienced "a scientific justification of faith in progress[,] was now being confronted by an accumulation of scientific evidence pointing to the reverse—the species doomed to extinction."[15] The only solution, he indicated, is not "an improvement of living conditions," as desirable as that might be; rather the inner pressures of history are the catalyst for evolution toward more being. We have reached the end of the expanding or "diversity" stage and are now entering the contracting or "unifying" stage. The human is on the threshold of a critical phase of superhumanization: the increasingly rapid growth in the human world of the forces of collectivization, the "super arrangement" or the megasynthesis.[16] At this point, Teilhard's theory runs counter to that of Darwin; the success of humanity's evolution will not be determined by "survival of the fittest" but by our own capacity to converge and unify.[17] The most important initial evolutionary leap of the convergence stage is the formation of what he called

---

12. "Cocreator" is a term put forth by Lutheran theologian Philip Hefner to describe his theology of the created cocreator. See Philip Hefner, *The Human Factor: Evolution, Culture, and Religion* (Minneapolis: Fortress, 1993), 208–9.

13. George Maloney, *The Cosmic Christ: From Paul to Teilhard* (New York: Sheed and Ward, 1968), 189.

14. Pierre Teilhard de Chardin, *The Future of Man*, trans. Norman Denny (New York: Harper and Row, 1964), 235.

15. Ibid., 298–303.

16. Archimedes Carag Articulo, "Towards an Ethics of Technology: Re-Exploring Teilhard de Chardin's Theory of Technology and Evolution," http://www.scribd.com/doc/16038038/Paper2-Technology.

17. Teilhard de Chardin, *Phenomenon of Man*, 243.

"the noosphere," a single thinking envelope so as to form a single vast grain of thought on the sidereal scale, the plurality of individual reflections grouping themselves together and reinforcing one another in the act of a single unanimous reflection.[18]

The noosphere is a psychosocial process, a planetary neo-envelope *essentially linked with the biosphere* in which it has its root, yet is distinguished from it—a sphere of collective consciousness which preserves and communicates everything precious, active, and progressive contained in this Earth's previous evolution. Teilhard's worldview is one of "deep catholicity," an emerging incarnational wholeness or cosmic personalization that is thoroughly christic. To live in deep catholicity is to be aware of this cosmotheandric process of evolution and to participate in it, cocreative of its future. How we think makes a difference to how we act and how we orient our lives for the future. For Teilhard, thinking is essential to evolution. To think is to unify, to gather the partials of experience into greater wholes. This type of catholicity or conscious wholemaking undergirds a deepening of love.

Interestingly, we find ourselves on the level of the noosphere today but without a sense of deep catholicity or meaning or purpose. It is a breathless, frenzied pace of technological achievement within a short span of human history, and we have not yet asked ourselves, What do we want with our technologies? Gordon Moore, cofounder of Intel Corporation, predicted that the computer chip would evolve exponentially; that is, the number of transistors in a dense integrated circuit would double approximately every two years. Today, we have more computing power in a cell phone than we had in the 1969 Apollo spacecraft. Computer scientist Bill Joy issued a warning in 2000 that "our most powerful 21st-century technologies—robotics, genetic engineering, and nanotech—are threatening to make humans an endangered species."[19] He suggests that humans should limit or even outright abandon their endeavors in these technologies in order to prevent the possibility of a catastrophe.

The fears of human extinction are less apocalyptic today than the emergence of a new posthuman species. Scholars indicate that the computer is giving rise to a new digitized human being who is more at home in the presence of artificial intelligence and virtual reality than among flowers and trees. Children are now spending their formative years online, and recent studies

---

18. Ibid., 251.
19. Bill Joy, "Why the Future Doesn't Need Us," *Wired* (April 2000), http://archive.wired.com/wired/archive/8.04/joy_pr.html.

show that excess computer usage is rewiring the brain.[20] Margaret Wertheim notes that artificial intelligence is spawning a philosophical shift, from reality constructed of matter and energy to reality constructed on information.[21] Philosopher Carl Mitcham, in his address to the American Philosophical Association, said, "A thousand or two thousand years ago the philosophical challenge was to think nature—and ourselves in the presence of nature. Today the great and the first philosophical challenge is to think technology and . . . ourselves in the presence of technology."[22] A new term, "cybergnosticism," has been coined to describe "belief that the physical world is impure or inefficient, and that existence in the form of pure information is better and should be pursued."[23] Michael Heim sees strong links between artificial intelligence and Platonic, Gnostic, and hermetic traditions, insofar as they emphasize the goodness of spiritual reality and corruption of material reality, an idea consonant with cyber life and posthumanism: "Suspended in computer space, the cybernaut leaves the prison of the body and emerges in a world of digital sensation."[24] Ray Kurzweil and others predict that a world of intelligent silicon-based creatures will soon coexist with carbon-based creatures, both growing and evolving together. Is the end of the human drawing to a close?

Although there are a wide range of technologies today to enhance human life, the term "transhumanism" refers to technologies that can improve mental and physical aspects of the human condition, such as suffering, disease, aging, and death. Transhumanism is "the belief that humans must wrest their biological destiny from evolution's blind process of random variation . . . favoring the use of science and technology to overcome biological limitations."[25] Transhumanists look to a postbiological future where superinformational beings will flourish and biological limits such as disease, aging, and death will be overcome. Kurzweil claims that machine-dependent humans will eventually transcend death, possibly by "neurochips" or simply by becom-

---

20. Dave Mosher, "High Wired: Does Addictive Internet Use Restructure the Brain?" *Scientific American* (June 17, 2011), http://www.scientificamerican.com/article/does-addictive-internet-use-restructure-brain/.

21. Stephen R. Garner, "Praying with Machines: Religious Dreaming in Cyberspace." *Stimulus* 12, no. 3 (2004): 20.

22. Carl Mitcham, "The Philosophical Challenge of Technology," *American Catholic Philosophical Association Proceedings* 40 (1996): 45.

23. Garner, "Praying with Machines," 20; D. O. Berger, "Cybergnosticism: Or, Who Needs a Body Anyway?" *Concordia Journal* 25 (1999): 340–45.

24. Michael Heim, *The Metaphysics of Virtual Reality* (New York: Oxford University Press, 1993), 89.

25. Articulo, "Towards an Ethics of Technology."

ing totally machine dependent. Robert Geraci states, "Our new selves will be infinitely replicable, allowing them to escape the finality of death."[26] The aims of transhumanism are framed by the Christian ideals of immortality, salvation from suffering, and personal happiness. In short, the transhumanist creed proclaims that technology will fulfill what religion promises.

Toward Ultrahumanity

Teilhard realized the power of computer technology and saw that it could usher in the next level of evolution if we use it toward this end. He used the language of ultrahumanity to emphasize the need for humanity to enter into a new phase of its own evolution. Man is psychically distinguished from all other animals, he wrote, by the fact that "*he not only knows, but knows that he knows.*"[27] He spoke of a new level of "*co-consciousness,*" a collective awareness brought about by the convergence of human beings (the noosphere) over the surface of the Earth. Teilhard predicted the evolution of the computer as the "brain" behind the noosphere and the catalyst for the next step of evolution. With the rise of technology he saw a forward movement of spiritual energy, a maximization of consciousness and a complexification of relationships. While technology can extend the outreach of human activity, it depends on a broader use of human activity and how humans control psychic, spiritual energy needs and powers.[28] Conscious evolution in and through technology demands a consciousness of the journey of the universe itself, our place in this journey, and the power of God drawing us into a new future. Teilhard did not anticipate the perfection of being through artificial means; rather, for him evolution is progression toward more being. He wrote, "It is not *well being* but a hunger for *more-being* which, of psychological necessity, can alone preserve the thinking Earth from the *taedium vitae.*"[29] He distinguished "more being" from "well-being" by saying that materialism can bring about well-being, but spirituality and an increase in psychic energy or consciousness bring about more being.[30] He imagined psychic energy in a continually more re-

---

26. Robert Geraci, "Spiritual Robots: Religion and Our Scientific View of the Natural World," *Theology and Science* 4, no. 3 (2006): 235.

27. Teilhard de Chardin, *Future of Man*, 293.

28. Joseph A. Grau, *Morality and the Human Future in the Thought of Teilhard de Chardin: A Critical Study* (Cranbury, NJ: Associated University Presses, 1976), 274.

29. Teilhard de Chardin, *Future of Man*, 317.

30. Grau, *Morality and the Human Future*, 275.

flective state, giving rise to ultrahumanity.[31] The noosphere is not the realm of the impersonal, but conversely it is the realm of the *deeply personal* through *convergence*. It is not simply a new level of global mind, but rather the new level of global mind is the emergence of Christ, "a new participation in God."[32] Through the convergence of consciousness and a deepening of love, God is being born from within; salvation is "becoming one with the universe."[33]

Teilhard awakens in us awareness of belonging to Omega and thus to a cosmotheandric whole: God, humanity, and cosmos deeply united in ongoing development. An embodied consciousness of presence is one of deep relatedness to Brother Sun, Sister Moon, Sister stars. Such consciousness is more than digitized random information, such as machine intelligence; rather, embodied presence entails encounter and response. Deep consciousness and deep relatedness mean apprehending the whole, an awareness of belonging to the whole. This "deep catholicity" radiated throughout Teilhard's life and work. His great love of matter and the dynamism of the universe impelled him to view technology within the context of cosmic evolution. He saw the power of computer technology to enhance personal relatedness and deepen it through shared consciousness and the emergence of global mind. But Teilhard did not advocate replacement or displacement of the human person by the machine. He advocated an "ultrahumanism," a deepening of personhood through shared consciousness and interrelatedness, an interpersonalization of cosmic evolution through the global mind and hence the global heart. Teilhard was keen to realize that what we become shapes the journey of the universe. The power of this journey is not mere impersonal information but the deeply personal energy of love.

---

31. Henry Kenny, *A Path through Teilhard's Phenomenon* (Dayton, OH: Pflaum Press, 1970), 105.

32. Philip Hefner, *Technology and Human Becoming* (Minneapolis: Fortress Press, 2003), 84.

33. Pierre Teilhard de Chardin, *How I Believe*, trans. René Hague (New York: Harper and Row, 1969), 81.

8

Teilhard and the Consecrating Universe

BEDE BENJAMIN BIDLACK

At the American Teilhard Association's annual conference in 2012 I present-ed some reflections on Teilhardian themes in *Journey of the Universe*.[1] At that time I spoke about the need for what Pierre Teilhard de Chardin called "zest for life" and the fantastic way in which *Journey*, as a work of art, communi-cated that. I emphasized three themes: wonder, power, and promise. In order to develop a zest for life, people must regain their capacity to look upon the cosmos with wonder, attend to it, and savor its magnificence. Particularly magnificent is Humankind[2] with its capacity to shape the future of Earth; this is the awesome power that people wield. If wonder and power are culti-vated in a disproportionate way, one can cower at the challenge set before us. What Teilhard teaches, however, is that the cosmos is moving with a purpose toward fulfillment in Christ Omega. People can bravely meet the challenge of the future because of the promise of the love of Christ Omega.

This volume is intended to honor the visionary Thomas Berry, whose 1978 essay "The New Story" was the inspiration for *Journey of the Universe*. His concluding remarks, titled "Transmitting Values," run parallel to the observations above. He insists on four points. First, people must obey the forces of Earth and not harm Earth processes as they unfold. How is this possible without a sense of wonder that admits that Humans are not in con-trol of these processes? Berry so often said that the Human is derivative of Earth. This being the case, to pretend that Humans, and not Earth, control

---

1. Bede Benjamin Bidlack, "'Zest for Life' in *Journey of the Universe*," American Teilhard Asso-ciation Annual Meeting, Union Theological Seminary, New York, April 21, 2012, http://teilhard-dechardin.org/mm_uploads/Bede_Benjamin_Bidlack.pdf.

2. English capitalizes proper nouns for the purpose of distinguishing the proper noun from other nouns. Doing so, authors draw attention to the proper noun so to affect in readers the atten-tiveness and respect authors believe the noun deserves. In addition, proper nouns often communi-cate an interior life, a consciousness possessed by the noun. Such are my intentions for capitalizing Human, Cosmos, and Universe—nouns that have been diminished due to exploitation, presump-tion, and a denial of their subjectivity.

Earth processes is to naively give credence to a kind of magic, here defined as "a belief and practice according to which men are convinced that they may directly affect natural powers and each other among themselves whether for good or evil by their own efforts in manipulating the superior powers."[3] Efforts to make the world more commodious for Human Beings has led Earth to correct itself through a changing climate, which science can neither predict nor control.

Berry's second point is that power is for the sake of building communion. Study of Earth processes is not to separate Humankind from Earth for the sake of dominating it, but for the purpose of growing in the appreciation of Human origins and for nurturing Human-Earth relations. Placing the Human in proper relationship with Earth and the Cosmos is not submission, but reconciliation. Reconciliation with Earth is an exciting, joyful endeavor, which will enhance one's zest for life. People may have to change their current lifestyle, but it is a change that will build communion and therefore enhance life, not detract from it.

Berry's third point is to have a mood of confidence. This runs parallel to my observation that people can look to the promise of the future and have confidence that they can meet the challenges reconciliation demands. Teilhard pointed out that to believe otherwise—to believe that war or environmental destruction will inevitably lead to doom—is to think that the 13.8-billion-year history of the universe is absurd. It is to think that somehow the Human's few hundred thousand years of existence can counter the eons of forces that brought it about. Laced with an unspoken pride, such a nihilistic attitude is the antithesis of the zest for life.

Berry's final point is that these prior three will provide the much larger context for Human existence. Bernard Lonergan argued that the questions a person asks will drive the answers, and the answers will shape that person's world. In the educational series of *Conversations* that accompany *Journey*, psychologist Sachiko Kawaura further states that the answers available to questions are limited by context.[4] That is, a small context will provide small answers; small answers will result in a small world. A small world is inconsistent with the science of the times; it is untrue. To live and act as if the world was small—as if it does not descend into the subatomic and expand light years to the edges of the Universe—is irrational, yet irrational "leadership"

---

3. Mariasusai Dhavamony, *Phenomenology of Religion, Documenta Missionalia* (Rome: Gregorian University Press, 1973), 31.

4. Mary Evelyn Tucker, *Journey of the Universe: Educational Series* (San Francisco: Northcutt Productions, 2011), disc 4. Later editions are titled *Journey of the Universe: Conversations.*

from the international community in response to climate change is sadly characteristic of the current age.

To quote one middle-aged climatologist, "Our generation has punted on the environment." The next generation must take the ball and run with it, but it can only do so if they inherit a zest for life founded on wonder-obedience, power-communion, promise-confidence, and context. The way people have communicated their essential context, the deepest truths from which the next generation must live, is through story. Mary Evelyn Tucker, Brian Swimme, and all of those involved in *Journey* use the powerful medium of film to tell the Universe Story and to transmit this larger context. The parallel between Teilhard and the creators of *Journey* as storytellers is the theme this chapter develops by looking at the art form of ritual as storytelling in which the participants not only transmit the story but also become the story itself. It does so by examining Teilhard's sacramental mysticism.

Teilhard as Storyteller

Much has been written on Teilhard's story as he put it in writing. This is Teilhard the scientist, theologian, and visionary. But one author, Thomas M. King, wrote of Teilhard as a priest in his final work, *Teilhard's Mass*.[5] As a priest, Teilhard used the canvas of the Divine Liturgy of the Eucharist to tell the Jesus Story, and importantly to demonstrate Humanity's role in the Universe Story. The Mass expressed Teilhard's sacramental theology.

Augustine defined a sacrament as "a visible sign of an invisible grace." Centuries later, the Council of Trent numbered the Sacraments at seven, and the Church is the dispenser of those Seven Sacraments. Since Vatican II in the mid-twentieth century, without diminishing the importance of the Seven Sacraments, theologians have emphasized that grace is not a thing or power to get from God, but is a constant activity of God by which God draws people into divine life. Grace is a verb. God graces us. Contemporary sacramental theology argues that God graces people through events, others, and Creation. In other words, God graces a person whenever he or she allows it.

This is the general outline of Teilhard's sacramental theology in the first half of the twentieth century, decades before Vatican II. Christ has a cosmic nature that inspires ("in-spirits") everything. People live and move in a divine milieu, the raw material of life.

---

5. Thomas M. King, *Teilhard's Mass: Approaches to "the Mass on the World"* (Mahwah, NJ: Paulist Press, 2005).

Rudolph Otto, by and large, defined the divine as *mysterium fascinans et tremendum*, a captivating and awesome mystery. People feel it when they have a sense of wonder. For Otto, the divine is wholly other, distant from the world. However, Teilhard and *Journey* argue that the divine is not separate, but coming forth from within the universe. Humanity, especially in its priestly role, plays an essential part in this process of divinization. Instrumentally bringing about the divinization of the world is the role of the priesthood. This is what Teilhard as a priest did at Mass, and this is what Vatican II urges upon the laity: to take up their identities as priests, by virtue of their baptism. *Sacrosanctum Concilium* states,

> Mother Church earnestly desires that all of the faithful should be led to that fully conscious and active participation in liturgical celebrations, which is demanded by the very nature of the liturgy. Such participation by the Christian people as "a chosen race, a royal priesthood, a holy nation, a redeemed people" (1 Peter 2:9; cf. 2:4–5) is their right and duty by reason of their baptism.[6]

Teilhard demonstrated that people should approach God with the Universe by offering up the experiences of their everyday lives. For example, in his 1955 summary of *The Divine Milieu*, "The Christic," he beautifully writes that the bread of offering is those aspects of life that quicken us, and the wine is those that diminish:

> And then there appears to the dazzled eyes of the believer the eucharistic mystery itself, extended infinitely into a veritable universal transubstantiation, in which the words of Consecration are applied not only to the sacrificial bread and wine, but . . . to the whole mass of joys and sufferings produced by the Convergence of the World as it progresses.[7]

Or again in "The Mass on the World":

> All things in the world to which this day will bring increase; all those that will diminish; all those too that will die: all of these, Lord, I try to

---

6. *Sacrosanctum Concilium* (Constitution on the Sacred Liturgy), 1963, http://www.vatican.va/archive/hist_councils/ii_vatican_council/documents/vat-ii_const_19631204_sacrosanctum-concilium_en.html, §14.

7. Pierre Teilhard de Chardin, *The Heart of Matter*, trans. René Hague (New York: Harcourt Brace Jovanovich, 1980), 94.

gather into my arms, so as to hold them out to you in offering. This is
the material of my sacrifice; the only material you desire.[8]

In other words, Christians are to be Christ-like by offering their lives to God.
Christ, in the person of the priest, transforms the particular elements of bread
and wine on the sacrificial altar into his divine Flesh and Blood. People, too,
are to say the words of consecration over their lives. Teilhard writes,

> That is why, in repeating over our lives the words the priest says over the
> bread and wine before the consecration, we should pray, each one of us,
> that the world may be transfigured for our use; "that they (the elements
> of our lives) may become for us the body and blood of our Lord Jesus
> Christ."[9]

In addition, the sanctification that takes place on the church altar does not
remain there, but, as Teilhard says, shimmers out, like a halo (Fr. *auréole*), into
the entire cosmos:

> The eucharistic transformation goes beyond and completes the transub-
> stantiation of the bread on the altar. Step by step it irresistibly invades
> the universe. It is the fire that sweeps over the heath; the stroke that
> vibrates through the bronze.[10]

What is well known is Teilhard's scientific mysticism: that the universe is
taken up into mind to be offered to God, most especially through research.[11] I
here draw attention to his sacramental mysticism. The celebration of the Eucharist has a divinizing effect upon the cosmos. For Teilhard, it is not only for the
redemption of sinners, but for the transfiguration of the universe. This is the
duty and privilege Christ gave us at the Last Supper: for the baptized to take up
their role as priests to sanctify Creation in the ongoing "Mass on the World."

To take this a step further, each person offering up his or her life serves as
the intermediary, the representative, of the universe, just as the priest represents Christ as the intermediary on behalf of Humanity. The Human is the
universe at prayer. *The Human Phenomenon* reads,

> Without metaphor, it is through the whole length, thickness, and depth
> of the world in movement that human beings see themselves capable of

---

8. Pierre Teilhard de Chardin, "Mass on the World," trans. in King, *Teilhard's Mass*, 146.

9. Pierre Teilhard de Chardin, *The Divine Milieu*, trans. Siôn Cowell (Portland, OR: Sussex Academic Press, 2004), 105.

10. Ibid., 87.

11. See Thomas M. King, *Teilhard's Mysticism of Knowing* (New York: Seabury Press, 1981).

experiencing and discovering their God. To be able literally to say to God, that we love him, not only with our whole body, our whole heart, and our whole soul, but with our whole universe.[12]

To summarize, traditionally speaking, only those consecrated with the charisms of the priesthood are priests, and only the Holy Eucharist is a Sacrament. However, each of the baptized is to be sacerdotal, and the entire cosmos is to be sacramental. God showers God's divinizing activity through people, events, and Creation. Humanity is to serve the universe by consecrating it through the offering of daily life to God. Teilhard told this story as a priest; the Mass as ritual transmits the story by involving everyone in it.

Conclusion

Journey of the Universe leaves the viewer with the exciting and challenging prospect that Humanity has the power to participate in the physical universe in a mutually enhancing way, which in turn affects the interior lives of people. These exterior and interior influences spiral into the future. This is the familiar interaction between radial (psychic) and tangential (physical) energy Teilhard presented in *The Human Phenomenon*. All of Humanity is involved in this process, and so *Journey* challenges all people to involve themselves in the new brokering of Human-Earth relations. But how? *Journey* ends on the hope-filled note, "Wonder will guide us."

In order to have the substance of the message appeal to the widest audience, the filmmakers did not overtly reference Teilhard and his Christian view of Christ pulling the cosmos from ahead. Therefore, this paper draws out Teilhard's sacramental mysticism, which is specifically Catholic and liturgical. The argument is that people need to renew their relationship with the sacred liturgy and respond to their sacerdotal calling as Vatican II insists. From the Catholic point of view, this is essential for moving ahead in our Human-Earth relations. No wonder that the Green Pope, Pope Benedict XVI, is a liturgical theologian.

To support the argument that the Body of Christ is "a royal priesthood," the paper references a document of Vatican II, *Sacrosanctum Concilium*. Another Vatican II document, *Nostra Aetate,* inspires another suggestion. *Nostra Aetate* unequivocally teaches that God reveals Godself through other religions. Since that is the case and since it is quite probable that God saves through other religions, could not one suppose that some activity analogous

---

12. Pierre Teilhard de Chardin, *The Human Phenomenon*, trans. Sarah Appleton-Weber (Brighton, UK: Sussex Academic Press, 2003), 213.

to the consecration of the Eucharist occurs in the ritual activity of other religions? Just as Catholics believe that Scripture and the Sacraments are the ordinary means by which Christ divinizes the world and that Christ reveals and saves in other religions, could there be some way that the ritual activity of other religions spiritualizes the world in a way suggested by Teilhard's sacramental mysticism? After all, Teilhard thought all relationship-enhancing activity—what he called activities of growth—divinized the universe. Similarly, Thomas Berry wrote of the Cosmic Liturgy, the work of the Cosmos in its movement toward ultimate fulfillment.

While the idea is provocative, perhaps it should remain only a theological possibility for now. The development of such a sacramental theology lies in the future, far in the future. Such speculation is in the realm of theologies of religions, and theologians like James Fredericks and Francis X. Clooney point out that Christianity must first get to know the religions before theorizing just what God is doing in them. Interreligious dialogue as it occurs today is a new endeavor. It took the Church five hundred years to come to some peace with Greek thought; Christians have only been getting to know the other religions for about fifty. Patience and study are needed.

That said, the best way to transmit the new story is through art, and arguably the best art form is ritual due to the high participation of all present. Ritual often involves its participants through a great deal of activity when other arts, with the exception of music, leave the audience at a distance to participate only minimally or passively. Humans are ritual beings, but the contemporary, developed world has set aside ritual in favor of the intellect. As it has done so, the other, nonintellectual capacities of Humanity are left behind. The result is a disembodied—that is, extraterrestrial—context that does not fully feel the urgency of our current moment in the Earth Story. Scientific evidence on graphs and spreadsheets has done little to influence daily lives or to provoke attention to Human-Earth symbiosis. Many people in the environmental movement are befuddled at this jeopardizing apathy. Numbers that appeal solely to the intellect only tell part of the story. This is why Thomas Berry urges us to transmit the New Story, the story told in *Journey of the Universe*. The power of transmitting a story across generations is something the religions of the world do very well, and why Mary Evelyn Tucker and John Grim have put their efforts into the Forum on Religion and Ecology at Yale. Religions embody their stories through their rituals, and the Catholic ritual par excellence is the Liturgy of the Eucharist. Teilhard lived the Universe Story when he was a priest celebrating Eucharist; the Church encourages the laity to do the same.

Part II

Dwelling in a Cosmos

A SACRAMENTAL UNIVERSE

❧

9

Living Cosmology and the Earth Community

Views of the Divine

JOHN CHRYSSAVGIS

It is in the nature of the universe to move forward between great tensions, between dynamic opposing forces. If the creative energies in the heart of the universe succeeded so brilliantly in the past, we have reason to hope that such creativity will inspire us and guide us into the future. In this way, our own generativity becomes woven into the vibrant communities that constitute the vast symphony of the universe.[1]

These are the closing words of *Journey of the Universe*, which takes us on a systematic voyage—actually, a sacred pilgrimage—through the creation of the sun and stars, the complexity of life and death, and the connection between animals and humankind. It is the same inspiring tension that has survived the conflict between sacred and secular, the clash between religion and science, as well as the confrontation between heaven and Earth.

We can now discern the relationship between divine essence and divine energies, as well as marvel at the connection between the living seed and the infinite universe. This is what Orthodox Christian thought likes to call a sacred—or sacramental—worldview, which is reflected in its mystical and prophetic literature, in both religious and philosophical circles.

---

1. Brian Thomas Swimme and Mary Evelyn Tucker, *Journey of the Universe* (New Haven, CT: Yale University Press, 2011), 118.

Cosmic Liturgy

It has always been a source of great comfort to me that Orthodox *spirituality* retains a *sacramental* view of the world, proclaiming a world imbued by God and a God involved in the world—a sacrament of communion. God is the Lord of the dance of creation, which is perceived as a voluntary overflow of divine gratuitousness and grace. Or as a seventh-century mystic, Maximus Confessor (580–662), puts it, the whole world is "a burning bush of divine energies," "a cosmic liturgy."[2]

This dimension of liturgy, of joyful praise in creation, is God's gift to the world and does not at all depend on any of our environmental efforts or awareness. So unless we willingly entertain and joyfully enter into this interdependence of all persons and all things, we certainly cannot hope to resolve issues of economy and ecology. For we should respond to nature with the same delicacy, sensitivity, and tenderness with which we respond to a person in a relationship. And our failure to do so is the fundamental source of pollution, a consequence of our inability to relate caringly toward the created world.

Such is the breadth and depth of the Orthodox Christian cosmic vision, one that is much larger than that of any one individual. I may be at the center of this vision or theophany, but I become aware that I am also but a detail of the living universe. Indeed the world ceases to be something that I observe objectively, or exploit selfishly, and instead becomes something of which I am a part, personally and actively. No longer then should I feel as a stranger, whether threatened or threatening, but as a compassionate friend in and of the world. For whenever we reduce life to ourselves (to our concerns and our desires), we neglect this enlarged, cosmic vision of creation. In fact, whenever we narrow even religious life to ourselves (to our concerns and our desires), we ignore our vocation to implore God for the renewal of the whole polluted cosmos.

"A Tale of Two Truths"

When medieval scholars maintained in their "tale of two truths" that science was called to observe the "Book of Nature" while religion was reflected in the "Book of Scripture," they were struggling with a long-standing debate regarding the inconsistency between science and religion as well as the incongruence between creation and church. In his book *Being as Communion*, Met-

---

2. Themes that run throughout the writings of Maximus.

ropolitan John [Zizioulas] of Pergamon, arguably the foremost Orthodox theologian today, compares these two different approaches and asserts that

> Science and theology for a long time seemed to be in search of different sorts of truth, as if there were not one truth in existence as a whole. This resulted in making truth subject to a dichotomy between the transcendent and the immanent.[3]

Indeed, this single truth about creation is proposed by many of the early church fathers—both Eastern and Western—in a variety of interpretations ranging from the fundamentally literal to the spiritually symbolic. Thus, in his exceptional treatise *On the Six Days of Creation*, Basil the Great (330–379) insists that the scriptural narrative is not a scientific explanation.[4] He strongly denounces those obsessed by the letter of the text, who overlook the spirit of scripture, and describes them as preferring "technology to theology." For St. Basil, the book of Genesis should be considered not as history but rather as metahistory.

Thus, the living universe invites us to an enlarged view of life, a more organic view of the world, not unlike that exposed with a wide-angle lens. By nature, it prevents us from using or abusing its resources; it prohibits a narrow, self-indulgent, self-serving way. Instead, the world becomes a celebration of the essential interconnection and interdependence of all things. What people conveniently overlook about the Genesis story is that the sixth day of creation is not dedicated exclusively to Adam[5] but shared with "living creatures of every kind; cattle and creeping things; and wild animals of every kind."[6] There is more that unites us than separates us, not only as human beings but also within the entire universe. This is a lesson we have only learned the hard way in recent decades.

The Universal Seed

Two theories or theologies about the origin of the universe are fairly well known. Almost two thousand years ago—long before the Big Bang theory was generated by physicists as a single-point, ever-expanding worldview—

---

3. John D. Zizioulas, *Being as Communion* (New York: St. Vladimir's Seminary Press, 1997), 119.

4. *Hexaemeron* 1.2, PG 29.8B; 1.11, PG 29.28B; 6.2, PG 29.120D; 9.1, PG 29.188D.

5. Genesis 1:26.

6. Genesis 1:24.

the classical concept of the germinative/generative principle or logos[7] and its far-reaching theological implications for ancient Greek philosophy and early Christian thought emerged. Logos was the logic observed in nature by the classical philosophers, especially Aristotle; and to the early Christian thinkers, the rational design of nature is clear evidence of a God who plans and implements the universe.

What most people may be less acquainted with, however, is *the notion of the universal seed*. The second verse of the book of Genesis relates how "the Spirit of God hovered over the waters."[8] St. Basil interprets this as the divine Spirit preparing the nature of water to produce all living beings, bracing the nature of water to produce the living universe—very much, as he observes, "like a bird broods on its eggs."[9] Similarly, in his *Third Commentary on Genesis*, St. John Chrysostom (349–407) refers to a fertile power that was active and alive in the waters and in the world, ultimately empowering all living things to emerge. Thus, all of life in the universe originates from a single ontological source or seed implanted by God in that original moment or "beginning." St. Basil continues,

> This short command ("Let the earth bring forth") is—in a moment— an elaborate system; so nature receives the impulse of this first command and follows its course without interruption until the consummation of all ages.[10]

Therefore, the world and the water were pregnant with every limitless variety of living species. Over the ensuing five days, God embellished the world, differentiating each creation by his commands and bringing forth hidden treasuries of forms stored within them. For St. Basil, this is precisely why the Greek (Septuagint) translation of the Hebrew text speaks of "one day" whereas the subsequent days are chronologically designated as "the second day," "the third day," and so on. Something unique and different happened on that seminal "day one," that original word, that first "bang."

So the church fathers shared a view of the world as containing a determinative force, through which God calls the immense variety of life forms to

---

7. In Latin: *ratio seminalis*. Adopted by the Sophists and Stoics, the phrase was incorporated into Jewish philosophy by Philo (20 BCE–50 CE) and into Christian thought by Justin Martyr (100–165).

8. Genesis 1:2.

9. *Hexaemeron* 2.6, PG 29.40f.

10. Ibid., 5.10. For the entire passage in English translation, see Philip Schaff and Henry Mace, eds., *Nicene and Post-Nicene Fathers*, Series 2, Volume 8 (Edinburgh: T&T Clark, reprinted 1996), 81.

unfold—from the elements through the plants and from the animals to the human beings. Such a dynamic view of nature is undoubtedly compatible with the scientific theory of biological evolution.

Essence and Energies: Unity and Diversity

However, there is yet another aspect of Orthodox cosmology or worldview that is worth highlighting here, namely that *creation is always perceived as an organic whole*. All living creatures are branches of the same tree, organic shoots of the same primordial seed, maintained in existence through the divine energies in which all life participates. Each January, on the Feast of Christ's baptism, Orthodox Christians proclaim, "The nature of waters is sanctified.... The earth is blessed.... The heavens are enlightened." And they pray that "by the elements of creation, by angels and human beings, by things visible and invisible, God's most holy name may be glorified."[11]

This binding unity is a direct result of the common evolution of all life, which shares the same elements (carbon, nitrogen, trace metals, etc.), the same processes (cell division, replication and repair of DNA, etc.), even the same genetic code. In this respect, all species share in unity even as they evolve in diversity. The *unity* helps us to perceive the relationship of all creatures as well as our relationship with the Earth itself. The *diversity* helps us appreciate the essential importance of all creatures, all life, and all environments for the sustainability and survival of our planet. No wonder, then, that the oldest surviving Christian liturgy prays, "Every material and spiritual creature proclaims the magnificence of God."[12] So the Green Patriarch, Ecumenical Patriarch Bartholomew, can assert,

> This connection is even detected in the galaxies, where the countless stars betray the same mystical grace and mathematical inter-connectedness. We do not need this worldview in order to believe in God or to prove God's existence. We need it to breathe; we need it simply to be.[13]

Unfortunately, there has been a cosmic shift and separation between God and man, heaven and Earth, cosmic liturgy and mathematical mechanics. However, the Eastern Christian tradition sought to modify the early Greek

---

11. Orthodox liturgy for the Feast of the Theophany.

12. The Liturgy of St. James is still celebrated twice a year in the Orthodox Church.

13. His All-Holiness Ecumenical Patriarch Bartholomew, "Faith and Environment: An Inspirational Perspective," Utrecht, the Netherlands, April 24, 2014.

concept of God as immobile *essence*, even while struggling to embrace its opposite—namely, the concept of a God understood as *becoming*. The Orthodox interpretation of scripture, liturgy, and spirituality reveals a God constantly reconciling all divisions by balancing the distinction between the immutability or stability of God with divine becoming or historicity—namely, God's intimate involvement in the created world and the human heart.

In this regard, Eastern theology prefers the model of a dynamic distinction between the *immutable essence* and the *uncreated energies* of God. The latter manifest the infinite possibilities and inexhaustible potentiality of the former. The divine energies—what the Hebrew scriptures call God's "glory"—charge the created world with reality and transparency, allowing it at once to reveal and to conceal the mystery of God. In the paradoxical language of Gregory Palamas (1296–1359),

> God is both existent and nonexistent; he is everywhere and nowhere; he has many names and cannot be named; he is ever-moving and unmoved; in short, he is everything and no-thing. . . . He remains wholly within himself and yet dwells wholly within us, causing us to participate not in his nature but in his glory and radiance.[14]

At the same time, God's essence remains totally transcendent—undefined and undetermined. But without divine energies there is no connection between God and the material universe, as affirmed in the twentieth century by one of the first Orthodox thinkers on the environment, Philip Sherrard (1922–1995):

> For if only the total transcendence of God is affirmed, then all created things, all that is in change and visible, must be regarded as without any real roots in the Divine, and hence as entirely negative and "illusory" in character; while if only the total immanence of God is affirmed, then creation must be looked upon as real in its own right, instead of as real only because it derives from and participates in the Divine; and the result must be a pantheism, and a worship of creation rather than of the

---

14. Cited by K. Ware, "God Immanent yet Transcendent," in *In Whom We Live and Move and Have Our Being: Panentheistic Reflections on God's Presence in a Scientific World*, ed. Philip Clayton and Arthur Peacocke (Grand Rapids: Eerdmans, 2004), 162. To quote Arthur Peacocke, "God would not be Creator unless the divine Being and the divine Becoming were facets of the same ultimate divine Reality." See his *Theology for a Scientific Age: Being and Becoming—Natural, Divine, and Human* (Minneapolis: Fortress Press, 1993), 185.

Creator, which must ultimately lead to the notion that God is superfluous, and hence to an entirely materialistic conception of things.[15]

One may infer from this passage, however, that because of the divine energies, nothing falls outside the embrace of God; everything is a reflection of the divine. The God contemplated by the Christian mystics of the medieval East was a God elusive yet familiar, both transcendent and immanent; it was a God who was afar and at the same time at hand.[16] This is the God worshiped in heaven while also venerated on Earth.

Conclusion: The Divine Seed

It is a tragedy that, in spite of the destruction and the suffering that we have inflicted on our planet, we have apparently not yet learned our lesson. The world remains for us a human-centered reality: we are still obsessed with ourselves, our problems, even our survival. And yet this is precisely what led us in the first place to this fateful predicament. Very little significance is attached to the reality that all things are coherent not just in their interrelatedness and interdependence but also in their relation to and dependence on God. For the fourth-century poet and theologian of Constantinople, Gregory Nazianzus (329–389), "All things dwell in God alone; all things swarm to him in haste. For God is the end of all things."[17]

The ancient Greeks had a similar worldview, recognizing the divine presence in all things. Thales (624–546 BCE) exclaimed, "Everything is full of God."[18] And Basil of Caesarea believed that even the slightest detail of creation bore the mark of the Creator:

> Look at a stone, and notice that even a stone carries some mark of the Creator. It is the same with an ant, a bee, a mosquito. The wisdom of the Creator is revealed in the smallest creatures. It is he who has spread out

---

15. Philip Sherrard, *The Greek East and the Latin West* (London: Oxford University Press, 1959), 35–36.

16. Western mystics confess the same worldview of God worshiped "in heaven as on earth." Thus, when Julian of Norwich (1342–1413) received "something small, the size of a hazelnut, in the palm of her hand," she was told that it resembled the whole world and "everything that was made." "I marveled how this could be," she conceded, "for it was so small that it might suddenly fall into nothingness." Whereupon a voice revealed, "It lasts and shall last forever; for God loves it. All things have their being in this way." See her *Revelations of Divine Love*, chapter 5.

17. Gregory Nazianzus, *Dogmatic Poems* 29, in PG 37.508.

18. *Fragment* 22.

the heavens and stretched out the immensity of the seas. It is he who has also made the tiny hollow shaft of the bee's sting.[19]

The same truth—discovered by science and discerned in theology—is poetically expressed outside of the theological world by the controversial twentieth-century Greek author Nikos Kazantzakis (1883–1957), whose work was regrettably misunderstood and maligned, even banned by the Vatican and condemned by the Church of Greece. Yet Kazantzakis retains a powerful religious worldview of the divine seed in the world—a view that critics might argue is a reinventing of Christianity. For Kazantzakis, created nature is the only premise and promise for either salvation or destruction; it is not a finished product, but a moving ground, a process of continuous self-transcendence and transformation. I close with his words:

Everything is an egg, and within it lies the seed of God, restlessly and sleeplessly active.... With the light of my mind and the fire of my heart, I beset God's watch—searching, testing, knocking to open the door in the stronghold of matter, and to create in that stronghold of matter, the door of God's heroic exodus.... For *we are not simply freeing God* in struggling with and ordering the visible world around us; *we are actually fashioning God*. Open your eyes, God is crying; I want to see! Be alert; I want to hear! ... For to save something [a rock or a seed] is to liberate God within it.... Every person has a particular circle of things, of trees, of animals, of people, of ideas—and the aim is to save that circle. No one else can do that. And if one doesn't save, one cannot be saved.... The seeds are calling out from inside the earth; God is calling out from inside the seeds. Set God free. A field awaits liberation from you, and a machine awaits its soul from you. And you can no longer be saved, if you don't save them.... The value of this transient world is immense and immeasurable: for it is on this world that God depends in order to reach us; it is in this world that God is nurtured and increased. ... *Matter is the bride of my God*: together they wrestle, together they laugh and together they mourn, crying through the nuptial chamber of the creation.[20]

---

19. Basil of Caesarea, *Commentary on Psalm 32*, 3, in PG 29.329.

20. Nikos Kazantzakis, *Ascetic Exercises*, 5th ed. (Athens, 1979), 85–89. Translation mine from the original Greek. This book is currently included in the series Kazantzakis Editions (Athens, 2009). For the English edition, see *The Saviors of God: Spiritual Exercises*, translated by Kimon Friar (New York, N.Y.: Simon and Schuster, 1960).

10

Between Creation and Apocalypse

CATHERINE KELLER

When I teach the Introduction to Theology course for my school's variegated international population of ecumenical mainliners, inner-city evangelicals, Asian Protestants, feminist posttraditionals, and spiritual ecoactivists, I work through several traditional doctrines. Each of these ancient symbols comes to life by way of certain nontraditional contrasts. The ancient contexts demand contemporary refractions. The doctrine of creation, for instance, would never let me teach it apart from some intensive work on our current planetary context. But now I find I can only do it with the help of *Journey of the Universe*. First we see the movie, then we interpret together the first chapter of Genesis. And this sets the whole semester's theological journey within the larger process of its only possible context.

What I am worried about is the sequel to the film. Watching it in September I had a premonition (not quite a preview) when in the beginning Brian Swimme points over from Samos to neighboring Patmos. I thought, *Uh-oh*. All too right. Is our civilization already headed back to that island? "Woe, woe, woe to the inhabitants of the earth," intoned John of Patmos two millennia ago.[1] Far from wonder at the universe and the intuition of divinity emergent from the awesome Universe Story, he confronts his readers with the other side of awe, the awful. Rudolph Otto prepared us for this doubleness with the *mysterium tremendum et fascinans*:[2] we may tremble in terror or attraction. But what combination of apocalyptically horrifying fact and spiritually transforming beauty will make the difference now?

When each of you teaches or preaches about climate change, you are no doubt trying to avoid paralyzing fear and energize motives for action. The motivating force of a magnificence in which we all participate is what the *Journey* project mobilizes as the boundlessly positive incentive for respon-

---

1. Revelation 8:13.

2. See Rudolph Otto, *The Idea of the Holy*, trans. John W. Harvey (Oxford: Oxford University Press, 1923), 1950. Originally published in German as *Das Heilige*, 1917.

sibility. But amid the current ecosystemic collapse and mass extinctions, the signs, as the filmmakers write, are unmistakable: "We are leaving behind the Holocene period of the past ten thousand years and entering the anthropocene, an era shaped primarily not by natural systems but by humans."[3] Perhaps we do not need a sequel. Perhaps we just need a day trip to Patmos.

You have, perhaps, each found fresh ways to convey wisely in your contexts not only the mounting effects of the irresponsibility but the millennial shift from an ecological language of *mitigation* of disastrous changes to one of *adaptation* to them. McKibben misspelled *Eaarth* for us in 2010, announcing, "We're running Genesis backwards, de-creating."[4] It seems that if we pay attention to the universe of Genesis we are sucked now through a wormhole right into eco-apocalypse. The data have become routine. At the point of writing this, the latest report of the Intergovernmental Panel on Climate Change (IPCC) offered along with its consensual road map to the situation little by way of hope. Instead the IPCC comes to the bleak conclusion that it may be too late—despite any efforts—to slow, let alone to reverse, climate change. It says that "even with adaptation, warming by the end of the twenty-first century will lead to very high risk of severe, widespread, and irreversible impacts globally."[5] The Patmos alarm is sounding all-too-eerily right.

The future is, of course, by definition open, full of unpredictabilities, incalculabilities, surprises of the worst and the best sort. But they do not erase the probability derived from the evidence. And it seems increasingly probable that we will do too little too late. So, then, is there a way nonetheless to teach that such difficult odds are no reason for despair? That improbable leaps forward—what Derrida calls "the impossible possibility of the im/possible"[6]— are par for the human course? I worry about this, wonder about this, work at this. Here I can only circle a bit within the dilemma.

Is there a way to acknowledge this lateness, this possible impossibility, our possible too-lateness, without sacrificing the awesome to the gruesome, the emergent creation story to a paranoid old apocalypse? In the ancient text, planetary human misery inflicted by the military economic dominion of the greedy whore-empire "Babylon" combines the four horsemen of hunger, pestilence, war and death with the collapse *avant la lettre* of Earth's ecosystems,

---

3. Brian Thomas Swimme and Mary Evelyn Tucker, *Journey of the Universe* (New Haven, CT: Yale University Press, 2011), 102.

4. Bill McKibben, *Eaarth: Making Life on a Tough New Planet* (New York: St. Martin's Griffin-MacMillan, 2011), 25.

5. http://www.ipcc.ch/report/ar5/syr/.

6. See Jacques Derrida, *On the Name*, trans. David Wood et al. (Stanford, CA: Stanford University Press, 1995), 43–44.

with the death of the seas, the great fires and floods, the deforestations, the droughts, the desertifications—vengeful, loveless misogynist John of Patmos may have been. But how can we avoid noticing that he, along with the whole legacy of Jewish apocalyptic literature, was intuiting something tragically true about the future course of history? Woe, woe, woe.

Whoa! Doesn't talking this way fuel a self-fulfilling prophecy of doom? So then it becomes important to announce that even John's Apocalypse does not mean the end of the world. "The end of the world," as it turns out, may be popular with fundamentalists. But it is not a biblical concept. There is catastrophic destruction, there is radical change. There is anticipated "a new heaven and earth."[7] Making new did not in its biblical texts ever signify a new creation out of mere annihilation, a *novo creatio ex nihilo*. Note also that "heaven" in Hebrew [*hashamayim*] can just as well be translated "sky" or "atmosphere." The Apocalypse was not predicting climate change. But if biblical symbols still haunt our imaginations, it may be worth repeating that for all the woeful destruction, he does not say it is too late, but rather, "Wake up, and strengthen what is at the point of death."[8]

What is now at the point of death is not Earth, is not the world—but *our* world, this stable ten thousand years of an Earth held to the average of a single degree Celsius of annual planetary temperature fluctuation. In this "sweet zone," civilization evolved. It may quite possibly be too late to save *this* world, Earth just as we have known it. But then may we say all the more awakeningly: it is not too late for a convivial future on Earth. It is not too late for a sustainable civilization. Not just a degraded survivalism but a new civil emergence does remain altogether possible. The New Jerusalem did once symbolize a just and inclusive new city, with its "tree of life" for "the healing of the nations" (as they still existed, implying a polis and political order, and so, evidently no absolute end). It is hard not to read the vision of this planet-sized artificial park, through the eyes of New Testament critic Stephen Moore, as "a megalopolis that is a continent-sized shopping mall, with a single stream and a single tree."[9]

Yet one can at the same time read there the crude and condensed dream of a new planetary polity, not a "return to nature," to be sure, but a realized cosmopolitanism. It may stir the utopian energies of a radically altered cosmos, a conviviality vibrant with compassion ("there will be no more tears"[10]). How-

---

7. Revelation 21:1.

8. Revelation 3:2.

9. Stephen D. Moore, *Untold Tales from the Book of Revelation: Sex and Gender, Empire and Ecology* (Atlanta: SBL Press, 2014), 225.

10. Revelation 21:4.

ever, any such dream can be abandoned to wishful thinking, fundamentalist second comings, or sci-fi postapocalypses if we do not frame it as our collective responsibility, our ability to respond—to "seize an alternative."[11] But it remains at once too grandiose and too pathetic to find ourselves at this point. Who are "we" to call ourselves to awaken, to make the difference between our extinction and our transformation?

The wake-up alarm is sounding. It has been sounding religiously for a couple of millennia; it is now science. James Lovelock—in the book written at eighty, the *Vanishing Face of Gaia*—offers a striking trope. He writes that it is as though an alarm clock is sounding as we sleep. He fears "that we still dream on and, rather than waking, we weave the sound of the alarm clock into our dreams."[12] Still—there is time. It is time, he writes, to wake up—"and realize that Gaia is no cozy mother that nurtures humans and can be propitiated by gestures such as carbon trading or sustainable development."[13] With the biologist Lynne Margulis, Lovelock had in 1974 advanced the "Gaia Hypothesis," trashed by many scientists in the last millennium, but now finding new currency. The leading continental science theorist, Bruno Latour—mindful of the scorn and misunderstanding heaped upon the Gaia metaphor—has recently written that it is "by facing Gaia, that wholly secularized and earthbound set of processes, that there is a dim possibility that we could 'let the Spirit renew the Face of the Earth.'" He goes on to call for a "multiplicity of engagements, of strategic assemblies of humans and nonhumans," as the only way "to cope with what the multiple loops traced by the instruments of science reveal of *the narrative complexity and entanglement of Gaia*."[14] "Wholly secularized"? Of course the Gaia theory must be protected against misplaced spiritualizations and well-meaning goddess worship. Gaia as a hypothesis is no deity. This was clear also to Rosemary Ruether in her prophetic 1992 *Gaia*

---

11. See the following websites: http://www.ctr4process.org/whitehead2015/s and http://www.pandopopulus.com.

12. James Lovelock, *The Vanishing Face of Gaia: A Final Warning* (New York: Basic Books, 2009), 29.

13. Ibid., 31.

14. Bruno Latour presented "Facing Gaia: Six Lectures on the Political Theology of Nature" as part of Edinburgh University's prestigious Gifford Lectures (February 18–28, 2013). He confronts the "controversial figure of Gaia" which is not Nature, nor a deity, but a form of power, best explored through the lens of political theology. Only once the new *geopolitics* of the Anthropocene are articulated can planetary boundaries be recognized as "*political* delin-eations" and the question of peace addressed. The lectures are available in video format from the University of Edinburgh (http://www.ed.ac.uk/schools-departments/humanities-soc-sci/news-events/lectures/gifford-lectures) and are forthcoming as a volume.

and God.[15] She did not read Gaia even as a nice Christian ecofeminist Earth goddess. But Ruether does render Gaia legible as one face of the divine matrix of life, and therefore as religiously meaningful. I see no deep contradiction here with Latour. Gaia's secularization makes possible new forms of coalitional complexity, across multiple religious and irreligious assemblies, building toward the Earth-wide and as Latour emphasizes "earthbound" chance of that pneumatological renewal: the prophetic impetus of every ancient proclamation of "the new heaven and earth."

An essay by Thomas Berry captures the vital insight that Gaia's homeostatic self-forming process of interrelation must also be embedded in a "Gaia cosmology."[16] "For the universe whence we emerged is constantly calling us back to itself. So too Earth is calling us back to itself, and not only to us but to all its components, calling them into an intimacy with one another and to the larger community within which all earthly realities have their existence."[17] That immense, perhaps infinite, intimacy is what he famously calls "the sacred community of the universe." Is this pantheism? Only if one rams an uninvited theism into his language. As one shaped by process pan*en*theism, I gladly embrace the sacred community of the universe as the very body of God—where God is a nickname not for an entity or a totality but for our personified relation to all that is.

What seems key for the community of creation spirituality, as for Whiteheadians, is that the awesome macrocosm is always folding in and out of the microcosm. The film does not just narrate, it cinematographically performs this folding: the unthinkable magnitude of the 100 billion galaxies each of 100 billion stars is enfolded in the intimacy of the narration. And so in response to the impersonal scales, the cosmic awe is able to activate rather than alienate. Rather than being made small and helpless, we get caught up in what is always larger than ourselves. And so we are ourselves enlarged. Its mystery does not mystify itself but magnifies our own capacities.

What about "God"? She/he/it is not named in the *Journey* film or in its primary resources, nor ever becomes the dominant subject of creation spirituality. God-speech will hinder some needed conversations even as it makes others—just as needed—possible. So, in the interest of the cosmopolitical solidarity we seek, ecotheology will and must continue its experiments in

---

15. Rosemary Radford Ruether, *Gaia and God: An Ecofeminist Theology of Earth Healing* (San Francisco: HarperSanFrancisco, 1992).

16. Thomas Berry, "The Gaia Hypothesis: Its Religious Implications," in *Sacred Universe: Earth, Spirituality, and Religion in the 21st Century*, ed. Mary Evelyn Tucker (New York: Columbia University Press, 2009).

17. Ibid., 14.

divinity, unsaying old notions of a transcendent disembodied sovereign him, uttering new kinds of divine ecology, matrix, milieu. I find also various old Christian names for God ecologically indispensable in varying moments— such as Cusa's *posse ipsum*, possibility itself. For the possibility of our planetary *convivencia* can only be actualized by the interdependent earthbound creatures of a boundless cosmos.[18] We may be helped through our apocalyptic angst not because God will fix our species' failure but because possibility itself does not cease to call: it even now offers the possibility that *we* will. Whether God "exists" or not, she/he/it insists upon our collective agency.[19]

I am helped, in other words, by reading Genesis (Greek for "becoming") as unfolding—even now—in and through the shifting entanglements that cross all space and time, down, way down at the quantum level, and on up and out—to an infinity dark with mystery. My favorite nickname for that infinity right now is the Supreme Entanglement. You can link it to Cusa's paraphrase of the divine infinity as the Enfolding (*complicatio*),[20] or more recently to O. Ogbonnaya's African rendition of "I am because we are" as "trinitarian communotheism."[21]

And you may or may not say that one may say that it is God who is calling, luring, even now, as Earth itself seems to call us to account for ourselves. That account shows ourselves inextricably and irresponsibly part of Earth that is calling. The call comes from beyond any human voice. It seems to call for a return to Earth, to a body that we have managed to abandon without ever leaving it, and so for a return that can only be a creative advance. And now we are called to a great advance, a great work—or else return is mere collapse: decreation. Here the very ending of the Holocene poses its frightening new beginning. If *apocalypse* means unveiling, and so disclosure rather than closure, then—as the old book says—the alpha and the omega are one[22] at any moment a call cuts across all space/time to each one of us, for what is possible, what we might embody together, beginning again now. Christians may call it Christ, whom I learned from John Cobb to define as creative transfor-

---

18. See Catherine Keller, *Cloud of the Impossible: Negative Theology and Planetary Entanglement* (New York: Columbia University Press, 2014).

19. The distinction between God's insistence and God's existence is the theme of John Caputo, *The Insistence of God: A Theology of Perhaps* (Indianapolis: Indiana University Press, 2013).

20. See Nicholas of Cusa, *De Docta Ignorantia* (*On Learned Ignorance*), bk 2. Originally published 1440.

21. A. Okechukwu Ogbonnaya, *On Communitarian Divinity: An African Interpretation of the Trinity* (New York: Paragon House, 1994).

22. Revelation 1.8.

mation. Of the environmental crisis Cobb wrote *Is It Too Late?* in 1971.[23] His answer is different now. But it does not lack the Christ-possibility: the chance to "seize an alternative." It *is* too late though (isn't it?) to make differences between potential eco-allies—between pantheism, panentheism, classical theism; between religions; between schools and movements and identity politics; between Whiteheadian and Teilhardian versions of cosmological theology; between religious and secular motivations—into squabbles that dissipate rather than collect communal energies for the adaptive creativity we are going to need. We—persons collected into the concrete contexts that together form the collectivity of a species of Earth—will need forms of resilience that we will more likely tap if we begin to cultivate them between us now, in the luxury of such forms of conversation as this. A privilege not to be squandered!

As I am here supposed to be speaking as a representative of Christianity, let me use that voice to declare: God doesn't care whether we believe in God—just that we *do* God. That we enact, that we actualize sacred community in its planetary *convivencia*. With the help of such new creation stories as the one told luminously by Thomas Berry, evolved by his cosmically gifted friends in this volume on *Journey of the Universe*, and fostered first and foremost by Mary Evelyn Tucker and John Grim, we just might. The New Jerusalem won't descend from above, but a new heaven and Earth, a renewed atmosphere and Earth, may yet arise through our work here below.

---

23. See John Cobb, *Is It Too Late? A Theology of Ecology* (Denton, TX: Environmental Ethics Books, 1995), originally published in 1971.

COSMOLOGICAL SPIRITUALITY AND ECOLOGICAL RITUAL

❧

11

"The Spirituality of the Earth"

Reflections on an Essay by Thomas Berry

KATHLEEN DEIGNAN

The journey of the universe has been a spiritual venture from the start—a mysterious, mind-boggling, heart-stopping wonder that our generation has come to know in its glorious unfolding more intimately and accurately than any other. Thanks to the great work of scientists, and the visionary poets who translate their magical and mostly indecipherable magisterium, a stunning narrative has begun to emerge. Because of such gifted transmitters, we are beneficiaries not simply of a new scientific cosmology but of a new cosmic story available to everyone, everywhere. Such is the multifaceted mediation of the cosmic adventure, *Journey of the Universe*, wonderfully conceived by Brian Thomas Swimme and Mary Evelyn Tucker, who by eloquently telling this great story have opened the peoples of the world to our own true and ever expanding nature.

Journey of the Universe—itself a monumental undertaking—is at once a cinematic project and an educational program; it is a traveling forum with global outreach inviting countless cohorts of students, teachers, preachers, scientists, policymakers, thinkers, and leaders of all stripes to awaken to the Great Story that lays beneath all others. Swimme and Tucker have made available an accessible foundational cosmology that can be translated and

114

integrated into any and every cultural and linguistic matrix. Specifically, *Journey* is an engaging event with particular address to the religious traditions of the world whose challenge is to evolve to their ecological phase by interfacing, interlacing, and integrating the ongoing revelations of cosmology, biology, and genetic science within their traditional wisdom legacies. This creative challenge comes with urgency in this critical and dangerous moment of Earth history, and the religions of the planet are perfectly poised to take it up for the life and healing of the world. As Thomas Berry so often said, the foundational scientific story of the universe awaits a "sacred telling," for only then will it have transformative power. This is the "Great Work" proposed and facilitated by the multivalent event that is the *Journey of the Universe*, which its creators and animators now call Christian communities to undertake in their labors to recover, reclaim, and reconstruct more viable theological, liturgical, catechetical, and spiritual paradigms required to bring our traditions to their ecological maturity.

In all of this we salute Thomas Berry, who ignited this fiery vision within Swimme and Tucker, who in turn are inviting all peoples to consciously participate in "the journey of the universe," and who now call Christians to ponder how an explicitly cosmological spirituality might emerge in witness and wonder within the Christian form. In honor of the centenary of the birth of Thomas Berry, I wish to enter this discussion by way of an article of his written in 1979, "The Spirituality of the Earth." Thomas immediately offers people of faith a radical shift of perspective in this summary initial statement:

> The spirituality of the Earth refers to a quality of the Earth itself, not a human spirituality with special reference to the planet Earth. Earth is the maternal principle out of which we are born and from which we derive all that we are and all that we have. We come into being in and through the Earth. Simply put, we are Earthlings.[1]

Thomas's rich and compact essay offers a unique perspective for Christians who would activate an Earth spirituality—a spirituality of Earth, for Earth—in light of contemporary cosmology. In it he proposes some recoveries within the Christian tradition that I found constructive and even somewhat surprising. However, in each instance of his critique and challenge to Christianity, Thomas repeatedly turns from his and our tradition to offer more compelling examples of his meaning from another. Therefore, in this brief set of provoca-

---

1. Thomas Berry, "The Spirituality of the Earth," in *The Sacred Universe: Earth, Spirituality, and Religion in the Twenty-First Century*, ed. Mary Evelyn Tucker (New York: Columbia University Press, 2009), 69.

tions for conversation, I would like to offer equally compelling resources from our own lineages, which need to be highlighted and creatively reintegrated into current Christian practice/praxis, so that we might accept his challenge to develop a thoroughly contemporary cosmic spirituality, a spirituality of Earth starting from our experience as Earthlings.

Berry's first prophetic challenge to Christianity regarding a spirituality of Earth confronts us with our Earth-blind sins and their sources. In the essay he focused most fiercely on the unamended attack on indigenous American peoples and their Earth mysticism by colonizing Euro-Christians, which he names one of the most barbaric moments in Christian history, a barbarism loosed also upon the American land with a destructive impact beyond calculation. He reminds us that while Christian scripture and liturgy have ever paid service to the natural world, the Earth process has been generally ignored by the religious-spiritual currents of the West, in preference to a redemption-from-the-world orientation, thereby withdrawing our sensibilities and conscience from an experience of Earth's sacrality.[2] How tragically ironic that this is the historical legacy of the only thoroughly incarnational religion on the planet.

Yet the denial of the sacredness of the natural world is not the whole story, nor is it constitutive of an authentic Christianity, as our recent recoveries of more original Christian Earth-centered spiritualities reveal. Indeed Thomas himself reiterates this more fundamental understanding throughout one of his last publications, *The Christian Future and the Fate of Earth*,[3] in which he insists that the dominant motif of the Christian event is not simply the redemption of the world by a messianic savior, but more essentially and originally, the long-awaited incendence of divinity into the cosmos, or in another metaphor, the flowering of divinity within the universe from the primordial Logos or seminal word of creation.

Even in Christianity's redemption focus, so critically noted by Berry, the early formators of our tradition actually iterate a story of liberation and rebirth of creation from the ravages of a spiritually disoriented human community who have refused the challenge of being collaborators in the historical unfolding of divine intention. The Christian story offers to this human predicament the promise of emergence of "a new creation," renewed and restored, perceived and received by those who would, like Jesus, their Christic exemplar, become a new kind of human—a new human-kind to cultivate and care for it. Indeed, the gospel of the "realm of God" springs up to reveal

---

    2. Berry, "Spirituality of the Earth."
    3. See Thomas Berry, *The Christian Future and the Fate of Earth,* ed. Mary Evelyn Tucker and John Grim (Maryknoll, NY: Orbis Books, 2009).

a dynamic terrestrial and cosmic field of grace proclaimed by a "reinvented humanity," to borrow Thomas's term. The developmental challenge of Christian life, then, was and remains the human labor to voluntarily let the "old hu/man" die, and take up the profound psychospiritual labor of regenerating the "new hu/man," a proleptic *anthropos-pneumatikos*: "the Christic." From Berry's perspective, the original Christian gospel was a summons to the Great Work of all Christian life and practice: to consciously, intentionally "reinvent the human at the species level."[4] How was this original and pivotal Earth-centered spirituality nurtured for the new Christic humanity in process? Our earliest Christian masters offer two modes of transformative practice: learning to read *The Book of Nature,* the primordial scripture, of which Thomas ever spoke, the work of *Theoria Physike*, or natural contemplation.

In 2002, rehearsing these ancient teachings, Pope John Paul II reminded Christians that creation is the original revelation, speaking clearly to us about the Creator and leading us ever more deeply into the mystery of God's love. He said, "For those who have attentive ears and open eyes, creation is like a first revelation that has its own eloquent language: it is almost another sacred book whose letters are represented by the multitude of created things present in the universe."[5] Earth-centered spirituality, therefore, is nothing new in Christianity, but is in fact its very heart and soul since Christianity celebrates the mystery of the incarnation of divinity into cosmic stuff, affirming creation as an unspeakable alphabet by which God continually spells out in inexhaustible creativity and splendor the terrible beauty and magic of a divine milieu. The scripture of creation and of the Bible are coextensive, all the early teachers will affirm. Given the right interpretive tools, one can read the divine design from nature back to the Bible and vice versa.

A cosmopoetics fashioned the Christian imagination from the start, summarized in St. Paul's word to the Romans that "ever since the creation of the world, God's invisible attributes of eternal power and divinity have been able to be understood and perceived in what he has made."[6] His letter to the Colossians reaches a truly cosmic vision of Christ seen not only as the center and meaning of history but also the epicenter of creation and all existing reality.[7] Later, John's Gospel announces a universe resplendent with divinity embodying everywhere from beginning to end, and especially in the mystery or sacrament of Jesus, setting Christian faith on its mystical

4. This motif runs all through the written and spoken legacy of Thomas Berry, and names the essential task of the "Great Work" of our time. See *The Great Work* (New York: Bell Tower, 1999).

5. Pope John Paul II, General Audience, January 30, 2002.

6. Romans 1:20.

7. Colossians 1:15–20.

course of seeing divinity manifesting throughout the universe. Irenaeus, a century later, echoes the dominant apostolic orthodoxy of such cosmotheology resounding in the burgeoning communities of faith: "The Universal Church, through the whole world, has received this understanding from the Apostles themselves: that creation reveals the One who formed it, and the very work made suggests the One Who made it, and the world manifests the One Who ordered it."[8] In the fourth century, Augustine of Hippo would put a bold exclamation point on the teaching. Some people, he noted, in order to discover God, read books: "But there is a great book: the very appearance of created things. Look above you! Look below you! Note it. Read it. God, whom you want to discover, never wrote that book with ink. Instead He set before your eyes the things that He had made. Can you ask for a louder voice than that? Why, heaven and Earth shout to you: 'God made me!'"[9]

This creation spirituality had many other articulations throughout the centuries of Christian history: in the monastic traditions of Benedict and Basil, in Celtic cosmocentric spirituality, and in the exquisite multimedia legacy of Hildegard of Bingen. This trajectory of incarnational Christianity reached its full flowering in Orthodox theology and practice and in the great Franciscan and Dominican schools of the Middle Ages. Later Meister Eckhart and the Beguines, Dame Julian of Norwich, Thérèse of Lisieux, and others lent their genius to the lineage. More recently Gerard Manley Hopkins, and in our own time, Pierre Teilhard de Chardin and Thomas Merton have revived this essential cosmo/Earth-centered spirituality for Western Christianity—and, of course, the one whom we celebrate during his centenary, Thomas Berry.

In his brief essay on Earth spirituality, Berry has delineated with insight and brevity the factors occasioning Christianity's derailment from this Gaiaphilic orientation: the catastrophes of lethal plague, the rupture of Christian unity, the philosophical ethos of enlightenment and secularization, the dawn of industrialization, the killing fields of the twentieth century, all climaxing in our current ecological dark night. Therefore, throughout his corpus, Berry challenges us to reinstate and reanimate an intensive Christian praxis of *lectio natura*—reading the Book of Nature—for the ongoing rebirth of the human species in its Christic form. Our concern, therefore, is how this essential engagement with and literacy of the primordial scripture of creation will be taught, by whom, where, and how. If in fact the cosmos is the first and very self-revelation of divinity, how shall we actualize this wondrous realization

---

8. Irenaeus, *Against Heresies* 2.9.1, in *Ante-Nicene Fathers*, vol. 1, ed. Alexander Roberts; available at http://gnosis.org/library/advh2.htm.

9. Augustine of Hippo, Sermon 126.6, in *The Essential Augustine*, ed. Vernon J. Bourke (Indianapolis: Hackett, 1978), 123.

noetically, aesthetically, ascetically, catechetically, liturgically, as an urgent mode of Christian therapy for the revival of our Earthling nature, for our own sanity, and for the life of the world?

If *lectio natura* is one way to revive Earth-centered Christian spirituality, another is the ancient practice of *theoria physike*, or natural contemplation, which is in process of retrieval in a soft form under the rubric of "creation spirituality." The more profound dimensions and demands of this teaching can be gleaned from Thomas Merton's lecture notes for his novices in Gethsemani during the 1960s. Here, Merton offers the young monks in training Maximus the Confessor's teaching on *theoria physike*, a stage in the journey of contemplative transformation (read "epistemological/psychic" development or evolution) in which the practitioner labors to sense—actually "to truly see by way of gazing"—the divine presence in the created universe.[10]

According to Merton, *theoria physike* is, then,

a) Reception of the mysterious, silent revelation of God in His cosmos and in its *oikonomia* [stewardship of all things, the structure of God's dispensation of the cosmos, God's providence and judgement], as well as in our own lives.

b) It is the knowledge of God that is natural to man, with God's help . . . what God intended for man in creating him . . . proper to him as a son of God, was his when in paradise. . . . We must be restored first of all to this "natural" contemplation of the cosmos before we can rise to perfect *theologia*.

c) This contemplation is demanded by the cosmos itself and by history. If man cannot know creatures by this spiritual gnosis, they will be frustrated by their end. If man cannot spiritually penetrate the meaning of the *oikonomia*, it runs the risk of being frustrated and souls will be lost.

d) Hence *theoria physike* is a most important part of man's cooperation in the spiritualization and restoration of the cosmos. It is by *theoria* that man helps Christ to redeem the *logoi* of things and restore them in Himself.

e) This *theoria* is inseparable from love and from a truly spiritual conduct of life. Man not only must see the inner meaning of things, but he must regulate his entire life and his use of time and of created beings

---

10. See *Merton and Hesychasm: The Eastern Church and the Prayer of the Heart*, ed. Bernadette Dieker and Jonathan Montaldo (Louisville, KY: Fons Vitae, 2003), 434–35. For a more in-depth analysis of the psychological transformations the early masters were evoking, see David Bradford, "Evagrius Ponticus and the Psychology of 'Natural Contemplation,'" *Studies in Spirituality* 22 (2012): 109–25.

according to the mysterious norms hidden in things by the Creator, or rather uttered by the Creator Himself in the bosom of His Creation.

f) The vision of *theoria physike* is essentially sophianic. Man by *theoria* is able to unite the hidden wisdom of God in things with the hidden light of wisdom in himself. The meeting and marriage of these two brings about a resplendent clarity within man himself, and this clarity is the presence of Divine Wisdom fully recognized and active in him. . . . At the same time he exercises a spiritualizing influence in the world by the work of his hands which is in accord with the creative wisdom of God in things and in history.[11]

Clearly few of our contemporary retrievals of *theoria physike* meet the depth of this early monastic contemplation, and this begs the question: How shall we accelerate the birth of the new human of which Berry so often spoke as the goal of the Great Work of our time if there is not this kind of intensive, lay meditative work? Again the challenges: How will this teaching be reappropriated for the evolution of the human sensorium? Who are its teachers, where are they trained, how shall they train, who will come for training? Remember Thomas's warning:

The deep psychic change needed to withdraw us from the fascination of the industrial world, and the deceptive gifts that it gives us, is too difficult for simply the avoidance of its difficulties or the attractions of its benefits. Eventually only our sense of the sacred will save us. Merton's gift . . . is this sense of the sacred throughout the entire range of the natural world.[12]

Another relevant item in his essay for our conversation about Christianity and the reactivation of cosmocentricity occasioned by *Journey of the Universe* is Thomas's rather startling proposal for the way we might speak of the "numinous maternal principle in and through which the total complex of Earth phenomena takes its shape."[13] He reminds us that the whole burden of modern Earth spirituality is to narrate the story of the birth of the human from our Mother the Earth, and how significant the title "Mother Earth" is for its intimacy, sense of relationality, and mutual care. Thomas ten-

11. Thomas Merton, *An Introduction to Christian Mysticism: Initiation into the Monastic Tradition*, bk. 3, ed. Patrick O'Connell (Kalamazoo, MI: Cistercian Publications, 2008), 121–136.

12. See Thomas Berry's Preface to Deignan's *When the Trees Say Nothing: Thomas Merton's Writings on Nature* (Notre Dame, IN: Sorin Books, 2003).

13. Berry, "Spirituality of the Earth," 75.

derly tells us, "Our long motherless period is coming to a close,"[14] and hope-fully with it the violation of Earth. But this will happen only if we radically mature beyond the "redemptive and transcendentally oriented spiritualities that have governed our own thoughts, attitudes, and actions" to this stage of our history. For "this new mode of Earth-human communion requires a profound spiritual context and a spirituality that is equal to this process."[15]

Then, in a startling proposal for its deep traditionalism, Thomas elects Mary, the mother of Jesus, the archetypal "madonna" of Western civilization, to carry the embryonic burden of the new Christic humanity, and indeed to bear the personification of Earth herself:

> The association of the Virgin Mother with the Earth may now be a con-dition to return Mary to a meaningful role. Her presence may also be a condition for overcoming our estrangement from the Earth. Because of our emphasis in the Western world on personhood, it is insufficient to see the Earth itself only as a universal mother. It must be identified with an historical person in and through whom Earth functions.[16]

Berry reminds us that few, if any, other civilizations were so deeply grounded in a feminine mystique as the medieval period of Western Chris-tendom, where references to Mary as the Earth are found throughout its reli-gious literature. Continuing to develop such Marian cosmopoetics is, to his mind, "a subject of utmost importance for our entire Earth-human venture. . . . Thus we cannot fail to unite in some manner these two realities: Earth and Mary."[17] Though our Christian poetics offers an abundance of such expres-sions, Thomas did not offer any specific examples of such ecstatic utterance, so allow me to share my adaptation of a hymn from a collection of prayers, poems, and songs of the Celtic world, *The Carmina Gadelica*, which sings of Mary this way:

> You are the fruit of the royal vineyard; you are the fruit of the boundless sea.
> O longed-for guest of the Visitation, come bless our homeland with your peace.
> O garden of virtues, O mansion of gladness, Mother of sadness and clemency . . .

---

14. Ibid., 76.
15. Ibid., 77.
16. Ibid.
17. Ibid.

You are the love song of all lovers; you are the garden of delight.
You are the fullness of Earth's desire; you are the radiance in the night.
O vessel of fullness, O chalice of wisdom, wellspring of health for all
    humankind,
Caress our heart and mind. . . .
You are the clear light of the heavens; you are the moonshine of the
    skies.
You wear a crown studded with the starlight; you are the radiance in
    our eyes.
O river of mercy, o wellspring of justice, fountain of grace, which re-
    news our lives, come live with us. . . .[18]

Thomas's third observation is a corollary of this Marian resurgence, that
in this moment of emergence of the new age of humankind, "Woman and
Earth, while differentiated, are inseparable. The fate of one is the fate of the
other. . . . Earth consciousness and woman consciousness go together. Both
play an essential role in the spirituality of the human as well as in the structure of civilizations."[19] However, I wish to note Berry's fourth observation
concerning our developing new capacity for "subjective communion with the
manifold presences that constitute the universe."[20] As we recover the more
primitive genius of humankind, we become more sensitive to the numinous
world that should be addressed in a responsive, reciprocal mood of affectionate concern. To illustrate this mature human capacity for receptivity and communion with the numinous subjectivity of every being, Thomas turns to the
Chinese tradition to underscore a form of feeling identity that is alive to the
vitality, the subjectivity, numinosity, and pain of all beings: "In Confucianism there is the universal law of compassion. As the early Confucian thinker
Mencius (372–289 BCE) suggested, this is especially observable in humankind, as every human has a heart 'that cannot bear witness to the suffering
of others.' When the objection was made to the neo-Confucian Wang Yang
Ming (1472–1529) that this law of compassion is evident only in human
relations, Wang replied by noting that even the frightened cry of a bird, or

18. Kathleen Deignan, "Ave Maria," inspired by a prayer from *Carmina Gadelica*. The original version is available at https://archive.org/stream/carminagadelicah30carm#page/134/
mode/2up/search/praise+of+mary. The sung version of this prayer can be found on the CD *Ave*,
available at http://www.ScholaMinistries.org.
    19. Berry, *Sacred Universe*, 78.
    20. Ibid.

the crushing of a plant, or the shattering of a tile, or the senseless breaking of a stone immediately and spontaneously caused pain in the human heart."[21]

To Berry's Asian references let us note equally compelling testimonies of our own faith-fellows who have richly lived in such a state of creation consciousness, even in the worst of times. So I give the last words to the abiding teachers of our one Christian legacy who await our recovery of their transformative wisdom and praxis. They await our creativity to transmit new soundings of their empowering voices telling and teaching us how to become cosmic Christians, Earth lovers, and defenders once again, as it was in our beginning. As it could be again. First a word from an ally who stands ready to aid us in this ecological dark night, to let us know there have been Earth-first prophets in the Christian community from the start, who leave us a testimony we could easily voice today, and from none less than the great teacher and bishop, Ambrose of Milan:

> Here are a people who find no delight in tapestries of purple or costly stage curtains. Their pleasure lies rather in their admiration of this most beautiful fabric of the world, this accord of unlike elements, this heaven that is spread out like a tent to dwell in to protect those who inhabit this world. They find their pleasure in the Earth allotted to them for their labors, in the ambient air, in the seas here enclosed in their bounds. In the people who are the instruments of the operations of God they hear music which echoes from the melodious sound of God's word, within which the Spirit of God works.[22]

> Why do injuries of nature delight you? The world has been created for all, while you rich are trying to keep it for yourselves. Not merely the possession of the Earth, but the very sky, air, and the sea are claimed for the use of the rich few. . . . Not from your own do you bestow on the poor man, but you make return from what is his. For what has been given are common for the use of all, you appropriate for yourself alone. The Earth belongs to all, not to the rich.[23]

---

21. Ibid., 79.

22. St. Ambrose, *Hexameron, Paradise, and Cain and Abel* in *Fathers of the Church: A New Translation*, vol. 42, trans. John Savage (Washington, DC: Catholic University of America Press, 2010), 69–70.

23. St. Ambrose, *De Nabuthe Jezraelita (On Naboth)* 3.11, http://hymnsandchants.com/Texts/Sermons/Ambrose/OnNaboth.htm.

Finally let us hear an echo of the "merciful heart" of Isaac the Syrian, who offers the passion of his ecological sensibilities and reminds us that he and a great cloud of witnesses have by faith buried such ecological genes in our Christian souls, for the life of the world to come: this world made new by those who become new.

> What is a merciful heart? It is a heart on fire for the whole of creation, for humanity, for the birds, for the animals, for demons, and for all that exists. By the recollection of them the eyes of a merciful person pour forth tears in abundance. By the strong and vehement mercy that grips such a person's heart, and by such great compassion, the heart is humbled and one cannot bear to hear or to see any injury or slight sorrow in any in creation. For this reason, such a person offers up tearful prayer continually even for irrational beasts, for the enemies of the truth, and for those who harm him, that they be protected and receive mercy . . . because of the great compassion that burns without measure in a heart that is in the likeness of God.[24]

---

24. St. Isaac the Syrian, *Homily* 81, in Metropolitan Hilarion, "St. Isaac of Nineveh and Syrian Mysticism," available at http://hilarion.ru/en/2010/02/25/1077.

12

A Spiritual Heart for the Ecological Age

CRISTINA VANIN

In *The Great Work: Our Way into the Future*, Thomas Berry reminds us that we do not choose the time in which we are born, the particular moment of history in which we find ourselves. Still, he says, "The nobility of our lives . . . depends upon the manner in which we come to understand and fulfill our assigned role."[1] And the role that befalls you and me is the Great Work of managing the "arduous transition" into the Ecozoic Era when human beings will be present to the Earth in a mutually enhancing manner. The Christian language that speaks to such transition or transformation is conversion.

On October 4, 2003, the feast of St. Francis of Assisi, the patron saint of ecology, the Social Affairs Commission of the Canadian Conference of Catholic Bishops issued a pastoral letter on what they called "The Christian Ecological Imperative." Then, in 2008, the United Nations' International Year of the Planet Earth, they issued another letter called "Our Relationship with the Environment: The Need for Conversion." In both documents, the bishops say that serious responses to the ecological crisis demand "that human beings change our thinking, relationships, and behaviors in order to recognize the interconnectedness of all creation."[2] What the bishops point to is that the conversion needed to address this crisis is moral and spiritual, as well as ecological.

In this chapter I focus on what it means to talk about ecological conversion. As well, if the crisis is moral and spiritual, then it is important to think about the type of spirituality that can contribute to this transformation. Not just any spirituality will do. If a spirituality is going to help humans move into the Ecological Age, it needs to be a cosmological spirituality, a spirituality that, like the *Journey to the Universe*, helps to orient and live our lives within the context of the immense complexity of the story of the universe. I

---

1. Thomas Berry, *The Great Work: Our Way into the Future* (New York: Bell Tower, 1999), 7.

2. Canadian Conference of Catholic Bishops, Social Affairs Commission, "Our Relationship with the Environment: The Need for Conversion" (2008), n. 14.

conclude with the description of a particular place, the Ignatius Jesuit Centre in Guelph, Ontario, whose work over the past fifteen years is a significant example of the impact of a cosmological spirituality on the development of a spiritual heart for the Ecological Age.

Why Do We Need Ecological Conversion?

In his book *Last Child in the Woods*, Richard Louv asks, "What happens when all the parts of childhood are soldered down, when the young no longer have the time or space to play in their family's garden, cycle home in the dark with the stars and moon illuminating their route, walk down through the woods to the river, lie on their backs on hot July days in the long grass, or watch cockleburs, lit by morning sun, like bumblebees quivering on harp wires? What then?"[3]

Thomas Berry also laments what has happened to our human children: "For children to live only in contact with concrete and steel and wires and wheels and machines and computers and plastics, to seldom experience any primordial reality or even to see the stars at night, is a soul deprivation that diminishes the deepest of their human experiences."[4]

Berry suggests that a primary cause of the devastation of the natural world is a consciousness in human beings that says that only we have value, that other beings have meaning and value not in and of themselves, but only when we use them. He describes in this fashion the alienation that results from such consciousness:

> While we have more scientific knowledge of the universe than any people ever had, it is not the type of knowledge that leads to an intimate presence within a meaningful universe.... Our world of human meaning is no longer coordinated with the meaning of our surroundings. ... Our children no longer learn how to read the great Book of Nature from their own direct experience. They seldom learn where their water comes from or where it goes. We no longer coordinate our human celebrations with the great liturgy of the heavens.... We no longer hear the

---

3. Richard Louv, *Last Child in the Woods: Saving Our Children from Nature-Deficit Disorder* (Chapel Hill, NC: Algonquin Books of Chapel Hill, 2005, 2008), 97.

4. Berry, *Great Work*, 82.

voice of the rivers, the mountains, or the sea. . . . The world about us has become an "it" rather than a "thou."[5]

He argues that, as a consequence of our alienation from the natural world, we end up teaching our children about our current economic systems that depend on the exploitation of life systems. If we want our children to have an attitude of exploitation, to think that the resources of the planet are there primarily for our human use, then what we need to do is make sure that they lose any feeling for the natural world, any relationship with it. This is not difficult to do because we, the human adults, have little sensitivity for the planet. We tend to regard the natural world as a backdrop to our human undertakings.

In other words, in our contemporary world, human persons do not truly live in a universe. We tend to live in cities and countries, in economic systems, in cultural and perhaps religious traditions. Our alienation from the natural world is so extensive that we are not even aware of it. Even the idea that we should have an integral and intimate relationship with the natural world lies so far outside our horizons that we cannot contemplate it. This is precisely why ecological conversion is needed, why we need to recover a capacity for being in communion with the natural world.

Bernard Lonergan on the Nature of Conversion

For a number of years I have been turning to the thought of Canadian philosopher and theologian Bernard Lonergan to help understand the nature and complexity of this kind of transformation. His work on conversion is especially relevant. Lonergan suggests that social breakdowns occur because of the cumulative effect of not acting as authentic human beings, because we violate the transcendental precepts that call us to be attentive, intelligent, reasonable, responsible, and loving human beings.[6] Operating as fully authentic subjects requires conversion, that is, a radical change in our horizons, in our self-understanding, in our way of operating in the world.

---

5. Ibid., 15, 17. See also "Everyone lives in a universe; but seldom do we have any real sense of living in a world of sunshine by day and under the stars at night. Seldom do we listen to the wind or feel the refreshing rain except as inconveniences to escape from as quickly as possible" (54).

6. Bernard Lonergan, *Method in Theology* (New York: Herder and Herder, 1972). With conversion, "It is as if one's eyes were opened and one's former world faded and fell away. There emerges something new" (ibid., 130). The living out of conversion affects all of our conscious operations, from what we attend to, to the way in which our understanding is enriched, our judgments are guided, and our decisions are reinforced.

When it comes, then, to ecological conversion, Lonergan helps us to realize that we need the intellectual conversion[7] that would open up the horizons of our inquiring and our decision-making. This aspect of conversion transforms us from a general bias that resists and blocks certain questions and insights. Ecologically speaking, we become willing to consider seriously the meaning and value of the Earth community in its integral wholeness and of all species in their interrelatedness.

In addition, we need the ecological moral conversion that shifts us from living our lives as if they have nothing to do with the natural world to living as if we will flourish only if the Earth flourishes. We need to shift from making decisions about development, energy, agriculture, oil, and so on, only in terms of the economic benefit to shareholders, to making decisions that take into consideration all stakeholders, including habitats, bioregions, particular species, and ecosystems. We need to become persons of integrity if we are going to collaborate to take corporate responsibility for the well-being of the planet.

At the psychic level, ecological conversion involves a shift from considering human beings as separate from the natural world to understanding humans as integral members of the whole Earth community of life, of the whole cosmos. It involves an ongoing process out of alienation from the natural world to a deepening intimate relationship with its rhythms and flows. To the degree that we can come to understand how intimate we are with the natural world, that difficult, often impenetrable, psychic barrier between humans and the natural world can be removed.

What Will Help to Nurture This Process of Conversion and Transformation?

Berry argues that "we need a spirituality that emerges out of a reality deeper than ourselves, a spirituality that is as deep as the Earth process itself, a spirituality that is born out of the solar system and even out of the heavens beyond the solar system."[8] Such a spirituality helps us to recover a capacity for being in communion with the Earth and understanding ourselves as integral with the universe process.

---

7. Lonergan distinguishes conversion as intellectual, moral, or religious, while recognizing that the three are interconnected in the human person. Robert Doran has added the notion of psychic conversion.

8. Thomas Berry, *The Sacred Universe: Earth, Spirituality, and Religion in the Twenty-First Century*, ed. Mary Evelyn Tucker (New York: Columbia University Press, 2009), 74.

Here is where the world's religious traditions have their role: they can help us appreciate that the story has a dimension to it that transcends the physical, and that the universe, from its beginning, is a psychic and spiritual as well as a physical reality. Christianity has the capacity to contribute to this arduous and liberating journey of rebirth into a larger, more comprehensive, and deeply spiritual realm of being.[9] It can help us to establish such a relationship of reciprocity and intimacy with the natural world if Christianity itself transforms into this new cosmological context.[10]

Lonergan talks about conversion as having a religious dimension; as such, it has to do with "being grasped by ultimate concern"[11] or "being in love with God."[12] With religious conversion, "A total being-in-love [becomes] the efficacious ground of all self-transcendence, whether in the pursuit of truth, or in the realization of human values, or in the orientation [human beings adopt] to the universe, its ground, and its goal."[13] Religious ecological conversion is the ongoing process of living out of the horizon of ultimate, divine loving, which is a loving of the whole cosmos.[14] Furthermore, any religion that helps human beings to develop their authenticity and self-transcendence "to the point, not merely of justice, but of self-sacrificing love,"[15] can help to sustain us through the sacrifices that will be required of us as individuals and as a species as we respond to the ecological crisis.

When we live within the horizon of the cosmos, we can begin to hear the voices of all creatures. We begin to listen to what the universe is articulating about itself in the stars, moon, mountains, forests, rivers, seas, meadows, songbirds, and insects. We learn of the integral relationship among all the members of the community of Earth and of the universe—that nothing is what it is without everything else. Berry insists that "we form a single sacred

---

9. Thomas Berry, *The Christian Future and the Fate of Earth*, ed. Mary Evelyn Tucker and John Grim (Maryknoll, NY: Orbis Books, 2009), 11: "Only religious forces can move human consciousness at the depth needed. Only religious forces can sustain the effort that will be required over the long period of time during which adjustment must be made. Only religion can measure the magnitude of what we are about."

10. Berry, *Sacred Universe*, 48. Berry warns that such a change "is not possible if we fail to appreciate the planet that provides us with a world abundant in volume and variety of food, a world exquisite in supplying beauty of form, sweetness of taste, delicate fragrances for our enjoyment, exciting challenges for us to overcome."

11. Lonergan, *Method*, 240.

12. Ibid., 105.

13. Ibid., 241.

14. Canadian Conference of Catholic Bishops, Social Affairs Commission, "You Love All That Exists.... All Things Are Yours, God, Lover of Life," Pastoral Letter on the Christian Ecological Imperative, October 4, 2003.

15. Lonergan, *Method*, 55.

society with every other member of the Earth community, with the mountains and rivers, valleys and grasslands, and with all the creatures that move over the land or fly through the heavens or swim through the sea."[16] If we could truly understand that our human story is integral with the story of the universe, "then we can see that this story of the universe is in a special manner our sacred story, a story that reveals the divine particularly to ourselves, in our times; it is the singular story that illumines every aspect of our lives—our religious and spiritual lives as well as our economic and imaginative lives."[17]

Knowing the universe as sacred, and knowing that our story and the stories of all other members of the Earth community are inseparable from the story of the universe, would help us to appreciate that everything shares a unity of origin. Berry's point is that everything that exists in the universe is genetically related to everything else. Also, community is at the heart of the nature of existence; there is a relationship of kinship of each being to every other being. "There is literally one family, one bonding, in the universe, because everything is descended from the same source. . . . On the planet earth . . . we are literally born as a community; the trees, the birds, and all living creatures are bonded together in a single community of life."[18] Within this horizon, we can find ourselves relating as subjects to subjects, no longer alienated from each other, but living in a relationship of communion with all.

Contemporary nature writers are important guides for teaching us how to develop our capacity for intimate communion by learning to be attentive to the natural world. Madeleine Bunting of the *Guardian* says that the point of this writing "is that nature is no longer something to be studied from a position of scientific detachment, but [it is] an experience, a relationship in which human beings are as much part of nature as any so-called wildlife."[19] She points out that "we need that attentiveness to nature to understand our humanity, and of how we fit, as just one species, into a vast reach of time and space."[20] Thomas Lowe Fleischner, editor of *The Way of Natural History*, says that "'natural history' is a practice of intentional, focused attentiveness and receptivity to the more-than-human world. . . . Attention is prerequisite to intimacy. Natural history, then, is a means of becoming intimate with the . . . world." Fleischner goes on to argue that attentiveness to nature matters

16. Berry, *Sacred Universe*, 85.

17. Ibid., 94.

18. Thomas Berry with Thomas Clarke, *Befriending the Earth: A Theology of Reconciliation between Humans and the Earth*, ed. Stephen Dunn and Anne Lonergan (Mystic, CT: Twenty-Third Publications, 1991), 14–15.

19. Madeleine Bunting, *The Guardian*, Monday, July 30, 2007.

20. Ibid.

because, "in a very fundamental sense, we are what we pay attention to. . . . Our attention is precious, and what we choose to focus it on has enormous consequences. What we choose to look at, and to listen to—these choices change the world."[21]

Lyanda Lynn Haupt, author of *Crow Planet: Essential Wisdom for Urban Wilderness*, asks how we are to attain intimacy with the natural world if we live at a remove from "nature," as most of us do in our urban and suburban homes. Do we need then to travel to far-off places, to participate in wildlife adventures? Haupt's answer is that this notion creates a disconnect between our daily living, which we think has nothing to do with nature, and wilderness that we regard as true nature. In contrast to this view, she argues, "It is in our everyday lives, in our everyday homes, that we eat, consume energy, run the faucet, compost, flush, learn, and *live*. It is here, *in our lives*, that we must come to know our essential connection to the wilder earth, because it is here, in the activity of our daily lives, that we most surely affect this earth, for good or for ill."[22] We are connected to the natural world in and through our everyday lives; our everyday lives are part of an emerging, evolving cosmic story. Truly coming to know this requires us to start walking the paths of our neighborhoods, to start knowing the breadth of all of our neighbors, human and nonhuman, "on and off the concrete, above and below the soil."[23] As we walk and develop our capacity to pay attention to the natural world, as we become more intimate and present to the members of the community of life, we can begin to know that our human story is integral with the story of the universe.

In 1999 James Profit, a Jesuit, invited me to become part of a working group that supported the work of what, at that time, was called the Jesuit Ecology Project, one of the social justice initiatives of the Jesuits in the English Canada province. The Ecology Project was housed at Ignatius Jesuit Centre in Guelph, Ontario. As part of the Working Group, I had the opportunity to work on a number of eight-day ecology retreats that we called Mysticism of Earth, retreats that were rooted in the discipline of the Spiritual Exercises of St. Ignatius of Loyola.

The Spiritual Exercises grew out of Ignatius's own experience of seeking to grow in union with God, and to discern God's will for him. All the notes, prayers, meditations, and reflections that Ignatius kept throughout his jour-

---

21. Thomas Lowe Fleischner, ed., *The Way of Natural History* (San Antonio, TX: Trinity University Press, 2011), as quoted by Mat McDermott, *Living/Culture*, August 18, 2011.

22. Lyanda Lynn Haupt, *Crow Planet: Essential Wisdom from the Urban Wilderness.* (New York: Little, Brown, 2009), 9.

23. Ibid., 13.

ney eventually were framed into a retreat. The four "weeks" of the Exercises
are structured around the life, death, and resurrection of Jesus Christ. In the
ecology retreats, we took this rich, extensively practiced, and meaningful
spiritual discipline and transformed it into the context of the comprehensive sacred universe. These retreats helped people to develop the "practice of
deep, intentional attention" to the natural world, to learn the sacredness of
the Earth community, and to develop the capacity to live out of the horizon
of ultimate, divine loving.

Each individual day invited retreatants to reflect on specific themes:

- Listening to what the voices of the community of life at Ignatius
  Jesuit Centre have to say to us about God's love for the whole of
  the universe, and for each and every living and nonliving being in
  the universe.
- Reflecting on the impact of our participation in ecological sin.
- Noticing how we experience the divine presence in the Earth's
  capacity for healing.
- Practicing a meditative walk that invites us to deep attentiveness,
  awareness, and intimacy of the natural world.
- Pondering what it means that God is intimately present in the pas-
  sion and suffering of the Earth.
- Celebrating the hope of the resurrection of the whole of creation.
- Contemplating how we are to respond in love and live our lives as
  members of this integral, comprehensive, and sacred community.

Each day, retreatants were invited to pray by walking the land, being attentive to the many voices to be heard on the land and to hearing the voice of
God in and through the voice of every creature. They were invited to relate
their prayer and reflection on the book of nature to what God speaks through
the books of scripture.

Each day also included the celebration of the sacrament of the Eucharist at
a specific location on the 240 hectares of property, a celebration that further
invited people to integrate the Christian story with the story of the Earth
and the story of the universe.

All of these aspects of the retreat were meant to help us to develop our
capacity for wonder and joy as well as critical, self-reflective awareness of our
responsibility for the future of the whole community of life: a capacity that
we can continue to develop in our everyday lives. Lyanda Lynn Haupt articulates the value of developing this capacity this way:

There is a way to face the current ecological crisis with our eyes open, with stringent scientific knowledge, with honest sorrow over the state of life on earth, with spiritual insight, and with practical commitment. Finding such a way is more essential now than it has even been in the history of the human species. But such work does not have to be dour (no matter how difficult) or accomplished only out of moral imperative (however real the obligation) or fear (though the reasons to fear are well founded). Our actions can rise instead from a sense of rootedness, connectedness, creativity, and delight.[24]

These retreats help to develop the sense of rootedness both in God and in the Universe Story, and to experience delight in choosing to live according to the ultimate meaning and value of this sacred story. They are an embodiment of the cosmological spirituality that Berry says must become the context for our contemporary spiritual journey of renewal and rebirth.

However, if an adequate spirituality for the ecological age is going to continue to emerge, humans need to articulate the story of the universe as fully, and in as many ways, as possible. Thomas Berry insists that it is our attentiveness to, and understanding of, the comprehensive, macrophase story of the cosmos that will provide us with the functional cosmology that can transform our alienated and separated consciousness. *Journey of the Universe* is one such significant articulation of the Universe Story for our time.

Ignatius Jesuit Centre decided to tell this comprehensive story through the development of a meditative walk called Stations of the Cosmos.[25] This is an additional way in which people can experience this new, deeper, and comprehensive spirituality, to learn to be more attentive and intimate with the community of life. The centre recognizes that, as Berry says, "our sense of who we are and what our role is must begin where the universe begins. Not only does our physical shaping and our spiritual perception begin with the origin of the universe, so too does the formation of every being in the universe."[26]

---

24. Ibid., 7–8.

25. In the 1980s, Holy Cross Centre for Ecology and Spirituality built such a walk on its Lake Erie property with eight stations that celebrated the emergence of the universe, Earth, life, the human, agriculture, culture and religion, science and technology, and the emerging Ecozoic Age. The Maryknoll Ecological Sanctuary in Baguio, Philippines, also developed such a walk with thirteen stations that integrate indigenous art and culture in the telling of the story.

26. Berry, *Great Work*, 162.

The walk comprises twenty-five stations that celebrate many of the significant moments of grace that are part of the story.[27] Each station is marked by images and photographs that help us situate our human story within this beautiful and incredible cosmological story. The opportunity to walk these stations is an opportunity to overcome our distance and nurture a relationship of profound presence and intimacy to the community of life.

So much of the impetus of Berry's work was his concern for *all* the children of the Earth—the child of the air, the sea, the soil, the flowers, meadows, trees, and human children. This concern resonates all the more deeply for me, now that I have two children of my own. Part of what I want to teach them is the importance of having deep relationships of love and justice with the whole of the community of life, with other human beings, and with God.

As St. Francis of Assisi's canticle about Brother Sun, Sister Moon, says: Ask the children, ask any of the Earth's children—the trees, the beasts, the fields, the rivers, the flowers, the worms—and they can teach us about God's love for the universe, that community of life of which each of us is an integral part.

We are then most true to ourselves when we are attentive to the community of life within which we live, when we strive to understand the nature and role of all members of the community, when we affirm the whole of the cosmos as the most comprehensive context of our being, and when we value the whole of the cosmos and take all of it into consideration as we make our choices. This is the essence of human authenticity and the basis of ecological conversion or transformation. It is the foundation of a cosmological spirituality that changes our hearts and minds, making it possible for us to become, with God, knowers, cohealers, and lovers of all that exists.

---

27. The significant moments that are celebrated include the initial flaring forth of the universe some 13-14 billion years ago; the development of galaxies; the formation of stars, including our sun; the amazing creativity of the Cenozoic era that emerged after the extinction of the dinosaurs; the arrival of the Homo sapiens species with its capacity for religious insight; the development of agriculture; the industrial revolution, with its increased human-induced extinctions; the profound shift in consciousness that occurred when humans saw the Earth from space for the first time; our present moment.

13

Being Church as If Earth Matters

A Response to *Journey of the Universe* from One Episcopalian's Perspective

STEPHEN BLACKMER

Journey of the Universe presents a magnificent story of the evolution of the universe from the unimaginable simplicity of a single point to the unimaginable complexity of all the myriad forms of matter and life. It opens our eyes to the wonder of this blossoming. It reveals the mystery that we all are—physically and spiritually—part of this extraordinary reality.

And it leads us to the critical question of how do we, how does each of us—how do I—live in harmony or right relationship with the rest of the blossoming universe? How do I recognize and inhabit the physical and the spiritual reality that flows from one source? How do I live in relationship with all my kin and neighbors, both human and nonhuman? How do I live in openness and receptivity to "the cosmic energies coursing through us" that *Journey of the Universe* suggests may renew the face of the Earth?

Human ability to manipulate the world has exploded in recent centuries. We have developed—or been given—godlike superpowers that enable us to create marvelous things . . . and to destroy so much.

With these powers comes enormous responsibility—responsibility we have often failed to honor. So the critical questions facing us are: How can we live in harmony with the rest of this extraordinary creation? The *Journey of the Universe* story itself doesn't attempt to answer these questions, but it does point toward this in the *Journey Conversations*, with rich examples of how some people are living into this new understanding.

From the perspective of the environmental community—that set of people, organizations, and actions at the forefront of the movement to live rightly in relationship with the Earth—the answers to these questions typically are to be found in science, better public policy, new ways of economic thinking, new forms of business, more advanced engineering, new technologies, and other modern marvels. And these all are essential. But they are nowhere near enough.

Even with all these, everything we know of human history strongly suggests that people will continue to wreak havoc on the Earth—and on each other. We always have. It is, in part, what we do. The Christian tradition has a name for it. But can we be humans who are conscious and fully alive in wisdom, compassion, and love—and who therefore use our superpowers to heal rather than to harm? This is the critical question.

Churching of the Green

Over the past half-dozen years, I have walked away from a very good life as an entirely unchurched environmental activist toward a new life as a baptized Christian and ordained priest in the Episcopal Church. As I have journeyed from eco-organizer to ecopriest, I have mused at length on what the Christian tradition offers to the movement to live in harmony with nature. It has been abundantly clear that my call to a life in Christ was not about abandoning my lifelong commitment to the Earth but about expressing it in a new way. But what new way? I honor those whose faith leads them to political action, but for me, the path has gone in the other direction. I didn't need Christian faith or church in order to be an environmental advocate and organizer; I had been that since I was sixteen years old. And I certainly didn't need the Bible or the church to teach me about biology and ecosystems, for it doesn't have much to say about them. But my activism—and my realization of its inadequacy—demanded something more.

What the Christian tradition gave me that the environmental movement does not—nor does politics or our economy or technology or science or mainstream culture—is a way and a tradition and a practice for becoming a human being fully alive, in the words of Irenaeus, second-century Bishop of Lyon.[1] Or, as the great ecologist Aldo Leopold wrote, "No important change in ethics was ever accomplished without an *internal* change in our intellectual emphasis, loyalties, affections, and convictions. The proof that conservation has not yet touched these foundations of conduct lies in the fact that

---

1. "For the glory of God is a living man; and the life of man consists in beholding God" (Irenaeus, *Against Heresies* 4.20.7, available at http://www.newadvent.org/fathers/0103420.htm).

philosophy and religion have not yet heard of it. In our attempt to make conservation easy, we have made it trivial."[2]

The inescapable—and difficult—conclusion must be that in order to "save the Earth," it is we who must change, who must be saved . . . which, of course, is what Christian faith and practice are about.

So six years ago, heart in my throat, I went to church for the first time in my life. I was baptized, I applied to seminary, I entered into discernment about becoming a priest, and finally—just over a year ago—I was ordained. I have no doubt that this is what I am called to do next in my call to heal the Earth. And yet, notwithstanding the rightness of all I have been doing, Christian practice, as I have learned it, does not provide a ready way to live out my lifelong and passionate love: to live body, mind, and soul in relationship with the Earth and all her creatures. The question I have carried all through seminary and, now, into ordination is: What would "church" be if Earth mattered? If we were to see and act as though the Earth is a sacrament—a living and tangible sign of God's endless grace—to be treated with awe and reverence as we would any sacrament?

Church of the Woods

Long before I walked into a "proper" church for the first time, I often told people that "my church is in the woods." Why, I wondered, would one want to go indoors on a Sunday when one could be in the woods, or on a mountain or river or lake, when it is so evident that the divine presence is there?

But now, my church really is in the woods. Starting in September, my church—the church where I serve as pastor and chaplain—is the Church of the Woods. We meet on a 106-acre patch of wild woods and wetlands in Canterbury, New Hampshire—where it is obvious that we are called to be in communion with all of creation. Our service is thoroughly recognizable—we gather, we sing, we read, we teach, we pray, we make offerings, we confess, we pass the peace, we share Communion. There's one major exception: the middle part of the service—right after we read the Gospel—consists of thirty to forty minutes of silent contemplation, sitting or walking, in the forest. So the book of nature as well as the book of scripture is our teacher, and people have

---

2. Aldo Leopold, *A Sand County Almanac* (San Francisco: Sierra Club–Ballantine Books, by arrangement with Oxford University Press, 1966), 246 (emphasis added).

the opportunity to experience and to learn a contemplative way of "doing church" as well as the exhortative way that we all know.[3]

Ecclesially, too, Church of the Woods is different. Though I am an Episcopal priest, the Episcopal Diocese of New Hampshire has been extraordinarily supportive, and our regular service is drawn from the Book of Common Prayer and the Anglican tradition, we are not a parish or congregation of the Episcopal Church but an ecumenical gathering supported by an independent nonprofit organization. We currently hold services on the second and fourth Sundays of each month in the Episcopal tradition, but the "church" is available for groups of any faith to use and we plan to offer hosting multifaith and ecological gatherings in the near future.

We look forward to connecting with others who are interested in being a "church of the land." Through this, we hope and believe we can help reunite love of, knowledge about, and care for the Earth with the ancient and powerful practices and teachings of the Christian tradition. We hope to show what "church" is like if the Earth and all that lives is sacramental.

Pilgrimage for Earth and Holy Week

If Church of the Woods is one example of what a church might be if our relationship with Earth and all her creatures was at the center, what might the central event of Christian faith—Holy Week—look like? Last June, in collaboration with St. James Episcopal Church in Woodstock, Vermont, we undertook a four-day, two-state Pilgrimage for Earth to explore this.

Maundy Thursday

What does it mean to "love one another as I have loved you," when Earth as well as Jesus is the lover and the beloved? In our Maundy Thursday liturgy at Church of the Woods, we remembered that our very lives flow from and are dependent upon communion with the Earth, as highlighted by Thomas Troeger's sermon, "Wind, Breath, Mud Creatures." At the same time, we recalled that we all are Judas—complicit in the betrayal of God's gifts through our destruction and desecration of nature.

Our service included transplanting white pine seedlings to clear them out of the path of the bulldozer that will be creating a driveway for our use, replanting them on a hillside in need of restoration from harmful logging by

---

3. More information and liturgies are available at http://kairosearth.org/church-of-the-woods/.

the previous owner. After the transplanting, while the sky let down the gentlest warm rain upon us all, we took water from a brook to wash each others' hands and feet—and we washed the feet of the seedlings as we watered them. We closed by singing our thanks and reenacting the Last Supper.

Good Friday

Our Good Friday liturgy was in two parts—a meditation on the Seven Last Words of Christ for the Earth, preached by Tom Troeger on the banks of the Connecticut River—the suture that forms the meeting place of New Hampshire and Vermont. This was followed by a Tenebrae service of lament at St. Barnabas Episcopal Church in Norwich, Vermont, with a *Litany of the Extinct*, chanted and sung in English and Latin.[4] Words cannot convey either the beauty or the heartbreak of hearing this ancient liturgy lamenting the extinction of species as the crucifixion of Christ.

Holy Saturday

With this backdrop, we moved to Mission Farm and Church of Our Saviour in Killington, Vermont, for a day of resting in grief and preparing for renewal. The liturgy was one of resting in stillness and silence as we recalled Jesus's death and the death of so much of the Earth around us. This was a day of asking, "How can we live in a world where we love so much—and where so much of what we love is dying?" We closed with a traditional celebration of the Easter Vigil at sunset, punctuated with African drum, Australian didgeridoo, and trumpet.

Easter Sunday

The Easter Sunday service, in the end, was the simplest of the liturgies to put together because, after all, Christians know how to do Easter, right? Yet at the end I found myself not quite satisfied. I was left with the question, What does resurrection look like in the face of mass extinctions, deforestation, widespread toxic chemicals, climate change, ocean acidification, and other ecological disasters that may never be reversed? While we take ultimate hope

---

4. This and the other elements of the pilgrimage liturgy can be accessed at http://kairosearth. org/church-of-the-woods/pilgrimage-for-earth/.

in the knowledge that "Christ has risen today," is that being manifested in the world? How?

Community

Perhaps the greatest impact of the pilgrimage was how it began to form a community of people joined by love of the Earth and commitment to following the way of Christ. And maybe this is the answer to the question that haunted me in the days after Pilgrimage Easter: What does resurrection look like when it is the Earth that is being crucified?

It looks just the same as it did when the disciples tried to come to grips with the death of Jesus—like a community of people banding together in hope and love to celebrate and rejoice in the eternal love that creates and binds all things. To sit in silence, offer prayers, sing songs, eat and drink together, mourn and celebrate, and perform rituals that bind them to one another, to the Earth we share, and to the God from whom everything flows and to whom everything returns. To be transformed by and in love. And then to go forth to take whatever action to heal the world we are able—which is diminishing ourselves and our destructive presence—giving enough space for nature's extraordinary healing to work.

Transformation

This, then, is my response to *Journey of the Universe*. It tells a magnificent story of an incredible reality—and it asks the question, what do we do?

We sit in silence, we offer prayers, we sing songs, we eat and drink together, we mourn and we celebrate, and we perform rituals that bind us to one another, to the Earth we share, and to the God from whom everything flows and to whom everything returns.

We ask to be transformed by and in love. And then we go forth to take whatever action to heal the world we are able—which means mostly getting out of the way.

For this, we need prayers, songs, and rituals that are connected to the Earth.

We need new expressions of our faith that are alive to the living and the dying of the natural world around us—and to our own complicity in that death.

We need to allow our hearts to be broken open.

We need to get out of the way so that God—and Earth, the first incarnation—is at the center.

We need to open ourselves to the transforming power of Christ (by whatever name), that we may become bearers of God's creativity and love.

What is your way?

PART III

Participating in a Living Cosmology

❧

14

Getting from Protestant Social Justice to Interfaith Creation Justice

What Does It Take?

Larry Rasmussen

"The Stone Age didn't end because we ran out of stones."[1] The update to this statement is, Will the Fossil Fuel Age end because we've left coal, oil, and gas in the ground? Or do we burn them? Max Weber feared the latter. He ended *The Protestant Ethic and the Spirit of Capitalism*, published in 1904, with the modern capitalist order pictured as an "iron cage" in which we've trapped ourselves. Bound to "the tremendous cosmos of the modern economic order," "the lives of all the individuals who are born into this mechanism [are determined] with irresistible force." "Perhaps [this order] will so determine them until the last ton of fossilized coal is burnt,"[2] he added. The year 1904 was a hundred and more years ago. Better than a century later, on the very same March day that the Intergovernmental Panel on Climate Change (IPCC) released its Fifth Assessment Report, strengthening yet again scientific consensus about human-induced warming, ExxonMobil, the most successful

---

1. Attributed to Sheik Ahmed Zaki Yamani, speaking as the Saudi oil minister to OPEC colleagues. Reported in Thomas L. Friedman, "Putin's Crimea Victory? Not So Fast," *New Mexican*, March 27, 2014, running Friedman's piece from the *New York Times*.

2. Max Weber, *The Protestant Ethic and the Spirit of Capitalism* (New York: Charles Scribner's Sons, 1958), 181. The original was published in German in 1904.

business in the history of money, issued a statement saying that it sees no end to the burning of oil and gas.

The difference between Weber in 1904 and the IPCC and ExxonMobil in 2014 is that now this "tremendous [economic] cosmos," thanks to those vast supplies of compact, stored energy in the form of fossil fuels, brings us the rolling apocalypse and mounting trauma of a new epoch on the only planet fine-tuned for us.

This is the setting for my question: What would it take to move from Protestant social justice into interfaith creation justice?

Let me signal the difference of 1904 and 2014 that matters most and then crawl back through some history that helps position us for new tasks.

The difference that sends pounding waves on all shores is this: a new Earth Age is upon us. The late Holocene has had a good run, from 11,700 years ago into the very recent past. Its distinctive mark has been climate stability sufficient to allow the triumph of life over and over again, and its importance for us is that it has hosted every human civilization to date, bar none. Say a prayer of thanks for the Holocene. It's been the time of our lives.

Now it appears the Anthropocene is upon us; and that means three realities that were not self-evident in 1904:

1. *Humankind collectively is now the single most decisive force of nature itself.* Most systems of the natural world are currently embedded as part of human systems, or profoundly affected by human systems—the high atmosphere, the ocean depths, the polar regions. It was never thus, but it is now, and for all foreseeable futures.

2. *Nature has changed course.* The scientists of the International Geosphere-Biosphere Programme claim that "evidence from several millennia shows that the magnitude and rates of human-driven changes to the global environment are in many cases unprecedented." "There is no previous analogue for the current operation of the Earth system."[3] Industrial humanity has brought on a nonanalogous moment, a unique geologic epoch.

3. *In contrast to the climate stability of the Holocene, the mark of the Anthropocene is climate volatility and uncertainty.*

What does this mean for social justice and creation justice? If our swollen powers are now exercised, to quote Willis Jenkins, "cumulatively across generational time, aggregately through ecological systems, and nonintentionally over evolutionary futures,"[4] then the time and space dimensions of human responsibility are stretched beyond anything humans have known. If no nat-

---

3. W. L. Steffen et al., *Global Change and the Earth System* (Berlin: Springer, 2004), v.

4. Willis Jenkins, *The Future of Ethics: Sustainability, Social Justice, and Religious Creativity* (Washington, DC: Georgetown University Press, 2013), 1.

ural terrain goes untouched by both human goodness and molestation, then everything turns on our action and choice, with consequences for present as well as distant generations of humankind, together with the other relatives aboard the ark.

Let me bare my soul: I find this swelling of human responsibility a frightening prospect, for two reasons. First, while the new reality means "the ascendency of ethics for our era, as an utterly practical affair,"[5] because everything turns on human choice and action, our canons and institutions of moral responsibility and justice don't begin to match the down-and-dirty consequences of actual human power. The neighbor we are to love as we love ourselves is no longer (only) the one at hand in time or space. We will cause Jericho road victims three generations hence because we handed them a diminished and dangerous planet. Put it this way: How are we to be Samaritans to neighbors we'll never see or know? What present justice takes you there?

But the question is not only about justice for deep time. It is also justice for the full community of life and its parental elements—earth (soil), air, fire (energy), and water. Yet our present canons and institutions of social justice are not competent, even conceptually, to gather in disappearing species and shrinking habitat, eroding soils, altered gene pools, collapsing fisheries, souring seas, environment-related disease, receding forests, melting glaciers, delta dead zones, migrating pests and diseases, rising sea levels, environmental refugees, biodiversity loss, changes in coastal zone structure, more greenhouse gases, surface temperature climbs, more intense storms and flooding, deeper drought, and climate volatility. Both deep time and a broader moral universe test social justice as we practice it.

Next question: If Anthropocene reality is morally awkward for social justice, what would it take to transform social justice to creation justice while still retaining the fire of social justice? That fire must *not* be lost because climate change makes social justice both *more difficult and more urgent*. More difficult because injustice has a further reach in space and time, on a planet likely bearing fewer resources, relative to the scale of the problems; and more difficult because those who contribute least to climate injustice suffer most from it. With climate injustice, "life [as] unfair" is on steroids.

In short, the puzzle is whether social justice can be transformed for a new vocation as creation justice.

To get a leg up on the answer, I turn to the story of why modern Protestant social justice emerged in the first place. It arose in response to the very kind

---

5. Larry Rasmussen, *Earth Community, Earth Ethics* (New York: Orbis Books, 1996), 5.

of economy that has since brought on the Anthropocene. It arose in response to what was identified early in the twentieth century as "the social question" or "the modern social problem." The phrases are those of Ernst Troeltsch, a contemporary of Max Weber's. Troeltsch wrote the following in 1911:

> This social problem is vast and complicated. It includes the problem of the capitalist economic period and of the industrial proletariat created by it; and of the growth of militaristic and bureaucratic giant states; of the enormous increase in population, which affects colonial and world policy; of the mechanical technique, which produces enormous masses of materials and links up and mobilizes the whole world for purposes of trade, but which also treats men and labour like machines.[6]

These were the consequences of the economy made famous in Adam Smith's work of 1776, *An Inquiry into the Nature and Causes of the Wealth of Nations*. Smith had identified its engine, soon named "capitalism" by French philosophers, and it was already a world-shaping power. Not that many years later, Karl Marx, shocked and awed by this capitalism, wrote in 1848:

> The bourgeoisie, during its rule of scarce one hundred years, has created more massive and more colossal productive forces than have all preceding generations together. Subjection of Nature's forces to man, machinery, application of chemistry to industry and agriculture, steam-navigation, railways, electric telegraph, clearing of whole continents for cultivation, canalization of rivers, whole populations conjured out of the ground—what earlier century had even a presentiment that such productive forces slumbered in the lap of social labor?[7]

The year 1848 was barely on the cusp of the "application of chemistry to industry and agriculture . . . clearing of whole continents for cultivation, canalization of rivers, whole populations conjured out of the ground." One hundred sixty–plus years later, all those factors and more belong to our earlier note that nature's systems are now embedded in human systems and profoundly influenced by them: earth, air, water, and the weather included. And while Marx was spectacularly wrong in his prophecy that the proletariat would become the gravediggers of the bourgeoisie and that the coming

---

6. Ernst Troeltsch, *The Social Teaching of the Christian Churches,* 2 vols. (Chicago: University of Chicago Press, 1981), 2:1010.

7. Karl Marx and Frederick Engels, *The Communist Manifesto* (New York: Henry Regnery, 1954), 23.

socialist revolution would upend capitalism, he was, like Troeltsch, dead on about the atomization of society and the exploitative nature of industrial orders that nonetheless captivated people with the lure of enormous productivity and mounting material prosperity. Had Troeltsch and Marx been present in 2000, they would likely not have been surprised at the staying power of the "social question" or the fact that it had gone global. The assault on settled, intact community still defines our world. The gap between the rich and the rest widens while institutions of family, community, and nation-state still struggle to stave off atomization amid unleashed economic forces, shifting identities, and unsure sovereignty. This might have saddened or angered these students of early capitalism, but it probably would not have startled them.

The social gospel of Walter Rauschenbusch and many others soon responded to "the modern social problem" across a broad front and on innumerable issues, including the right to unionize, a minimum wage, decent and safe working conditions, housing that went beyond hovels, the franchise, child labor laws and an eight-hour day, legal recourse to discrimination in hiring and firing, protests against obscene wealth for the robber barons but meager rewards for those receiving slave wages, and some beginning provisions for child care and health care.

Does the twentieth century report anything else? Yes, and it went largely unnoticed by Rauschenbusch and Troeltsch, less so by Marx.

The "ecological question" joined the "social question." It, too, is the direct outcome and downside of the organization, habits, and exacting requirements of modern industrial-technological society and its expanding extractive economy. It manifests itself as the unending transformation of nature, a parallel to the unending transformation of society, both of them in ragged pursuit of mammon. We cited its warning signs earlier: shrinking habitat and disappearing species, eroding soils, altered gene pools, glacial melting, rising seas, and so forth. Like the social question, the ecological question has also gone global.

I have crawled back through this history so that we notice what has largely gone missing in Protestant social justice—namely, the natural world as worthy of any reverence such that it might make moral claims upon us or itself be due justice. Protestant social justice tacitly affirmed the "tremendous [economic] cosmos" of the industrial paradigm, whether as industrial socialism or industrial capitalism. The quest of social justice, and the glory of its considerable achievement, was to render the consequences of the new economy far *fairer* in the lives of those determined by it. But this was justice captured by an economic cosmos tone deaf to the needs of the natural world.

Differently said, social justice as it emerged here, while admirably driven by fair play and fair outcome for human communities, is justice shorn of any doctrine of creation except creation as the décor for the human and a storehouse of resources for human benefit. The ecological question has little presence in either the formation or execution of this justice. This is justice that assumes that the basic unit of human survival itself is human society. It is not. It is planetary creation comprehensively, with all the primal elements—earth (soil), air, fire (energy), water—truly primary. The human common good is not possible without care for the goods of the planetary commons.

My criticism is not a criticism of what Protestant social justice has included and accomplished. It has not only taken on the vital issues we mentioned. It has been spot-on in developing the means by which to do race, class, gender, and cultural analysis, thereby demonstrating in detail the gross injustice of severely maldistributed benefits and burdens. Liberation theologies refined these analytical tools and, like the Hebrew prophets, made the locus and play of power a centerpiece of theological and social ethical method itself. That focus on power and that laying bare of injustice, both near the center of Christian reflection itself, are precious gifts that now belong to both social and environmental justice. Yet the humanity/nature dualism of most justice traditions worked like blinders on a horse. Attention straight-ahead left nature off to the side and external. The growing and unsustainable footprint of humanity in every natural domain that mattered—the carbon footprint, the water footprint, the biodiversity footprint, the ecological footprint (land and energy footprints), and the material footprint, meaning the measure of resources used.

The conclusion is this: For social justice to arrive at the threshold of creation justice, a deep theological and moral transformation is required. Certainly, one of the keys to this transformation is a different cosmology, some viable alternative to the cosmology of industrialized humanity and its economy. *Journey of the Universe* is an instance of that. What follows is a starter list of requirements for the cosmology of creation justice.

In keeping with the ancient understanding of *oikos*, the Greek root of "ecology," "economics," and "ecumenics," economics and ecology merge to become "eco-nomics." Eco-nomics embeds all economic activity within the ecological limits of nature's economy and pursues the three-part agenda of production, relatively equitable distribution, and ecological regenerativity. Growth as a good is not precluded, provided it is ecologically sustainable and regenerative for the long term, reduces rather than increases the instability that large wealth and income gaps generate, and bolsters rather than undermines the capacity of local and regional communities and cultures to nurture and draw wisely upon their cultural and biological diversity. In all events, "the first law of economics must be the preservation of the Earth economy."[8]

A parallel exists for energy policy. Anthropocene energy use is markedly changing planetary processes. Yet most justice attention to energy is about energy resources and use: Do we have enough to continue to grow the economy to meet human needs? Are we energy-independent? How will energy be distributed fairly? These discussions go on without *first* asking what energy sources and uses are mandated by the planet's climate-energy system. They assume that human energy use is primary, then we'll deal with environmental effects. This is exactly backward. The first law of energy is preservation of the *planet's* climate-energy system as conducive to life; human energy use is necessarily *derivative of the planet's*. This is the energy parallel to Thomas Berry's maxim that the first law of the human economy is the preservation of nature's economy.

Yet getting economic and energy policy right, vital though it be, is likely not sufficient for creation justice, even when other key sectors, such as water and air, are added and primacy is granted the atmosphere's dynamics, or the hydrological system's. All of these rightly reframe social justice by internalizing nature's requirements for its own regeneration and renewal on its own terms. They observe the necessary order in Berry's maxim that "planetary health is primary and human well-being is derivative."[9]

Principles and policies have little traction, however, apart from virtue. A cosmology for creation justice nurtures a soul-deep *feel* for creation, a gut connection that is Earth-honoring and Earth-healing as well as God-praising and God-fearing. This entails the religious conviction that all creation is sacred, not only human life. All creation is worthy of wonder and awe, respect and reverence (the substance of ecological virtues). The sheer unearned gift of life bears a value far beyond the stark utility accorded it by the industrial paradigm. In our economic cosmos nature's value is market value and little more. Unlike the cosmology of *Journey of the Universe*, here there is no place for the sacred, the mystical, and the numinous. Or, to summon Max Weber again, creation justice and its cosmology counter the "disenchantment" of nature effected by our economic cosmos with a "reenchantment" that restores to nature what it has been—the bearer of mystery and wonder, the medium of the sacred and the presence of the divine, the ethos of the cosmos itself.[10] Alice

8. Thomas Berry, "Conditions for Entering the Ecozoic Era," *Ecozoic Reader* 2, no. 2 (Winter 2002): 10.

9. Thomas Berry, *Evening Thoughts: Reflecting on Earth as Sacred Community*, ed. Mary Evelyn Tucker (San Francisco: Sierra Club, 2006), 19.

10. Weber, *Protestant Ethic*, 180–82.

Walker understands reenchantment: "The more I wonder, the more I love."[11] Market value does not.

Creation as sacred, bearing moral claims upon us for planetary health, and asking justice of us for its well-being together with ours, leads to a second element of creation justice—namely, *good God-talk.* Worthy God-talk gathers in all of Earth's voices to sing the hymn of creation or to reflect creation's "shine of the holy."[12] God-talk that does not encompass all 13.8 billion years of the universe's pilgrimage to date and the immense wheeling of 100 billion galaxies, each swimming with billions of stars and who knows how many planets; God-talk that does not gather in all species come and gone, as well as those leaving as we speak; and God-talk that does not embrace the whole drama of life in all its misery and grandeur is unworthy. Shorn of the universe, the worship of God is worship of a human species idol. It is God rendered in our own diminished image. Worthy God-talk, then, is about the mystery of matter and its drama—all of it, past, present, and future. It is an invitation to "sing with all the people of God and join in the hymn of all creation"[13] so as to give voice, however partially and inadequately, to the carnal presence of the "uncontained God."[14] The uncontained God is the God of creation justice.

The last element of creation justice is an *upgraded understanding of ourselves,* an improved anthropology, if you will. Our present segregated sense of ourselves as a master species is a miserably shrunken grasp of who we are in the scheme of things. From both a scientific and a religious point of view we are "fearfully and wonderfully made," in the phrase of the psalmist. We are the handsome fruit of two wombs: our mother's and Mother Earth's. So the singer sings of God:

> For it was you who formed my inward parts;
> you knit me together in my mother's womb.
> I praise you, for I am fearfully and wonderfully made.
> Wonderful are your works; that I know very well.

---

11. Alice Walker, *The Color Purple* (New York: Pocket Books, 1982), 290. The full quotation, in a letter from Celie, is, "I think us here to wonder, myself. To wonder. To ast. And that in wondering bout the big things and asting bout the big things, you learn about the little ones, almost by accident. But you never know nothing more about the big things than you start out with. The more I wonder, he say, the more I love."

12. Joseph Sittler, "A Theology of Earth," in Richard C. Foltz, ed., *Worldviews, Religion, and the Environment: A Global Anthology* (Belmont, CA: Wadsworth/Thomson, 2003), 17.

13. From the Lutheran Eucharist liturgy, *Lutheran Book of Worship* (Minneapolis: Augsburg, 1979), 88.

14. Denise Levertov, from the poem "Annunciation," in *The Door in the Hive* (New York: New Directions, 1984), 85.

My frame was not hidden from you,
when I was being made in secret,
intricately woven in the depths of the earth.
Your eyes beheld my unformed substance.
In your book were written all the days that were formed for me,
when none of them as yet existed.[15]

To be "fearfully and wonderfully made" across the stop-and-start eons of evolutionary life, to belong to life's drama and grandeur and have a perch of our own in the great Tree of Life, is our glory. "I just try to act my age," quips Joanna Macy. "My atoms are 14 billion years old."[16] When justice is creation justice, it takes on such transcendence as this.

These three elements, then, composted together with deep changes in economic and energy policy, transform social justice to creation justice in a way that can be affirmed widely on an interfaith basis—creation deemed sacred, God-talk worthy of the uncontained God, and a "fearfully and wonderfully made" understanding of ourselves. The "tremendous [economic] cosmos" of "the last ton of fossilized coal"[17] has more than met its match—the wonder of the real cosmos and its economy.

---

15. Psalm 139.
16. Cited from "Letters," a feature found at www.joannamacy.net.
17. Weber, *Protestant Ethic*, 181.

15

Evolutionary Cosmology and Ecological Ethics

WILLIS JENKINS

Journey of the Universe puts in cosmic perspective the remarkable emergence of human creativity and, in our own pinprick of time, the emergence of human influence over Earth. How can the cosmic perspective help humans exercise their influence responsibly? How does cosmology matter for ethics?

Like many readers and viewers, *Journey of the Universe* made me pause in wonder at the movements and expanses of the universe. Wonder at the deep sources and improbable emergence of life always deepens my affiliation with the thin green pulse of Earth's biosphere. There seems to me some connection between wondering at the stars and loving the Earth, between opening ourselves to the question of our place in the universe and deepening our sense of place on Earth.

More than half of humanity now lives beneath the orange haze of urban lights, and many of the rest of us orbit around artificial lights at night. Like the migrating songbirds who circle bright buildings all night long, unable to see the stars by which their ancestors learned to orient themselves, most humans cannot see the stars by which our ancestors learned to wonder. In the same historical moment that humanity began to learn the expanse of the universe, we also began to cut ourselves off from the night skies that provoked our wonder in the first place.

The stars remain, of course, for those who still seek darkness, but dark skies can be difficult to find and the stars easier than ever to ignore. For many people, the universe described by astrophysicists seems impossibly alien from the experience of daily life, which moves in orbits of artificial light. No longer looking into the dark night sky, we lack the experience of being questioned by the stars and of wondering at great spinning worlds not our own.

Meanwhile, the orienting traditions of human meaning, whose stories were surely composed and told beneath the questioning stars, seem to have lost their connection to the universe. Perhaps our ancient traditions stagger before the knowledge of contemporary cosmology: not just an overhead canopy of stars, but billions of unseen galaxies, moving in mind-bending ways in an expanding universe, which itself may be just one of a multiverse. The science of astrophysics seems to confound the meaning-making capacities of humanity's great creation stories and orienting myths. The result, so far, has been that traditions go on telling the old stories in uneasy dissonance with the sciences of the stars, with the consequence that the great stories of human meaning disorient us from the universe.

Brian Swimme and Mary Evelyn Tucker aim to heal that narrative disorientation. By rendering evolutionary dynamics of the universe within an unfolding story, they restore the stars to humanity's meaning-making—and at just the moment that this evolved capacity for storytelling has begun to matter for the future evolution of the species and the planet. If evolutionary cosmology will have any part in human self-understanding, it must be made part of the stories by which humans interpret themselves. *Journey of the Universe* is a remarkable cultural accomplishment in this regard, for it makes a narrative from the nearly unaccountable complexity of astrophysics, and thereby presents evolutionary cosmology within human meaning-making. *Journey* exemplifies its thesis, that in humanity the universe becomes self-conscious.

Evolution Underdetermines Ethics

So how much practical difference should we expect such mythic healing to make to ecological problems like diminished migrations and the extinction of bird species? If humans learn to better interpret the question of our own meaning within evolutionary context, would we become more responsible for migratory birds? Would we discover an obligation to dim the lights of our building and restore the stars to avian navigation? Clearly the story told by *Journey of the Universe* establishes a mood of wonder that supports delight in and respect for nonhuman life. It also celebrates how the universe has become exuberantly self-conscious in a species of primate whose niche capacity for symbolic thought seems to have endowed its minds with the universe. So does the role of that symbolic brilliance in the story (as the presupposition of storytelling at all) entail norms for how the human species should exercise its powers? In this case, are there norms for how humans should modulate the brilliance of its cities?

The iconic image associated with discourse of the Anthropocene depicts Earth at night, with the planet lit up by urban networks of human brilliance. Does the story of life's emergence from the stars carry practical norms for how humans should modulate the literal brilliance of their cities? More generally, does narrating the emergence of life from the stars help uncover something like natural laws for how humans should host other forms of life, now that humans have such influence over the Earth's habitats?

It is not obvious. Cosmic narrative is a difficult genre in which to find specific norms or natural laws. Evolutionary cosmology can help put in context practical questions of what to do about something like species extinctions— but then that context seems to overwhelm the question. It is a truism that everything that exists is a product of the processes preceding it; how might appreciation of those processes reform specific exercises of power?

But maybe norms or laws seek too much moral confidence; maybe the moral of cosmology is found in humility. The astrophysicist Neil de Grasse Tyson has suggested that a perspective informed by contemplating the universe can make wars over oil seem silly and—if imagining intelligent visitors from another planet—even embarrassing.[1] So perhaps the decentering perspective required to appreciate evolutionary cosmology should humble human projects before the life around them.

Still, the humility of a decentered perspective only goes so far in informing discrete judgments of appropriate exercises of power. I am amazed by the evolutionary history of birds, these descendants of dinosaurs who navigate by the stars from which they also originate, which makes me ashamed of the undeclared war against nature by which humanity is driving so many bird species to extinction. I, too, appreciate that the Ebola virus originates from the same stars that I do, and yet I have no objection to the World Health Organization's "war" against it—and might find its ultimate extinction justified. I agree that recent American wars over oil have been embarrassing for their small-mindedness. But not all political violence is silly just because it appears small in the perspective of the universe. Armed protection of refugees, for example, finds thin warrant in evolutionary cosmology, but some contexts, I would argue, require measured violence to protect human rights. What about armed protection of nonhuman migrations or militarized conservation efforts? Ethical norms for exercising, restraining, and judging human power require more determined warrants.

My point here is simply to caution against one kind of simplistic enthusiasm for evolutionary cosmology by underscoring the basic point that evolution underdetermines ethics. Obviously the processes by which everything came to be make up the conditional possibilities for any scene of human freedom. But wonder at those processes does not go very far in helping humans

---

1. Neil DeGrasse Tyson, *Living on Earth*, National Public Radio, October 18, 2013.

decide what to do with the unanticipated emergence of their own planetary powers. *Journey* offers a narrative that, by depicting the emergence of consciousness and its significance for the future of Earth, brings readers up to the cusp of ethics, on which the rest of the story now depends.

"We live on a different planet now," write Swimme and Tucker, "where not biology but symbolic consciousness is the determining factor for evolution."[2] Some readers would like to find in *Journey of the Universe* a kind of natural law—a normative pattern of the unfolding universe—that could guide the appropriate exercise of human power in the Anthropocene (or let us call it instead the "Ecozoic," as Thomas Berry named the era marked by a spasm of anthropogenic extinction events and by the need for civilizations to reorient themselves to the ecological rhythms of the planet they now shape.[3] Reckoning with the challenge that the planetary problems of the Ecozoic present to ways of thinking marked by separate domains of the human and the natural, ecological ethics has just begun trying to think about responsibility without nature (or at least without the idea of nature as objective standard for human behavior). It would not aid deliberations over obligations to migrating birds, nor to any other hybridizing human/environmental system, to attempt to reinstate natural law by way of an evolutionary story. But we might consider the role of moral moods.

Wonder as Moral Mood

Having issued a caution for enthusiasts to be at least cautious about its ethical implications, let me return to the moral significance of telling stories in wonder at the stars. The stars do not demand any particular way of making meaning from them; astrophysics does not require any particular existential mood. But telling stories and using interpretive symbols to set moods seem innate to our species, and the interpretive dispositions to reality sustained by those stories and cultivated in those moods may have pervasive ethical implications. The dispositions and moods set the context for what will count as acting well or poorly.

Those implicit accounts of good action are not at all required by the science that seems inevitably to give rise to them. Robert Bellah's *Religion in Human Evolution* observes the tension: evolution is the modern creation myth, grounded in disinterested science, but it inevitably gives rise to extra-

---

2. Brian Thomas Swimme and Mary Evelyn Tucker, *Journey of the Universe* (New Haven, CT: Yale University Press, 2011), 101.

3. Thomas Berry, *The Great Work: Our Way into the Future* (New York: Bell Tower, 1999).

scientific ideas, shaped by the interests of story-formed creatures. "When it comes to telling big stories about the order of existence, then, even if they are scientific stories, they will have religious implications."[4] Science takes on a different function when it is used to satisfy humanity's craving for meaning. Bellah contrasts the pessimistic feeling communicated in Steven Weinberg's statement that the more he studies the universe the more it seems pointless, with Eric Chaisson's more optimistic account of evolution as basis for cultural meaning. Either way, evolution is glossed in a way that seems religious, in the sense that recounting histories of nature cultivate interpretive dispositions to reality. Those dispositions to reality in turn lend some kinds of actions more intelligibility than others.

So while there is not a direct correlation between the causal processes that lead to stars and the moral practices of humans (no direct correlation between the starry skies above and the moral law within), there is a connection between how humans make sense of the causal processes that lead to stars and how humans make sense of possibilities of action. There is no direct correlation between how the cosmos evolved and whether humans should alter their atmosphere, but wonder at the history of the stars may give rise to kinds of symbolic play through which humans will conceptualize appropriate climate action.

Bellah follows evolutionary theories arguing that play allows organisms to create an environment that supports new developmental and evolutionary directions. Play seems to create a relaxed field of competition that can support the development of capacities that make organisms and perhaps species more successful. Bellah then suggests that ritual might be thought of as a kind of cultural play, allowing symbol-making creatures a relaxed field in which to pursue new possibilities. Religion might then be considered this kind of functional nonfunctionalism by supporting ventures of symbolic transcendence that allow humans to interpret the problems of their existence in new ways. That is one way to think about the axial age of cultural development.[5] New biocultural possibilities for civilizations may be opened or at least made intelligible by symbolic play that, in effect, uses culturally approved exercises of wonder to rethink everything about a culture. *Journey of the Universe* may be symbolic play, in that sense, and may well help open new biocultural possibilities from its response to findings of evolutionary cosmology. It takes leave

---

4. Robert Bellah, *Religion in Human Evolution* (Cambridge, MA: Harvard University Press, 2011), 45–46.

5. Ibid., see chapter 2 and Conclusion.

from evolutionary science, but its symbolic play has axial implications and is therefore rightly considered religious.[6]

Bellah closes his massive work on the deep history of religion by explaining what it can do for humanity's current moral situation. Deep history puts in perspective humanity's impact on Earth's life as sudden destruction on the order of war. Extinguishing bird species by drowning them in light appears a turn as tragic as accidentally starting a world war. Deep history also allows its students to accept deep religious pluralism, since many human traditions of meaning, each explicable on its own terms, is just what one would expect from a long and open field of symbolic play. Deep history lets scholars give up the search for one absolutely true myth by which to measure the others. The science of evolution is not "true myth" but rather the modern condition of myth-making, and only sometimes also its subject. Bellah closes his book with the implication of those two findings: "If we could see that we are all in this, with our theories, yes, but with our practices and stories, together . . . we might just make a bit more likely . . . a world civil society that could at last restrain the violence of state-organized societies toward each other and the environment."[7]

Evolutionary cosmology matters for ethics, then, when wonder at origins animates forms of cultural play that suspend belief in the competitions that ordinarily order human life. Consider, as both example and theory, the Buddhist teaching of dependent origination. It holds that appreciating the complex and contingent relationships on which our reality depends can help persons overcome the illusion of independent identity. The Dalai Lama writes, "When we come to see that everything we perceive and experience arises as a result of an indefinite series of interrelated causes and conditions, our whole perspective changes. We begin to see that the universe can be understood in terms of a living organism where each cell works in balanced cooperation with every other cell to sustain the whole."[8] Thinking from Bellah, we might say that for the Dalai Lama, the teaching itself becomes play, and it opens possibilities of compassion. In appreciation for our own emergence within a story of elegant contingencies, the Dalai Lama says, we may begin to see that our self-interest is bound up with the flourishing of the whole, and so begin to question our notions of "self-interest." The very exercise of questioning our notions of self and the interests we think it has can diminish violence against other humans, against the birds, and against Earth.

---

6. This is consonant with Thomas Berry's sense of the religious dimensions to the great cultural work of our time. See Berry, *Great Work.*

7. Bellah, *Religion in Human Evolution*, 606.

8. Dalai Lama, *Ethics for a New Millennium* (New York: Riverhead Books, 1999), 41.

If we interpret the universe as the Dalai Lama does, it is not because science demands it. It is because his way of playing with reality permits us to see the interpretive possibility of compassion as the most fitting or elegant way to make sense of the facts and problems of our existence. Practices of compassion make the story true.

Conclusion

So, how should humans live brilliantly in the dark? Responsibility for other creatures begins in recognition that human dwelling is increasingly becoming the habitat for all other life on Earth. Right now, that recognition should make us look upon our forms of dwelling with some shame, for as habitat they have little character, hospitality, or elegance. As habitat for birds our cities are (for the most part and not without some heartening exceptions) hostile. The story told in *Journey of the Universe* cannot produce a practical ethic of responsibility or care for migrating birds, but it should make readers and viewers question how the infrastructural hostility to birds fits with the story.

The many nonintentional ways we cut ourselves off from wonder may have debilitating effects on our capacities of cultural play (our "religious" capacities, in that sense), with the result that we lack the imagination to rethink the narratives that lend contingent meanings to our lives—the narratives that make us proud of yardage or of compassion. Songbirds circling our empty lit skyscrapers betoken a worse tragedy than the individual harms to those birds. Into what story of Earth does that sort of thing fit, and can humans recognize themselves within it? Restoring the stars to the birds may indeed require restoring the stars to the human imagination, to a role in the stories we tell of ourselves.

Received that way, as a restoration of the universe to the imagination of a species, *Journey of the Universe* does indeed matter for ecological ethics. Received within a mode of symbolic play, evolutionary cosmology can open critical and reparative possibilities for the stories humanity lives by. Told in the context of mass species extinctions, stories made from evolutionary cosmology can make humans reflect on the stories that will be told from the actions taken in our generation. Will the commitments that matter most to us shape the story that future generations tell of us? Abrahamic people of faith may ask, What story will God eventually tell of humanity's tenure in creation? In play, others might ask, What story would Earth tell of its latest upstart species? What story would the universe tell?

How humans cope with planetary problems like species extinction will make the meaning of the universe stories that we tell ourselves. The emergence of the universe is indeterminate for ecological ethics, which awaits the (contingent, fallible) interpretations given evolutionary cosmology by the wondering freedom of humans. Those interpretations happen as humanity uses its cultural and biological inheritances to meet the challenges of its existence. The interpretations are good and true ones just insofar as they are used to sustain fitting, even graceful, ways of inhabiting the cosmos and navigating by its stars.

16

The Nature and Limits of a Contemporary Evolutionary Cosmology's Ethical Significance

FREDERICK SIMMONS

We have learned so much about our universe over the past century that remarking upon this fact has become trite. Yet what, if anything, has our knowledge of the starry heavens above to do with the moral law within us?[1] Given the celebrated distinctions between fact and value and is and ought, it may appear that astronomy and ethics have no connection at all. However, our growing awareness of the size, duration, complexity, precision, mystery, beauty, and serendipity of the universe evokes an awe and admiration that do not seem so separate from ethical sentiments. Brian Thomas Swimme and Mary Evelyn Tucker's *Journey of the Universe* is a recent rendition of contemporary evolutionary cosmology that deftly summons just such responses and is intended to advance urgent ethical aspirations.[2] Nonetheless, Swimme and Tucker meticulously forebear from deriving much moral guidance from their cosmology, and I shall argue that they are correct to remain so circumspect. I proceed by considering *Journey*'s possible connections with Christianity and thus implicitly examine a facet of the religion and ecology movement. While I conclude that *Journey*, per se, hardly provides an ethic, I contend that it can inform cardinal Christian ethical commitments in consequential ways and thereby signals a principal reason to explore the potential relationships between religions and ecology.

---

1. Cf. Immanuel Kant, *Critique of Practical Reason* (Cambridge: Cambridge University Press, 1997), 133 [5:161 in German Academy of Sciences edition of Kant's *Collected Works*].

2. Brian Thomas Swimme and Mary Evelyn Tucker, *Journey of the Universe* (New Haven, CT: Yale University Press, 2011).

Journey of the Universe and Christian Theology

Journey of the Universe is a wondrous story splendidly told, and thus appreciation is a fitting response to it—both by Christians and others. Indeed, *Journey* beguilingly suggests that appreciation is a fitting response to the universe itself, and not simply this elegant story about it, and I think such appreciation succinctly conveys both the nature and the limits of *Journey's* significance for Christian ethics. In order to contextualize this ethical aspect of its meaning, I begin with a few observations about *Journey's* potential implications for Christian thought more broadly.

Of course, *Journey* arrestingly attests to not just the connections but also the congruence between life on Earth and the rest of material reality.[3] For example, *Journey* observes that like organisms, matter itself is a self-organizing process[4] that ignites creativity in new forms whenever possible[5] and ultimately gives rise to astonishing complexity.[6] At the same time, *Journey* discerns a deep ambiguity threaded throughout the universe,[7] for destruction is inextricably entwined with the process of bringing forth ever more complexity[8]—whether in supernovae[9] or evolution by natural selection and the self-consumptive dynamics of ecological cycles that allow finite ecosystems to sustain themselves indefinitely. Either way, such congruence between life on Earth and the rest of material reality suggests that attention to inorganic phenomena can offer insight into the organic, that nature displays realities evident in culture, and that matter discloses truths about the spirit. If so, in addition to orienting religious energies to address ever more global environmental challenges, cosmology may conduce to religious knowledge.

Of course, *Journey* does not present itself as natural theology, and I think it is right to refrain. Even assuming Christian convictions, *Journey's* theological import is contestable, for although Christianity explicitly concerns theology, Christians have disagreed about whether—and if so, to what extent— God may be known through nature apart from special revelation. Moreover, Christians who accept natural theology have generally understood it to corroborate, and perhaps enrich, what they have learned of God and faith by other means rather than constitute a self-sufficient source of religious wis-

---

3. Ibid., 48.
4. Ibid., 27.
5. Ibid., 24.
6. Ibid., 13.
7. Ibid., 32.
8. Ibid., 68.
9. Ibid., 34.

dom.[10] Thus, while reality, per se, may prove revelatory given Christian convictions, Christians usually insist that the book of nature remains incomplete and indeterminate without the book of scripture, and I argue momentarily that the same is true for *Journey* with respect to environmental ethics, mutatis mutandis.

Nevertheless, I believe the relatively new story that *Journey* tells ought to matter to Christian thought, for cosmology does not simply affect Christian convictions but is a dimension of Christian faith itself. After all, Christians affirm that God created everything that is not God, remains involved in the unfolding of God's creation, and will fulfill God's purposes for it. *Journey* helps us understand what we have come to know about the reality that Christians confess God to have made, and so can help Christians appreciate what they take to be God's creation. It also suggests some profound parallels between fundamental Christian beliefs and the material world. I briefly mention two.

First, *Journey* contends that astronomy, geology, and biology all discern dynamism as inherent to matter and life, and therefore disclose that material reality is not incidentally but essentially historical.[11] Temporality matters in the universe, for new things emerge and physical properties change. Given their doctrine of incarnation and its interpretation of God's fullest self-disclosure as an event, Christians likewise maintain that truth is not simply propositional or expressible in creeds but is finally historical.[12] Second, *Journey* describes the sun as transforming its essence into the energy on which Earthly life depends. Such self-consumptive self-giving that sustains others is reminiscent of the Johannine interpretation of Christian discipleship rooted in the crucifixion of Jesus, wherein the faithful are known by loving as Jesus loved, namely, by laying down his life for others.[13] More broadly, I have noted *Journey*'s insistence that life on Earth manifests the deep patterning of the universe itself,[14] and indeed ecological dynamics nourish life perennially by consuming all of the organisms they support. Christians have reason to

---

10. Cf., e.g., David Fergusson, "Types of Natural Theology," in *The Evolution of Rationality: Interdisciplinary Essays in Honor of J. Wentzel van Huyssteen,* ed. F. LeRon Shults (Grand Rapids: Eerdmans, 2006), 380–93.

11. Swimme and Tucker, *Journey*, 2, 40.

12. Cf., e.g., Walter Rauschenbusch, "The Influence of Historical Studies on Theology," *American Journal of Theology* 11, no. 1 (January 1907): 127. "When God revealed himself, it was not by communicating abstract propositions or systems of doctrine. The fundamental fact in the Christian revelation was that the Word became flesh. Therewith Truth became History."

13. See, e.g., John 13:34–35; 15:12–13; 1 John 3:16.

14. Swimme and Tucker, *Journey*, 48.

extend this continuity further still, and interpret the deep patterning of life on Earth and in the universe as a whole to manifest fundamental spiritual truths—for example, about the significance of temporality and the role of self-giving in sustaining life.

Incidentally, *Journey* may offer Christians unexpected resources for theodicy as well. Modern natural sciences intensify the challenge that natural disasters pose for Christian attempts to reconcile God's goodness, wisdom, and power with the world in which we live since these sciences make it less plausible to suppose that such disasters are ultimately due to human sin.[15] The Lisbon earthquake of 1755 is perhaps the most famous instance in modern Europe, but other destructive consequences of the tectonic system have caused grave suffering on a massive scale and led many to question Christian faith. However, *Journey* observes that the tectonic system has also produced new minerals and polymers found nowhere else in the solar system,[16] and contributed decisively to the formation of the atmosphere that allowed life to emerge on the planet. Hence, rather than simply random or otherwise uselessly ruinous, tectonic forces appear integral to making Earth fit for organisms.

Yet just as *Journey* is not natural theology, it is not theodicy, and does not purport to be. *Journey* does not claim the tectonic system's contributions to life on Earth outweigh its harms, or attempt to explain why the God of Abrahamic faith would rely upon that system to help create the conditions for human flourishing. *Journey* does not proffer such justifications, nor should it, for *Journey* aims to tell the what and how of the universe, rather than the why.[17] Christians and others who seek to answer that latter question, however, may rightly draw upon *Journey* and the rest of our knowledge of the world as they attempt to do so. Indeed, theodicies that do not appreciate the tectonic system's role in making the Earth inhabitable miss data that may be relevant to consistently combining this system's violence with a wise, powerful, and loving God.

---

15. Of course, modern natural sciences also intensify the challenge that life's development and diversification pose for Christian theodicy; cf., e.g., Christopher Southgate, *The Groaning of Creation: God, Evolution, and the Problem of Evil* (Louisville, KY: Westminster John Knox Press, 2008).

16. Swimme and Tucker, *Journey*, 41.

17. To be sure, *Journey* posits a role for various aspects of the universe (e.g., 112)—including human beings, as I discuss in the following section. Nevertheless, *Journey* does not ascribe a purpose to Earth or the universe as a whole.

Journey and Christian Ethics

I believe *Journey* has similar significance for Christian ethics, namely that it is directly relevant to Christian ethical thought, but insufficient to deliver determinate moral mandates on its own. Normative ethical theories—whether Christian or otherwise—combine doctrines of value and obligation to organize and justify commitments concerning the substantive content of morality.[18] With its elegant narrative of this vast, dynamic, complex, creative, and finely tuned universe, and the astonishing intricacy, variety, and capacity of the life it has come to support, *Journey* fosters wonder and appreciation for all that is. As such, *Journey* inculcates a sense that material reality and the processes that have brought it into existence are valuable. This is a major contribution to ethical reflection, particularly in a context where many people and social practices mindlessly destroy the products of these processes and damage the processes themselves. However, this sweeping sense of the value of what has emerged and the way that it developed does not pretend to provide a definition of goodness, much less attempt to specify the relative value of the multiple good things that have come to be. In short, *Journey* does not fashion or imply an axiology discriminating enough to clarify many of our applied moral choices.

Moreover, since ethics concern the right as well as the good, normative commitments are necessary in order for *Journey* to mean something ethically and not just evaluatively. Nevertheless, as a story of the universe's development and diversification derived from contemporary natural science, *Journey* appropriately adumbrates no more than vague and occasional normative principles. For example, rather than attempt to cull specific ethical injunctions from their wondrous story, Swimme and Tucker scrupulously avoid overreaching and instead conclude by asking whether "our destiny has something to do with [the] desire to journey and to experience the depths of things . . . [with] knowing how we belong and where we belong so that we enhance the flourishing of the Earth community."[19] Only after thus tentatively posing the prospect of a human telos do Swimme and Tucker venture this modest normative invitation. Their restraint is all the more remarkable given the ethical zeal that animates their interest in cosmology.[20] Still, since this evo-

---

18. Cf., e.g., Shelly Kagan, *Normative Ethics* (Boulder, CO: Westview Press, 1997), 2.

19. Swimme and Tucker, *Journey*, 112–13.

20. For example, John Grim and Mary Evelyn Tucker end their book *Ecology and Religion* (Washington, DC: Island Press, 2014) by effectively answering *Journey*'s concluding questions affirmatively and eliciting ethical orientation from these affirmations: "This is our collective challenge: how to respond with a more inclusive ethical understanding and more comprehensive eco-

lutionary cosmology alone engenders at most conditional and exceedingly formal moral obligations, *Journey*, per se, lacks the resources needed to furnish determinate moral direction, let alone a theory of the right. As a result, because it affords such a diffuse ascription of value and can scarcely delineate our moral obligations, in itself *Journey* does not supply much of an ethic, environmental or otherwise.

Yet *Journey* remains ethically significant, even if its own prescriptions are often too general to illumine—much less identify—what we should do in a given instance, since *Journey* can have more determinate moral meaning when coupled with normative commitments and sensibilities adduced from other realms.[21] For many people, religion is of course a primary source of such commitments and sensibilities, and hence religion naturally complements *Journey*. Thus, just as a robust Christian natural theology requires the book of scripture as well as the book of nature, so *Journey* requires religions or other normative traditions to supplement its cosmological story in order to offer practical ethical mandates.

The field of religion and ecology recognizes and addresses precisely this need. Emerging from concern for the growing environmental crisis, the field of religion and ecology appreciates that cosmology and even ecology do not themselves supply concrete moral counsel, but may powerfully shape and galvanize moral sensitivity and resolve when united with religious commitments.[22] In fact, by essentially embracing Walter Rauschenbusch's conviction that a great faith is required to meet the tremendous social challenge of our time, the field of religion and ecology proves a contemporary, ecumenical manifestation of the social gospel, "explor[ing] religions as vehicles encouraging change of attitudes and values regarding the environment."[23] Yet the

---

logical knowledge to the diminishment of planetary ecosystems. In this way we can create the grounds for the flourishing of life. Revaluing nature is at the heart of this discussion. . . . If we recognize humanity as part of complex evolving ecosystems, we may become more beneficial participants in the planetary ecosphere. . . . In so doing we may fulfill our deepest role as humans within the Earth community" (ibid., 170).

21. In *Ecology and Religion*, Grim and Tucker explain that *Journey* "draws on the perspectives of scientists, religious thinkers, and environmentalists to deepen our understanding of the evolutionary process and to outline new directions for the flourishing of life. *Journey of the Universe* was inspired by Thomas Berry's understanding of the need for a new comprehensive story integrating the sciences and the humanities in relation to the ecological and social challenges we are facing" (ibid., 9).

22. Ibid., 6; cf., e.g., Larry Rassmussen, *Earth Honoring Faith: Religious Ethics in a New Key* (New York: Oxford, 2013), 6.

23. Grim and Tucker, *Ecology and Religion*, 11. Cf. Walter Rauschenbusch, *Christianizing the Social Order* (New York: MacMillan, 1915), 42. See also Grim and Tucker, *Ecology and Religion*, 26, 87.

converse is equally true, for religious traditions likewise require ecology and other contemporary natural sciences to realize their potential contributions to the environmental movement, since these sciences help us recognize the nature of our environmental problems and some of the ways we might manage them.[24]

Further, the conjunction of religion and ecology brings more than such pragmatic benefits. For instance, I have noted that religions frequently involve doctrines of creation or conceptions of the natural world, and thus scientific cosmological findings—whether at the astronomic or ecological levels—may enrich these religious convictions.[25] Similarly, religion and ecology is not simply a social movement but also an emerging field of interdisciplinary academic inquiry that is stimulating multiple insights.[26] Perhaps most pivotal, the field of religion and ecology has decisively advanced previous study of the relationships between religion and science that sought to explore the compatibility of these domains at this generic level.[27] Rather than assume that differences between religious traditions and among the natural sciences make no difference to the relation between religion and science, from its inception the field of religion and ecology has investigated connections between specific religious communities and a particular natural science.[28]

---

24. Grim and Tucker, *Ecology and Religion*, 6.

25. Grim and Tucker identify religious cosmologies as religious traditions' narratives of the origins and unfolding of the universe, Earth, and life in ibid., 63; cf. 2, 43.

26. Cf. ibid., 9, 220n6.

27. For example, Ian Barbour, a leading figure in late-twentieth-century reflection on religion and science, took such a generic approach, especially with respect to religion; among many books; cf., e.g., Ian Barbour, *When Science Meets Religion: Enemies, Strangers, or Partners* (New York: Harper Collins, 2000). Similarly, despite criticizing Barbour on precisely this point and repeatedly emphasizing both scientific and religious pluralism as well as the importance of contextual specificity to scientific and theological rationality, J. Wentzel van Huyssteen also ultimately addressed the relationship between religion and science generically and instead sought to overcome these challenges to establishing a basis for interdisciplinary reflection; cf., e.g., J. Wentzel van Huyssteen, *Duet or Duel? Theology and Science in a Postmodern World* (Harrisburg, PA: Trinity Press International, 1998). Of course, concentration on connections between specific religious traditions and particular branches of natural science is not confined to the field of religion and ecology. For example, Stephen Pope's criticism of epistemic reductionism underscores the importance for Christian ethical engagement with contemporary biology of distinguishing between the natural sciences; see Stephen Pope, *Human Evolution and Christian Ethics* (Cambridge: Cambridge University Press, 2007), 61–69.

28. Cf., e.g., the conference series on World Religions and Ecology organized by Tucker and Grim and the accompanying nine edited volumes published by the Harvard Center for the Study of World Religions that each approach the connections between ecology and religion from the perspective of principal branches of a given religious tradition.

Journey, Christianity, and Environmental Ethics

So what does *Journey* mean ethically, given a particular religious tradition and a specific science? This remains a vast topic despite these delimitations, in part because ethics is such an immense subject, and further because Christianity is so diverse. Indeed, I have devoted a monograph to ecology's potential significance for one dominant strand of Christian ethics and faith, and I will not venture to summarize its findings here.[29] Instead, I focus my consideration of *Journey's* implications for Christian morality on the faith's environmental ethics, and offer two general observations about both the good and the right. Concerning the good, I suggest *Journey* compellingly fosters Christians' appreciation for the world they are committed to valuing, and helps them to recognize that the processes that produce and sustain the world that they value ineluctably involve disvalue for all organisms, including human beings.[30] Concerning the right, I propose that *Journey's* contribution to Christians' conception of the world's value encourages them to respect higher levels of biological organization in nature rather than attempt to diminish these levels' disvalue, and to prioritize systemic and holistic factors when our duty to respect higher levels of biological organization in nature requires us to redress—or if that is not wholly possible, mitigate—the damage we have impermissibly done.[31]

Undoubtedly, these are still very broad principles. Religious traditions offer a matrix of normative commitments that may hone *Journey's* amorphous moral meanings, but given Christianity's ethical variety it would be necessary to examine particular branches of the faith and the more specific ethical commitments that they generate to narrow these principles even with *Journey's* insights. Cosmology can inform and enrich ethical reflection, and since Christian faith involves cosmological claims, Christians ought to allow *Journey* to inform and enrich theirs. At the same time, cosmology underdetermines ethics, even for Christians and members of other religious traditions, since these faiths have multiple ethical meanings and thus interpret the moral significance of a given cosmology differently. Yet although Christianity and *Journey* per se do not collectively yield a discrete environmental ethic or provide express moral instruction on particular environmental problems, they concur about the propriety and importance of relating to the environment ethically. Environmental ethics surely involve much more than this.

---

29. Frederick Vernon Simmons, "Life and Value: What Ecology May Mean for Christian Ethics and Faith" (PhD diss., Yale University, 2010).

30. Swimme and Tucker, *Journey*, 68.

31. Ibid., 113.

Nonetheless, consistently relating to the environment ethically would be a significant shift in itself and constitute another step in the universe's journey on planet Earth.

Integrating Ecology, Justice, Race, and Gender

❧

17

The Universe Story and Social Justice

Growing Healthy, Just, and Sustainable Communities in an Age of Global Warming

Carl Anthony and M. Paloma Pavel

Don't ask what the world needs. Ask what makes you come alive, and go do it. Because what the world needs is people who have come alive.[1]

We are living in an historic moment. We are each called to take part in a great transformation. Our survival as a species is threatened by global warming, economic meltdown, and an ever-increasing gap between rich and poor. Yet these threats offer an opportunity to awaken as an interconnected and beloved community.[2]

The Universe Story, with its focus on deep time beginning with the Big Bang, offers a crucial framework for placing our life and work in the larger context of creation. First articulated by Thomas Berry, this story has been elaborated and expanded in *Journey of the Universe,* a powerful trilogy that

---

1. Howard Thurman quoted in Gil Bailie, *Violence Unveiled: Humanity at the Crossroads* (New York: Crossroad, 1996), xv.

2. Desmond Tutu, from the foreword to Anne Herbert and M. Paloma Pavel, *Random Kindness and Senseless Acts of Beauty* (New York: New Village Press, 2014).

includes a book,[3] an Emmy Award–winning film,[4] and an educational series of conversations[5] with a diverse circle of scholars and practitioners who are generating this profound perspective for our time. The Universe Story is a call to each of us to tell our own stories in a new way, orienting and linking them to the processes that bring forth the stars and galaxies, our sun, solar system, planet Earth, the oceans and continents, mountains, rivers, and every living thing. This story expands the imagination and encourages a deeper, more spiritually grounded approach to our relationship to the Earth and our ecological resources. In the opening years of the twenty-first century, this urgent message of radical inclusion and interconnection is a clarion call to diverse communities that have been marginalized and excluded, to recast our understanding of our place and role in the order of things and forge a new story for our time.

The Universe Story provides a profound invitation to communities of color to frame our issues in new ways, to connect our struggles for social justice to the urgency of the climate movement, and to lead the way in creating a just and sustainable world. This work embodies Thomas Berry's concept of the "Great Work," in which humanity must evolve from our current exploitative model of living,[6] not only through caring for our planet and its resources, but also through reinventing our deep-seated power structures that distribute opportunities and advantages along lines of race, class, gender, sex, faith, and many other social divides. Faith-based organizing groups who are grounded in the leadership of African American communities and other communities of color are a powerful site of energy, discussion, and action for marginalized groups, and a hub for community leaders and activists to organize with great success.

After several decades of participation and research with these groups, we have seen the power of deepening and reimagining personal and community stories as a vital step in community leadership development. In this light, we see that faith-based communities of color provide a compelling contemporary context for an application of the cosmology of the Universe Story, revealing the potential to inspire the next generation of community climate

---

3. Brian Thomas Swimme and Mary Evelyn Tucker, *Journey of the Universe* (New Haven, CT: Yale University Press. 2011).

4. Brian Thomas Swimme and Mary Evelyn Tucker, *Journey of the Universe*, DVD directed by David Kennard and Patsy Northcutt (2011; New York: Shelter Island, 2013).

5. Mary Evelyn Tucker, *Journey of the Universe: Conversations*, DVD directed by Patsy Northcutt and Adam Loften (New York: Shelter Island, 2013). Originally released by Northcutt Productions, 2011, under the title *Journey of the Universe Educational DVD Series*.

6. Thomas Berry, *The Great Work: Our Way into the Future* (New York: Bell Tower. 1999), 3.

justice leaders to step into the privilege and responsibility of caring for our planet Earth.

For nearly two decades we have been working collaboratively on multi-racial, multiclass leadership development and coalition-building in communities of color, frequently organized in faith-based settings—including ten years leading the Sustainable Metropolitan Communities Initiative (SMCI) at the Ford Foundation. Over this time, and in twenty different SMCI sites across the United States, we noted a pattern—that communities of color who were able to innovate at new levels of collective impact had begun by deepening their understanding of their historical and community context and consciously creating a new story about their situation.[7] This process of reinventing our story allows for an expansion of the imagination, opens up the possibility of new roles and new alliances, and contributes to ground-breaking victories for marginalized communities. We have been exchanging these strategies with our community organizing partners who are using the power of story in grassroots leadership training. Based on the success of these re-storying strategies, we see potential for the Universe Story to galvanize community leadership in the climate justice movement, by helping us to redefine our role as caretakers of the planet and forge unprecedented alliances across social and geographic boundaries.

Telling a New Story

The Universe Story includes the story of the cosmos, the story of planet Earth, the story of human development, and the story of Now—the Ecozoic era with its current climate challenges.[8] In creating a new story, we see the need to reexamine some conventional assumptions around the story of human development. Using the African American community as a case in point, below we provide an illustration of how telling the story of human development from the perspective of people of color illuminates a different set of details, and increases the value and viability of the Universe Story to communities of color.

The human race was born in Africa. Two hundred thousand years ago, after 6 million years of hominid evolution, Homo sapiens emerged out of the

---

7. See M. Paloma Pavel, ed., *Breakthrough Communities: Sustainability and Justice in the Next American Metropolis* (Cambridge, MA: MIT Press, 2009), 153–238.

8. Brian Swimme and Thomas Berry, *The Universe Story: From the Primordial Flaring Forth to the Ecozoic Era—A Celebration of the Unfolding of the Cosmos* (New York: HarperOne, 1994), 3–15.

cataclysmic processes of planet Earth.[9] About fifty thousand years ago, some Homo sapiens began migrating to many places throughout the globe. Over time they learned to speak many languages, invented tools, began to farm, and built relatively autonomous villages and towns, a few overland trade routes, and a few large cities. The inhabitants of that Afro-Eurasian continent eventually came to populate both North and South America.[10] By 1492 CE, only the oceans prevented communication between the American and Afro-Eurasian continents.

In 1500, European and African expansion began. Between 1500 and 1800 CE, a million and a half Europeans had migrated to the new world, and seven and a half million Africans had migrated to the new world as slaves. The contact between the American populations and the European and African populations resulted in the death of up to 30 million people. In those years between 1500 and 1800 CE, the world experienced a major convergence in which many populations of the world became connected, with a clear divergence in wealth, power, freedom, and access to resources and global learning between some European people and African and Native American people.

While the year 1800 saw only 2 to 3 percent of the human population living in cities, two centuries later, over 50 percent of the world lives in cities. Many of our major US cities contain a hidden history, shaped by the surge in wealth generated during historic moments of excessive abuse—slave trade, colonization, genocide—which developed in tandem with humanity's unsustainable relationship to the environment. The story of how these injustices shaped our metropolitan regions challenges the dominant narrative of urban development, which begins with the Fertile Crescent, then focuses on Europe and its colonies, then fails to include the influence of people of color. Port cities of the United States—New York, Philadelphia, Baltimore, Charleston, Savannah, and New Orleans—were all initiated between 1500 and 1800, shaped by the dominance of the triangular slave trade. These origins persist in our metropolitan geography today, in the stratification by race of the cities and the suburbs, and the school-to-prison pipeline in low-income neighborhoods of color. This joint history—the shaping of our cities, the institutionalization of racial divisions, and the large-scale ecological shift following the industrial revolution—gives some clues as to how solutions for sustainable and just metropolitan regions might be developed. Now that we are in a place of reconsidering our relationship to the environment, we are

---

9. Noel T. Boaz, *Eco Homo: How the Human Being Emerged from the Cataclysmic Story of the Earth* (New York: Basic Books, 1997), 23–50.

10. Cynthia Stokes Brown, *Big History: From the Big Bang to the Present* (New York: New Press, 2007), 38–56.

also given the opportunity to reconfigure the legacy of racism and build cities with justice for everyone.

It is essential for justice that our world's most marginalized communities take the lead in reimagining our cities and reinventing our relationship to Earth and to one another. A new story that can make change this big will need the representation of people at every level of society. Thus, to build a balanced, holistic vision for the future, it is crucial that we raise the voices of those who are least likely to be heard. Vulnerable communities on the front lines of climate change are already experiencing its first and worst effects, along with centuries of precipitating environmental and social injustices. These frontline (most at-risk) communities are the biggest stakeholders experiencing climate change firsthand, and their perspective is invaluable to understanding what is at stake. On the other hand, people in positions of privilege (even the well-meaning ones) will often find a way of rationalizing and reproducing business as usual. If our most vulnerable communities are left out of the process of creating this new story, the end result will surely continue to reinstate the same inequities and disadvantages that ultimately result in global collapse.

Our experience with communities working at the intersection of environment and social justice demonstrates how new voices emerging from communities of color become a valuable asset in articulating environmental issues.[11] Their advocacy illuminates the pitfall of a mainstream environmentalism that ignores the stratification of society and the differential impacts of pollution depending on social location, and reveals the vitality and agency of communities of color and marginalized communities in generating new solutions.[12] The Universe Story has the potential to offer our communities a liberatory arc of time, freeing us from narrowly defined identities that are limited by constrictive notions of class and race. In this framework, a marginalized, criminalized, or disempowered person can discover that he or she is in fact made of stars, and connected to the regenerative possibility of all life. We can reclaim the power of the whole universe, embedded in our own cellular structure.

Faith-based community groups are a vital hub of organizing and energy for vulnerable communities, and so they are an important place to focus on

---

11. Richard Marcantonio and Alex Karner, "Disadvantaged Communities Teach Regional Planners a Lesson in Equitable and Sustainable Development," *Poverty and Race Journal* 23, no. 1 (2014): 1.

12. Federal Reserve Bank of San Francisco and the Urban Institute, eds., *What Counts: Harnessing Data for America's Communities* (San Francisco: Federal Reserve Bank of San Francisco, 2014), 29–31.

when we talk about community engagement. Moreover, faith-based groups stand on a liberatory history that also offers new potential for bringing the Universe Story into action. For African American communities that were carried to the new world in slave ships named for Christian saints, there can be an understandable ambivalence around Christianity—but we have seen the power of the Universe Story approach to Christian cosmology to reconcile these tensions. Interfaith community organizing, which welcomes faith approaches that embrace the root of all traditions, offers especially fertile ground for this application. The Universe Story creates a bigger tent in which to pray, worship, and act on behalf of vulnerable communities in the best of Christian faith traditions.

The Potential of a Bigger Story for Faith-Based Community Organizing

One of our most admired and cherished partners in faith-based community organizing is Mary Gonzales, founder of the Gamaliel network, who inspires and emboldens us with her use of re-storying in leadership development and coalition building. We are presenting Gamaliel's successes with these strategies in order to point toward the potential of using the Universe Story for the same purpose, to empower individuals from marginalized communities and enable new kinds of global, interfaith, multiracial climate justice coalitions.

Mary Gonzales is a bilingual and bicultural professional organizer who, over thirty-five years, cofounded the Gamaliel Foundation, a network of forty-three community-led organizations in sixteen US states, with chapters in South Africa and Great Britain. She also founded *Ntosake*, a national training program for women, and the Arc of California, a statewide network of twenty-three agencies serving people with disabilities. Throughout a lifetime of community organizing work, she has found great success in community-driven impact for environmental and social change.

Gamaliel builds power for interfaith, cross-sector, intergenerational, multiracial, and multicultural groups in the most vulnerable communities as a pathway to faith and leadership development, using strategies that train community leaders to see their stories in a larger context. A foundational principle of their leadership training is that change-making does not happen in isolation; one must develop a "public self" to win changes in the larger society from the standpoint of a community leader. Gamaliel's transformative process includes rewriting one's own biography as a hero's journey, embedded in the story of one's community. Mary describes this process as "very intense but very exciting. People come out of there feeling like it's a whole new world."[13]

Mary's conviction in this process was fueled by her own initial experiences being trained as an organizer, which allowed her to see the story of herself in a new way:

> I always thought I was just a crazy inner-city person, angry about many things that I didn't understand, feeling bad that I was angry, feeling like I had to be silent, feeling just inept, inadequate, all kinds of nasty feelings about myself, and through this training everything that I had experienced was held up. . . . There was a career that supported what I wanted to do, there was a real language that was acceptable and that supported what I believed.[14]

Along with re-storying techniques, Gamaliel trains its leaders to build coalitions with new allies from diverse backgrounds, including low-income communities and communities of color, academics, policy advocates, public-interest lawyers, student groups, labor unions, business leaders, education officials, and environmentalists. With these strong leadership-building and coalition-building strategies, Gamaliel networks have seen remarkable successes at the intersection of environment and social justice. Recent successes in California include winning back over $70 million in transportation funds toward affordable public transit in Oakland,[15] and winning a free Eco Bus Pass for all eighteen-and-under youth in Alameda County, simultaneously reducing greenhouse gases while giving vulnerable youth from marginalized communities better access to jobs, quality education, healthy grocery stores, and other opportunities.

Gamaliel and other faith-based community groups that are building leadership and power in vulnerable communities have seen great success in strategies that reimagine individual and community stories in a larger context. This reimagining empowers the exploration of new capacities and skills and the forging of new alliances beyond the boundaries of geography, race, class, sector, and faith, resulting in real victories for vulnerable communities. Based on these success stories we see a great potential for the Universe Story framework to galvanize vulnerable community members, especially in faith-based

---

13. Excerpted from a 2013 interview with Mary Gonzales, in M. Paloma Pavel, ed., *Climate Justice: Frontline Stories from Groundbreaking Coalitions in California*, forthcoming.

14. Ibid.

15. See http://genesisca.org/Documents/Statement_about_Civil_Rights_victory.pdf.

settings, to reach new levels of climate justice organizing at this most critical moment in history.

Going Global: We Need a New Story

Since Berry first published *The Dream of the Earth* in 1988, nearly three decades ago, we have been experimenting with the potential of community stories in the context of the Universe Story to fuel our activism. At our most recent Universe Story event, called the Gathering Our Stories Charrette hosted at the California Endowment, we invited our climate-justice coalition partners (including some community partners from Gamaliel) to post personal, community, and universal trials and triumphs on a twenty-foot-long collaborative timeline that spanned the wall of the conference room, beginning with the Big Bang. Paloma Pavel facilitated the day, introducing the deep time map, its center point representing the present moment of our immediate opportunities and actions for Climate Justice, while two funnels designating the past and future fanned out on either side. On the far left, in the deep time past, Paloma drew the great flaring forth in vibrant colors, while Carl Anthony narrated the Universe journey: the initiation of the galaxies, the elements, and the laws of physics; the formation of planet Earth; the emergence of life; and the development of our atmosphere. We guided our journey forward through deep time, Carl describing the first appearance of humankind in Africa, Paloma catalyzing the group to share their own cultural heritage as she drew these diverse emerging strands on the time map, converging on the present. Together as a group, we filled in key moments in the social justice and environmental movements, and identified our activist predecessors whose shoulders we stand on—from ancient times to the Civil War and Reconstruction, to the Civil Rights Act, the Montgomery bus boycott, and the freedom rides.

Next, we facilitated community leaders to come forward and place Post-it Notes signifying key moments in their lives that had inspired their activism, as well as recent key victories in our climate justice work. Finally, in the funnel flowing out from the present moment into the future section of our deep time map, Paloma led the gathering and marking of hopes and dreams for the future, and envisioning of the world we hope to leave for generations to come.

This story-gathering exercise provided a renewal and regeneration in movement building and collective visioning. We were surprised and encouraged by the hidden sources of passion and fuel that this day tapped. The result was invigorating—individuals saw themselves not only as a product of history, but as making history.

To make big changes, we need to keep the big picture in mind. Keeping up with regional politics can keep us in a shortsighted and reactive frame, but while we must indeed act quickly and locally, we must also think globally—in time as well as in space. While seizing pivotal opportunities, we can use a Universe Story framework to keep our actions aligned with larger goals. *This Universe Story has the power to bring together a global Earth community, to cut across social distinctions of race, place, religion, and class, highlighting our universal heritage as the descendants of stars, capturing our place in the creative development of matter, life, and consciousness.* Yet we cannot neglect our community stories. Each community narrative reveals solutions and resources found within our own traditions that can be employed to solve global problems. The story of the African American experience, from prehistory to the present, is one of a multiplicity of community narratives needed to counterbalance the dominant European narrative of conquest and bring about an understanding of the forces shaping our world. With an eye to our fundamental interconnectedness, we must ground ourselves by honoring our community and individual stories, recognizing the historical factors that have brought us to where we are.

We need something this big to harness the courage, resourcefulness, and imagination necessary for the challenges to come. This is how we will find the kind of imagination and innovation that our ancestors did, when they chose to leave the trees and walk upright in the savannah. We are facing a similar leadership challenge now: to leave the shelters of our familiar ways of thinking, activating latent technical and emotional capacities, and working together under a bigger sky. The arc toward justice may be long, yet we see its resilient bend when we take a longer perspective.

We have seen community groups take strength in reimagining their stories to see the potential of movement building and innovative city-making. We have seen the success of creating new stories that open the possibility of linking arms across boundaries, creating radical community partnerships between frontline communities, policy experts, academics, philanthropic organizations, and government institutions. Now we must connect struggles for social justice and environmental sustainability with the bigger picture—the journey of the universe, from the Big Bang to our emergence as conscious beings, to our role in managing the precious resources of life on planet Earth—to create a new story for our time. The global breakdown of our economy and environment thus also provides an unparalleled opportunity for a breakthrough—a reimagining of our purpose and a comprehensive coming home to our place in the universe.

18

Religious Resources for Survival

Ecofeminism and Earth Community

MARY E. HUNT

The impact of climate change and the rapid deterioration of living conditions for many species, in particular for people who are made poor, require resources for survival. Ecofeminist religious thinking and acting is one such resource that holds real promise in its theoretical creativity, practicality, and coherence. These terms also describe the foundational work of Thomas Berry that excites many imaginations and undergirds the film *Journey of the Universe*.[1] Their common foci on religion as a cultural expression of what binds (*religare*, "to bind fast") a diverse cosmos hints at more commonality than is obvious at first glance. Thomas Berry's one hundredth birthday celebration is a delightful occasion to bring together elements of both strands in a mutual strengthening for maximum strategic impact.

As an ecofeminist theologian I focus primarily on resources by women, queer people, and people of color. So Thomas Berry's work, while familiar to me in the grand scheme of things religious, had not been an explicit object of study. Moreover, I found some people who use his work, whose good praxis is sparked by his writing, to be insufficiently critical of his content. Insofar as he made grand claims without much reference to the specificities that have come to characterize contemporary work in religion, especially issues of sex, gender, race, and class, I have long been reluctant to embrace his content. I prefer not to reinforce a seemingly hegemonic way of thinking and acting when so many emerging approaches remain to be taken seriously.

So imagine my surprise when reading *Thomas Berry: Selected Writings on the Earth Community* that I began to sense I had read something quite sim-

---

1. Brian Thomas Swimme and Mary Evelyn Tucker, *Journey of the Universe*, DVD directed by David Kennard and Patsy Northcutt (2011; New York: Shelter Island, 2013).

ilar before.[2] It was not that the content was so familiar, but the dynamics were—broad strokes, creative insights, unique syntheses, courageous thought experiments, scant regard for the negative reactions of those less able to step up and step out. *Aha,* I thought, *this man has a lot of similarities to feminist philosopher and theologian Mary Daly.*

A look at the two of them together furthers my goal: to find resources that deepen ties and help build coalitions of like-minded but differently situated people and groups. The urgency of dealing with what I term the "cosmic conundrum"—how life gives life, or not—makes pairing intellectual odd fellows like Thomas Berry and Mary Daly a strategic innovation. If these two seemingly disparate thinkers have points of coincidence, imagine how many more overlaps exist among the thinking and actions of other people. Beginning with this case of two quite unusual figures paves the way to bring lots of people into creative conversation and collaboration.

In this chapter I outline the surprising parallels that I have begun to discern between the life and work of historian/geologian Thomas Berry and the life and work of feminist philosopher/theologian Mary Daly, both Catholic outliers. I think of them both as champions of the cosmos, intellectuals who trained their considerable talents on the whole of reality, not simply a tiny piece of it. Their lives and work suggest some implications for ecofeminist work in religion that I find important in my own efforts.

Many ecofeminists in religion have found Berry's work useful, and one might hope that Berry and followers find the work of ecofeminists helpful as well. To date, I can report more embrace of Berry's ideas by ecofeminists than even acknowledgment of feminism by many ecologists and theologians who claim to be doing ecological theology. I hope to even the balance. I am struck by the largely unnoticed similarities between the two. Berry offered the outline of a story capable of communicating the historical scaffolding of the universe. The film *Journey of the Universe* is a powerful expression of that. Daly provided the early and strong foundations for ecofeminist religious thinking. Both did so with a commitment to cosmic survival and thriving that has become even more urgent after their natural deaths.

I trust that a comparison will help bring those influenced by Thomas Berry to a deeper understanding of feminist work and those influenced by Mary Daly to a fuller cosmological vision. It has been thus for me. Taken together, their two intersecting intellectual projects shed light on and provide tools to address the pressing needs of our gorgeous—but threatened—geo-home.

---

2. Mary Evelyn Tucker and John Grim, eds., *Thomas Berry: Selected Writings on the Earth Community* (Maryknoll, NY: Orbis Books, 2014).

Twentieth-Century Catholic Outliers

Thomas Berry (1914–2009) and Mary Daly (1928–2010) were both cradle Catholics who lived during a century of profound changes in their religious home, some of which they created. They were born into a church that was expanded by immigrants and controlled by Rome. Berry, one of thirteen children, and Daly, an only child, had early mystical experiences in nature that they claimed shaped them profoundly. Thomas encountered creation in a grassy meadow with white lilies, while Mary found "be-ing" in a clover, and later in a hedge. They were hardly garden-variety Catholic kids! That they turned into quasi-mystics later in life is no surprise. To pass over this mystical component, is, in my view, to miss the why of their prodigious efforts to explain and embrace the world.

Both studied theology and philosophy. Thomas's route took him through priestly ordination in the Passionist Community, though he engaged more in the work of Teilhard de Chardin than in ordinary pastoral ministry. Mary found her way through the thickets of sexist discrimination that prohibited her from studying philosophy in Catholic undergraduate and master's programs where she did degrees in English. They both earned doctorates, Berry in history and Daly (who actually earned three doctorates) in theology and philosophy. It would belabor the obvious to say that their genders conditioned their lives.

They went on to teach in Jesuit institutions, Professor Berry at Fordham University and Professor Daly at Boston College, where they were outsiders of sorts. As a Passionist cleric and a laywoman, they did not have the built-in access and leg up, a kind of affirmative action that their Jesuit colleagues enjoyed in those days. Rather, they were free and at the same time required by lack of easy entrée to do their work in much broader contexts. It showed in their remarkable creativity and scope. I daresay they were the most accomplished scholars of their generation in their respective institutions. At least in Mary's case, the prophet was not appreciated in her own land. I suspect the same may have applied to Thomas.

Thomas and Mary were both smitten with Thomas Aquinas. The rigor of his thinking and his systematic way of working through arguments stood them both in good stead. They had studied Aristotle, but with Aquinas they arrived at the conclusion that matter matters. Both resisted the lure of neo-Thomism (as well as transcendental Thomism) in their day, preferring to engage in Berry's case with other world religions, and in Daly's case with Paul Tillich and process thought. Not even those fields were sufficient for their enormous reach, so they went on to create whole new discourses. I can imagine that they were enigmas to their colleagues and perhaps even the object of professional jealousies for their prolific production and huge impact. Their clear and compelling writing left no doubt as to where they stood, even in those instances where they stood alone, better, ahead of the rest of us.

They also differed sharply in many ways that bear mention. Mary was aggressively antichurch after Vatican II, given the ways women were oppressed and excluded by institutional Roman Catholicism. All of her training notwithstanding, there was finally no place for her in a male church. Thomas was heavily imbued with the influences of world religions, especially Confucianism and Buddhism, so his focus was not on in-house church issues either. As a Passionist, he was simply not part of the diocesan structures of power, nor was he a natural fit in the Jesuit-heavy intellectual circles that shaped policy and polity.

I do not know if they ever met, though they may have crossed paths at professional meetings. Both attended the American Academy of Religion annual meetings on occasion, but I leave it to Berry and Daly biographers to say whether they ever actually had a conversation. What stands out for me is that while both were shaped in the Catholic ambience of the mid-twentieth century, neither of them was mired in it intellectually or religiously. Instead, they used their solid training in languages, ideas, history, and speculation as a springboard to a much broader arena, nothing less than the cosmos. Still, their radical work had sufficient echoes of Catholic thinking and acting that it resonated with many who have pushed toward an ecoconsciousness, and the "feministization" of religion from that starting point as much as from others.

Champions of the Cosmos

When I think of Berry and Daly, it is not grandiose to imagine them as champions of the cosmos. While others had their causes, Thomas and Mary were quite concerned with the Cause of Causes, Final Cause, and ideas of that magnitude. They thought big and bold about things, seeing both the forest and the trees, while others around them often saw neither. Certain similarities obtain that are more interesting than their differences.

First, they both had certain central insights, perhaps the fruit of those early mystical encounters. For Thomas Berry, the universe is unique in its status insofar as everything else is described in relation to it. It has its own story in which human beings play an important role. "Differentiation," "subjectivity," and "communion," he explained, characterize the unfolding of the universe;

"the ecological age must now activate these principles in a universal context if the human venture is to continue."[3] So we humans have our assignment, should we choose to accept it.

Mary Daly understood God as a verb. She is popularly known for saying, "If God is male, then the male is God," which sparked an already nascent feminist revolution in religion.[4] But her insight into the nature of human oppression was but one element of a much larger project. "At the very core of the philosophy of *Beyond God the Father* is the Realization that the women's revolution is about participation in Be-ing," she wrote. "That is, it is an onto-logical movement. As outsiders, or extra-environmentals, women are espe-cially equipped to confront the structured evil of patriarchy."[5]

Even patriarchy proved too small a canvas on which to paint "the blossom-ing of the intuition of be-ing which had been Foretold in the Be-Speaking of the clover blossom."[6] Instead she dealt with time in Spiral Galaxies, conflat-ing future and past, memory with moment. She encompassed animals and water, "Spinning," as she put it, her way through Earth travel and intergalac-tic thinking with equal determination. It will take generations to sort out her thoughts, but many have already begun to act on her insights.

Second, the proof of their genius and creativity (surely some colleagues thought they were mad, crazy Catholic curmudgeons perhaps!) was that both Thomas and Mary needed new words to describe what they meant. Berry coined "Ecozoic" to signal the end of the geological age and the beginning of a time when serious attention needs to be paid to the future of the cosmos.

He described himself not simply as a historian or expert on world reli-gions, not as a philosopher or theologian, but as a "geologian," one who was a scholar of Earth. Nothing else was quite right. He wrote, "A new language, an Ecozoic language is needed. Our Cenozoic language is radically inadequate. A new dictionary should be compiled with new definitions of existing words and introductions of new words for the new modes of being and of function-ing that are emerging."[7]

How convenient that Mary Daly wrote just such a book, her *Websters' First New Intergalactic Wickedary of the English Language* (with Jane Caputi)![8]

3. Thomas Berry, *The Dream of the Earth* (San Francisco: Sierra Club, 2006), 44.
4. Mary Daly, *Beyond God the Father* (Boston: Beacon Press, 1973), 19.
5. Mary Daly, *Outercourse* (New York: HarperCollins, 1992), 159.
6. Ibid., 160.
7. Remarks at the Center for Reflection on the Second Law, Raleigh, NC, quoted in Tucker and Grim, *Thomas Berry*, 144.
8. Mary Daly and Jane Caputi, *Websters' First New Intergalactic Wickedary of the English Lan-guage* (Boston: Beacon Press, 1987).

She redefined some old words, but also made up "New Words" and "Other New Words," as she called them, in order to communicate her insights. "The purpose of the *Wickedary* was the freeing of words from the cages and prisons of patriarchal patterns," she wrote.[9] What is remarkable is her insight into the need for new words coupled with her willingness to write the wordbook herself.

Her new volume was necessary to express "Mystery," later, "Metamystery," in place of "history" in order to grasp the enormity of rethinking the cosmos.[10] Words like "Hag," and "Crone," long used negatively about women, came back as honorifics in the emerging cosmic context. She coined words like "Gyn/Ecology" to describe a reality that encompasses women's well-being. Indeed Daly did in her *Wickedary* what Berry intuited was necessary in order to create what she called "New Space" and "New Time." Her ambitious imagination translated ordinary words into powerful tools for social change.

A third aspect of their cosmos championing can be found in Berry's *The Great Work* and Daly's *Outercourse* conclusion, "The Great Summation," which bear more than a little resemblance.

In *The Great Work*, Thomas Berry writes,

> The Great Work now, as we move into the new millennium, is to carry out the transition from a period of human devastation of the Earth to a period when humans would be present to the Earth in a mutually beneficial manner. . . . Such a transition has no historical parallel since the geobiological transition that took place 67 million years ago when the period of the dinosaurs was terminated and a new biological age begun. So now we awaken to a period of extensive disarray in the biological structure and functioning of the planet.[11]

In "The Great Summation," Mary Daly writes,

> We must not think of our powers to change the world merely by "summing up" our numbers. . . . Rather we should focus on our enormous Diversity within our Unity. We are not merely four thousand, or forty thousand, or even forty million. We cannot think that way, because of the enormous variations among us. We are not merely, say, one hundred thousand members within the same "species." We are more like a hundred thousand species. And the more we recognize and develop our

9. Daly, *Outercourse*, 294.
10. Ibid., 294–95.
11. Thomas Berry, *The Great Work: Our Way into the Future* (New York: Bell Tower, 1999), 3.

unique Elemental powers, the more powerfully we can communicate with each other and Act in the world.[12]

She went on to say, "Because of our Diverse and Varied talents, our Biophilic Energy can increase exponentially. Each variable 'factor' depends upon other variable 'factors.' . . . The Great Summation is following our Final Cause, Realizing our participation in Be-ing. It is cumulative affirmation and celebration of Life."[13]

These immodest treatises were just what the cosmos's doctors ordered. Little comments about small topics would never be adequate to the magnitude of their projects. We can excuse the grandiosity of their titles, indeed evaluate even those as perhaps too limited to contain their boundless and boundary-less content.

A fourth way to relate these two original thinkers is to analyze the people who read, heard, and were influenced by them. After all, they/we are their legacy. Thomas Berry's impact is assured in the veritable cottage industry that has developed on the basis of his thinking. Both the scholarly production for which thanks are due to Mary Evelyn Tucker and John Grim, and especially the film *Journey of the Universe* that Mary Evelyn and Brian Swimme made, provide marvelous access to Berry's work for a wide audience. It is impossible to quantify such impact. Dazzling images and scintillating ideas about things most humans can scarcely grasp provide a real service that translates concepts into action.

People's actions inspired by Thomas Berry's work are as impressive as the writings. They demonstrate concrete ways in which seemingly abstract concepts are rooted in and give shape to everyday life. One major cohort of practitioners influenced by Thomas Berry is the so-called green sisters or eco sisters. They live on farmlands and run centers like Genesis Farm, Crystal Spring, and other earth-focused places in the United States and Canada. Visitors can learn practical skills as well as engage in theoretical, theological, and spiritual search, much of it rooted in Berry's writings. These communities, many developed by Catholic women religious, embody the confluence of feminist and ecological streams.

Another practical application of Berry's work is in the emerging field of Earth jurisprudence. Lawyers utilize Berry's insights into cosmic connections as a resource for developing a new approach to legal theory and practice based on the needs of the cosmos, not simply on the wants of human beings. New

---

12. Daly, *Outercourse*, 415.
13. Ibid

thinking about humans as a responsible not a privileged species is beginning to give shape to field. One can only imagine what will emerge from those efforts, but it is sure that Earth will benefit and humans will find our rightful place in the panoply of creation.

Mary Daly's influence was equally abundant and practical. While there is nothing akin to Berry's colleagues' production, a Mary Daly reader is in preparation that will reintroduce her work to a new audience. Scholarly symposia, dissertations in the works, and countless references to her work in feminist theory assure that her intellectual impact endures.[14]

As with Berry, the practical impact of Mary Daly's work is, though harder to pinpoint, equally abundant. There was a bookstore in Cambridge, Massachusetts, named New Words. Croning ceremonies are common ways that older women celebrate their maturity. The massive movement away from patriarchal Christianity—people seeing and moving "beyond God the father" in many denominations—has been spurred by Mary Daly and nurtured by other feminist scholars.

There are not, as far as I know, physical places where people embody her thought as such. But she wanted, and began to create late in her life, "Hedge Schools," where those who were marginalized by educational power structures (as happened with the Irish who started them in the eighteenth and nineteenth centuries to teach their own and not British culture) could gather to educate and learn. Moreover, her presence is palpable among many feminist scholars of religion who realize how much New Space and New Time she created for the rest of us. Imitating her would be antithetical to her own method. But joining her in commitments to life abundant is consistent with her approach.

Ecofeminist Religious Work

Thomas Berry and Mary Daly may seem to be strange intellectual bedfellows. But a reason to hold them together is to demonstrate how broad the coalition of thinkers must be to confront the urgency of the moment. The speed with which the environment is losing its capacity to support life justly leaves no time to dwell on small differences. The Ebola outbreak only dramatizes the interconnectedness of all beings, something that we resist learning at our peril until the spread of infection wakes us up with a start. I take three con-

---

14. "Special Section in Memory of Mary Daly," *Journal of Feminist Studies in Religion* 28, no. 2 (2012): 89–117.

crete insights from these Catholic outliers who championed the cosmos that now inform my own ecofeminist work in religion.

First, even the most seemingly disparate approaches to ecojustice, ecosanity, even eco-eros, as I like to think of it, have some overlapping and mutually beneficial elements. Just as Berry and Daly dovetail on their cosmic concerns, so, too, do ecologists and ecofeminists. However, the need to be explicit about the commonalities is important. Gone are the days when ecologists could claim that their issue was primary, without which all else is doomed. And "over," too, are the days when ecofeminists could ignore the devastating environmental impact on the poorest of the poor that include men. Those caricatures do not help. Rather, both groups need to consult the new dictionaries and begin to speak a common language, including one another's vocabularies of action. This is an effort worthy of the signal thinkers' approaches.

Second, the role of religion in ecological enhancement is pivotal. But it helps no one to bring twentieth-century religion into conversation with twenty-first-century needs. The important changes in religion that have come about because of antiracist, antisexist, antiableist, postcolonial, queer, and other differentiated ways of reshaping religion need to be reflected in the discourse. Likewise, the people who represent various religious traditions in international, interreligious conversations need to represent those new realities. For example, it would be inadequate to bring US Roman Catholic bishops into conversation with other religious leaders when it is clear that they do not represent the overwhelmingly lay constituency of Catholicism. Most of them have no clue about cosmology, much less ecofeminism. Other people can do the job better. Breaking with the models of the past because religions have already changed into something new is a strategic challenge that requires interreligious cooperation.

Third, the liturgies, rituals, and art forms that carry the content of ecofeminist work in religion need to reflect equally differentiated sources. It serves no purpose to use the same language and imagery, the same hymns and texts, the same frameworks and roles to try to express a new reality. For instance, worship that employs exclusive or anthropocentric language about the divine, that includes texts uncorrected for racial and gender bias, or that is led by a presider who is cast in a role above or beyond that of other participants contradicts cosmos-friendly approaches. There are plenty of new resources to use for meaningful celebrations that are coherent with contemporary cosmic concerns.

Much remains to be done to operationalize ecofeminist religious work. But outliers of many stripes and champions of the cosmos of various sorts are finding common cause. In the spirit of Thomas Berry and Mary Daly, the work can and must be done with mystical faith, vivid diversity, and tireless energy before it is too late. The very doing together across differences is its own joy.

19

Planetary Journeys and Ecojustice

The Geography of Violence

WHITNEY A. BAUMAN[1]

In order to explore the implications of *Journey of the Universe* for ecojustice, I have chosen to focus on the first word of that title: "journey."[2] Indeed, "journey" has been a part of the more mystical/spiritual side of most religions, mythologies, literature, and even some philosophies over the span of recorded history. It is both metaphorical and literal in its manifestations, and it is no mere coincidence that "journey" is used as the lead word for the title of a documentary film discussing the implications of contemporary cosmology for human meaning-making practices. What I want to argue here is that the concept of "journey" is an apt metaphor for thinking about ourselves as planetary meaning-making creatures, and that the notion of journey has much to say about issues of planetary justice (a term I use here to include both ecojustice and environmental justice).[3] In particular, there are three components of journey writ large as a metaphor for cosmology that are relevant for discussions of planetary justice: uncertainty, multiple possibilities, and the focus on the terrains or geographies of knowledge vs. the temporal evolution of knowledge and ideas. I'll take the rest of this chapter to discuss each of these three concepts, but first let me describe in a bit more detail the metaphor of journey.

---

1. My thanks go to Lisa Stenmark for commenting on an earlier version of this chapter.

2. See, e.g., Whitney Bauman, "Journeying," in *Vocabulary for the Study of Religion*, ed. Kocku von Stuckrad and Robert Segal (Leiden, Netherlands: Brill, 2015).

3. On "planetarity," see, e.g., Gayatri Spivak, *Death of a Discipline* (New York: Columbia University Press, 2003). See also Stephen D. Moore and Mayra Rivera, eds., *Planetary Loves: Spivak, Postcoloniality, and Theology* (New York: Fordham University Press, 2009).

The Context of Journey

Journey, as discussed above, has been a part of multiple religious traditions: pilgrimages (a specific type of journey), the *Hajj*, walkabouts, dreamtime, the exodus, and many other journey stories are central to religious traditions. Furthermore, the idea that life is a journey "along the way" permeates many religions. Sometimes this idea of journey can actually cheapen the value of this life (as in this is one stage along the way to something greater), but in many ways journey can help us to trouble or blur certain assumptions that can structure the world in violent ways. Let me explain a bit further what I mean.

Journey is at least in part inimical to singularity and its epistemological cousin, certainty. In other words, journey involves always and already a bit of unknowing and openness toward the "not yet."[4] This is an important point in terms of discussing planetary justice in the context of the "journey of the universe." While we may be discussing the context of the universe in which we live, such discussion should never preclude the multiple journeys that can and do take place within that universe, nor the possibility of a multiverse.[5] Would the ethical implications of the journey of the universe be different, in other words, from different begging points and paths within that larger journey, or if there were multiple universes? Philosophers and theologians from many different thought-traditions have been thinking about this point for centuries. The importance and need for pluralism has also been central to the Forum on Religion and Ecology, and to the *Journey of the Universe* project, which includes engagements with multiple religious traditions and communities. For instance, in the "conversations" associated with *Journey* we hear from people such as Carl Anthony, who talks about the practical implications of a new story of the universe for issues of environmental justice. Nancy Maryboy and David Begay share how many of the components of relationality and kinship are compatible with Navajo understandings of the world.[6] In other words, the story is really about multiple tellings and journeys and how these journeys come together to form a collage-style view of cosmos we inhabit. This multiperspectivalism and the proliferation of possibilities

---

4. Ernst Bloch, *The Principle of Hope*, 3 vols. (Boston: MIT Press, 1986).

5. On the historical development of the idea of the multiverse and its implications for thinking about contemporary cosmology, see Mary Jane Rubenstein, *Worlds without End: The Many Lives of the Multiverse* (New York: Columbia University Press, 2014).

6. See Mary Evelyn Tucker's four-part DVD series, *Journey of the Universe: Conversations*, directed by Patsy Northcutt and Adam Loften (2011; New York: Shelter Island, 2013).

that arise from multiple tellings make a huge difference ethically when we are talking about cosmology.

The insistence on multiple possibilities is not just important for academic ease or for some politically correct need to be inclusive. Rather, I would argue, along with many others, that recognizing and cultivating multiple possibilities and the unknowing that is entailed in this are both central to understanding issues of planetary justice. In light of "wicked" problems, such as global climate change and globalization, we need to have a multigenerational, multiperspectival approach toward thinking about any ecosocial problem that we face.[7] In other words, not a single solution but rather every solution that we can think of creates further problems, and it is possible that our solutions do not matter any more than on a local (and precisely not a universal) scale. These located, contextual stories, then, together provide a richer account of the journeys, but can never exhaustively explain or describe the whole journey. Finally, our solutions if they are to matter must pay attention to their multigenerational implications for the future. This type of long history and long-term view is precisely what *Journey* (and many of the world's religious traditions) aim to cultivate.

Having said all of this, there is at least one sense in which attaching our actions to a universal journey—which at least metaphorically is connected to Joseph Campbell's idea of the individual hero's journey—might allow us to go "away" from where we are rather than to stay focused on the specifics of how each and every action reverberates through the multiple lives we influence in the here and now.[8] Though entirely sympathetic to Joseph Campbell's "hero's journey," couching journey as merely an existential condition tends to create a situation in which the entire universe becomes a stage for individual self-actualization. *Journey of the Universe*, in many ways, aims to keep this existential tendency in check: for it is about the journey of the universe and all the journeys therein, rather than the individual's universal journey.[9] Still, there may be an anthropocentric connotation with "journey," and perhaps the "palaver" of our universe might be a better way of naming the journey of the universe in which we find ourselves. Palavers, like conversations, can meander, shift, abruptly end, and are precisely not attached to any one end

---

7. See, e.g., Willis Jenkins, *The Future of Ethics* (Washington, DC: Georgetown University Press, 2013), 111–89.

8. Bruno Latour, "Thou Shalt Not Freeze-Frame," in James Proctor, ed. *Science, Religion, and the Human Experience* (Oxford: Oxford University Press, 2005), 36.

9. Joseph Campbell, *The Hero with a Thousand Faces* (Princeton, NJ: Princeton University Press, 1972).

or outcome because the outcome is profoundly unknown.[10] In other words, the palaver focuses us on the current context along with its unknowns and multiple possible outcomes, rather than allowing us to be swept away by the drama of single-trajectory solutions or the certainty of a destination.

What might be more important than defining a "common universe" in which we live is asking how our multiple journeys affect one another here and now, and for generations to come. This is at heart what I think *Journey of the Universe* and its conversations aim toward. There are multiple conclusions about what the sciences are telling us about the universe, after all. If we agree with the majority of current scientists that the sun will eventually burn our planet into a cinder and the universe will one day fizzle out, isn't it just as easy to read chaos, destruction, and a "live as if there is no tomorrow" attitude into the "journey" of the universe as it is any sort of environmental ethic? Put another way, there is no foundation in reality for ethics; here I agree with problems that fall under the rubric of the naturalistic fallacy, namely, that one can't derive an is from an ought. However, its opposite, the supernaturalistic fallacy, in which we think we (humans/individuals) can impose order on the world willy-nilly, is equally problematic. Rather than siding with either of these extreme positions regarding the relationship between ethics and value to reality, I argue that there is only coconstruction of meaning and that there is a proliferation of possible paths down which we might walk. These paths or "lines of flight" lead to unknown territories, or at least no place in particular.[11] Thus, what becomes important is how we act/react and cocreate these multiple paths along the way to the not yet known, and how these actions, reactions, and cocreations affect earth bodies. The reason for siding with this open-ended, meandering approach to reality has to do with the relationship between certainty, uncertainty, and violence.

The Openness of Uncertainty vs. the Violence of Certainty

Michael Warner suggests that queer theory, among other things, needs to pay attention to and challenge "the hierarchies of respectability that saturate the world."[12] Though I won't dive into queer theory here, what I do want to

---

10. Isabelle Stengers, "The Cosmopolitical Proposal," in *Making Things Public*, ed. Bruno Latour and Peter Weibel (Cambridge, MA: MIT Press, 2005), 1003.

11. Gilles Deleuze and Felix Guattari, *A Thousand Plateaus: Capitalism and Schizophrenia* (Minneapolis: University of Minnesota Press, 1987).

12. Michael Warner, "Queer and Then," *Chronicle of Higher Education* (January 1, 2012), http://chronicle.com/article/QueerThen-/130161/.

focus on are these "hierarchies of respectability" and their tendencies to create norms, categories of good and bad, and labels of natural and unnatural, and to structure ideas of progress and development. In other words, these projected certainties (where there are none) are what I am claiming often result in the violence. We are all familiar with these types of hierarchies of respectability: two common examples are patriarchy and the great chain of being. They have sources in Platonic reason; in the *imago Dei* tradition; in release from *samsara*, which was historically largely thought to be possible only from the incarnated space of a male human; and even in models of modern science that place the entire world as "standing reserve" toward human ends.[13] Of course, each of these founding myths come from multiple contestations, as so many are writing about today.[14] Nonetheless, these hierarchies of respectability tend to function in such a way that by virtue of one's embodiment (race, gender, sex, sexuality, class, religion), one is more or less privileged in the world according to them. The more privileged one is, the more he or she is able to "background" one's multiplicity and dependency on multiple human/earth others.[15] Such backgrounding is also bolstered by the projected certainty of the privileged center of a given ordering of the world. This certainty no doubt does not like the messiness of hybridity, of multiple possibilities, nor the pace of palaver and journey; rather, certainty prefers linear progress and singularity to such ambiguities. And it is this projection of progress and certainty that creates violence for living, evolving bodies. One only needs to recall the violence toward earth bodies committed in the name of development, progress, civilizing, and enlightenment.

The journey of the universe thwarts our attempts at certainty in many ways and forces us to live with unknowing at the edges of known reality: namely, the mystery of what is beyond the universe. Is there a multiverse? Does the universe rest on an infinite stack of turtles? Though science continues to describe the mysteries at both the quantum, physical level and the cosmological level, these new descriptions lead to more and more mysteries. This has, in fact, been the trajectory of the history of science: a theory is posited based upon how it can order and deal with factual information and it holds until the anomalies that it cannot explain build up around its edges and eventually burst the theory open toward a new way of thinking. Copernicus,

---

13. Martin Heidegger, *The Question Concerning Technologies and Other Essays* (New York: Harper and Row, 1977), 3–35.

14. See, for example, Rubenstein, *Worlds Without End*; and Catherine Keller and Laurel Schneider, *Polydoxy: Theology of Multiplicity and Relation* (New York: Routledge, 2010).

15. On "backgrounding," see Val Plumwood, *Environmental Culture: The Ecological Crisis of Reason* (New York: Routledge, 2002).

Darwin, and Einstein all had problems with the implications of their own findings, in part because science (like religion) also settles on certain ways of making meaning out of the world. Einstein even threw out mathematic equations that suggested the universe was expanding at an accelerated pace because he couldn't deal with the implications of that finding. Only when Hubble detected the expansion did Einstein go back to correct his faulty math, calling this mistake his greatest blunder.[16] The point is that the lure of certainty is strong, but the costs of certainty cannot be ignored.

The ultimate nomadic nature of the journey of the universe is such that certain mappings will continuously be revised and even thrown out. This forces at least some amount of uncertainty into our understandings of the cosmos, of the future of life, and of one another. Religion and science both have uncertainty built in, and the journey of the universe helps us to embody this uncertainty as we draw from religious and scientific sources to make meaning out of our lives. Though we may have common contexts as humans, earthlings, mammals, and stardust, such common contexts need not necessarily lead to any singular, certain meaning or future.[17]

Multiple Possibilities for Becoming

Oddly enough, a story about the universe precisely cannot become *the* universal story. One result of the unknowing at the edges of our own understanding of the world is an opening onto the contexts of knowledge. In other words, we become located, planetary creatures (among many other creatures) in an ever-expanding universe. Possibilities for extraterrestrial creatures abound, but possibilities for terrestrial, hybrid becomings and experiences also abound. We become aware of our contextual locatedness on a planet and of the many other "stories" that exist on this single planet.

If, as Karen Barad (among others) has suggested, performative agency is at work "all the way down" to the quantum level, then the universe may indeed be indeterminate, as Niels Bohr claimed.[18] If this is the case, and if we can extrapolate a metaphor for understanding the world based upon indeterminancy, then perhaps truth ought to be reconsidered as regimes or habits of

---

16. Rubenstein, *Worlds without End*, 39.

17. This is at least what I argue in part in *Religion and Ecology: Developing a Planetary Ethic* (New York: Columbia University Press, 2014).

18. Karen Barad, *Meeting the Universe Halfway* (Durham, NC: Duke University Press, 2007), 3–38, 97–131.

truth[19]—that is, as ethically, aesthetically, and politically constructed over time, rather than as having to do with ontology and metaphysics. Such thinking allows us to ask not what is true for all times and all places, but into what truth regimes we ought to live. How do our current truth regimes affect other bodies and the becoming of our planetary future?

I am not suggesting here that we create willy-nilly our own realities, which would require omniscience and omnipotence as if we were acontextual, all-powerful agents creating worlds out of no context at all. Rather, I am suggesting that we are cocreators among a universe of cocreators and that we are also cocreated (by many different entities on a daily basis).[20] It truly does take a village of bacteria, chemicals, atoms, ideas, languages, proteins, genes, and animals—of various combinations of earth, air, fire, and water—to make me "me," from moment to moment. I have some choice in that, but I'm not the arbiter of my own existence, much less any others. We are born into and live as part of various coconstructions, and we have some wiggle room as to how we want to coconstruct the future, but not total control over it.

Rather than mistake my own ideas about what are some good vs. bad ways of becoming with reality and progress toward something that should be evident to all peoples, places, and entities in this giant, journeying universe, I'd suggest a bit of humble pie, and talk rather of political persuasions toward certain ways of becoming. In other words, rather than abstracting my idea of the true and good and mistaking it for what is True and Good (regardless of context), what is true and good becomes much more about orthopraxis than orthodoxy. This shift in focus turns our attention to the geography of our meaning-making practices and how these practices affect various bodies within the planetary community.

The Geography of Knowledge vs. the Evolution of Ideas

In his book *Slow Violence*, Rob Nixon suggests that we begin to look at the geography of daily actions rather than placing our actions into some temporality of progress.[21] In other words, focusing on "progress" at times allows us to get caught up in a narrative where we move from one problem to another,

---

19. Michel Foucault, *Power/Knowledge: Selected Interviews and Other Essays (1972–1977)* (New York: Random House, 1980), 237–58.

20. I adapt this idea from Philip Hefner, *The Human Factor: Evolution, Culture, and Religion* (Minneapolis: Fortress, 1993).

21. Rob Nixon, *Slow Violence and the Environmentalism of the Poor* (Cambridge, MA: Harvard University Press, 2013), 45–46.

addressing specific incidents along the way. It is easier to stand where we are, look back into history from our perspective, and focus on the evolution of an idea that leads to the ground on which we stand rather than explore the geographical distribution of an idea and how that idea takes shape and affects multiple places, bodies, and geographical areas. In other words, a geography of knowledge must look at how various ecosystems, places, and topographies are affected by an idea, whereas the evolution of an idea allows for solipsism in that it confirms one's own idea of progress. Many of the issues that fall under the rubric of environmental justice are issues that are dispersed over vast geographies and generations: hence Nixon's trope, "slow violence." How, for instance, does our daily bread create vast networks of violence across the planet? How does my writing of this chapter create networks of violence due to the construction and waste of the computer on which I type it? What is the ecological footprint of my long-distance romance? More broadly, what would a geopolitical mapping of my own way of living and becoming in the world look like from the perspective of other planetary creatures? Rather than assuming that our knowledge of the world and universe is somehow progressing or evolving—as if we live in the most enlightened time and all past others were in the dark and in need of "education" or "development" and one day we will be proven by future thinkers to have been backward— why not consider that we are living under different truth regimes? The truth regimes of one thousand years ago and one thousand years from now were not and will not be necessarily any more or less "true"; rather they cocreate the world in different ways. As Mary Jane Rubenstein says in her latest book, "The shape, number, and character of the cosmos might well depend on the question we ask it."[22]

This means that "better" and "worse" ideas materialize in the world that create less or more violence toward planetary bodies, respectively. Given various truth regimes—the modern scientific, the capitalist, the environmental, for example—we can begin to judge how meaning-making practices take shape in the world for various planetary bodies. What are the ecological, social, and other ills and benefits, and how are these ills and benefits distributed geographically? Such an analysis forces us to take responsibility for the truths and meaning-making practices that we live by; our knowledge becomes planetary, embodied, ethical, and political.[23]

---

22. Rubenstein, *Worlds without End*, 235.

23. See, e.g., Bruno Latour, *Politics of Nature: How to Bring the Sciences into Democracy* (Cambridge, MA: Harvard University Press, 2004); William Connoly, *A World of Becoming* (Durham, NC: Duke University Press, 2010); and Jane Bennett, *Vibrant Matter: A Political Ecology of Things* (Durham, NC: Duke University Press, 2010).

The Sacred Dimensions of Land, Food, and Water

∾

20

Unless Contemplatives Return to the Land…

Chris Loughlin

Located at the mouth of the Eagle Hill River in Ipswich, Massachusetts, where it opens out onto Plum Island Sound are seventy acres of land permanently protected. The site lies within a wetland complex with a broad sweep of coastal bluffs and islands, open waters, and intertidal areas. The expanse is a nesting area for shorebirds and water fowl, for land birds and raptors. In season this heritage landscape, called the Great Marsh, becomes a resting, feeding, and staging area for birds migrating along the Atlantic flyway.

Far to the south on the shores of Buzzard's Bay, at the west end of the Cape Cod Canal, is an undisturbed parcel of coastline of great ecological importance. A heart-determined decision to preserve 110 acres of privately owned land inspired neighbors to join in the action. Collectively, these lands at the tip of Great Neck protect over three hundred acres of contiguous habitat, forested uplands and freshwater wetlands, coastal beach and shellfish beds.

Far inland, the Petersham Preserve in Massachusetts includes the biologically rich Tom Swamp area that folds in with the mature woodlands of the Harvard Forest and the vital Quabbin Reservoir. Added, now, to this stretch of state- and privately protected land is another two hundred acres, abundant with plant and animal life. This broad expanse has become a critical link for wildlife corridors and human recreation.

Along a stretch of the Connecticut River, a tilled tract of furrowed rows stands out from tangled underbrush and shade trees. Land of Providence is less than a mile from the center of the city of Holyoke in Massachusetts. On this twenty-six-acre reservation, Nuestras Raices rents small incubator plots to local people who want to learn commercial farming. The program helps fledgling farmers develop business plans that will allow them to move out into the community on their own. Land of Providence is a living example of what is becoming common practice—partnerships between religious orders and land trust foundations, and in this case an additional nonprofit organization supporting community farmers. None of these narratives could happen without the gift of the land.

Located on rocky coasts, tucked into wooded hillsides, situated on scenic riverfronts and open farmlands are monasteries, retreat houses, seminaries, schools, and hospitals belonging to religious orders of women and men. We arrived with the immigrant populations from Europe or were founded in this country to serve the needs of the peoples arriving on North American shores. We were gifted with or purchased at minimal cost some of the most diverse and ecologically precious lands across this continent. Today these lands have become an incredible resource to place at the service of other communities, particularly the communities of the most ancient voices, some diminished or gone forever, that long to tell us the story of the journey from whence we have arrived.

Just a mile from the center of Attleboro, Massachusetts—a small manufacturing city—lies the National Shrine of Our Lady of La Salette. Built on the site of a historic spring, the shrine actually covers little of the 120-plus-acre property. The remaining acreage has been untouched for years, allowing it to return to its natural state of woodlands and fields, ponds and streams. Much of the site is designated by the state's Natural Heritage and Endangered Species Program. On the day of celebration to mark the opening of what is now called Attleboro Springs Sanctuary, the mayor spoke of the 117-acre preserve as a "beautiful, pristine area filled with opportunity for nature study, walking, and contemplation." The land specialist from Mass Audubon didn't hesitate to explain that while the mission of the religious order in the past had been reconciliation with God and with neighbor, the program of the new alliance of stakeholders would be reconciliation with land and its bounteous expressions at Attleboro Springs.

Not one of these conservation projects could have happened without a new alchemy, the chemistry creating new alliances between religious orders and public land trusts, and most often a third partner whose participation creates, maintains, and fosters a new relationship with people and land in its multivalent forms. Religious orders cannot "save nature" by our good intentions and our service mode of the past. We have neither the influence to effect societal change nor the power to activate the environmental efforts required. We need to join our effort with those who, long before we recognized the necessity of such diligence, were caring for, conserving, and setting in place the legal process to protect the living systems—the flora and fauna, the flying ones and crawling ones, the budding ones and fruiting ones—within a bioregion. The new alliances build fluid structures where self-nourishing, self-governing, self-fulfilling functions activate spontaneously and often with amazing and unexpected recovery, a recovery that permeates through the whole biological field, even the human spirit!

Just beyond my small office at our Earth literacy center—Crystal Spring in Plainville, Massachusetts—are the offices of Red Tomato, a community-supported agricultural endeavor. In former years, icons of saints hung on the walls, and statues of the blessed stood on pedestals. Now a gallery of New England's farmers pictured in orchards, row crops, and farm stands speak even to the casual observer: *Good people working together to do good things—that's farming at its best!* If you listen closely to conversations or observe the intensity and focus of the Tomato staff, the rooms resound with the vibrations of that earlier era—a novitiate, a place of new learnings and spiritual practices to imbibe the disciplines required for a worthy and needed response to the times.

Vision actualizes now as a new generation shapes the work. The vision to deliver the diverse bounty of our region's farms directly to our region's retailers and institutions multiple times a week all season long with quality-intensive customer support is our local discipline in "our return to the land." The land here, once a gift to the order, has again been gifted. A conservation restriction protects thirty-two acres that in time may simply leave an ecological footprint that a religious order once walked this way.

As we listen to the voices that differ, the voices speaking with great urgency—the wetlands and shorebirds, the soils and maple swamp—we awaken to the great task that lies before us. These are the voices that call us to a new relationship with Earth, and within that relationship we discover the deepest currents of spiritual energy urging us toward a future.

Father Vincent McNabb was an Irish Dominican who lived at the beginning of the twentieth century. He was a strong presence in London as it was embarking on its industrial future. Alarmed at the decline of family farms and appalled at the dehumanizing conditions of the cities, Father McNabb cited that it was impossible for Christian values to permeate the social and economic fabric of life without a rootedness in the Earth and the natural world. He said, "If there is one truth more than another which life and thought have made us admit, against our prejudices, and even against our will, it is that there is little hope of saving civilization or religion except by the return of contemplatives to the land."[1]

For Father McNabb the recovery of the human spirit was commensurate with the functioning of family life folding in with the seasonal cycles and particularly the liturgical celebrations that marked the wondrous rhythms of birth-death-and-rebirth. However, by the end of the twentieth century, a new cosmology—a new worldview—had reshaped our understanding of how things came to be. The cosmos is the primary referent, and it calls us to collaborate with its agency.

We have at last begun to pay attention to other perspectives and acknowledge our too-narrow view of the human story. The language of the anguished cry from the voices of the disappearing penetrates our sensitivities.

To be among the first generations to learn the story of an unfolding universe is sheer gift, but even more an opportunity for great risk. To comprehend the breath of dwelling within a cosmogenesis, we must surrender to the fullness of the experience. Over the past few decades, some have learned to tell the story. With clarity and confidence our words describe the birth of the cosmos, the coming of galaxies, the death of a star, the forming of Earth, the marvel of early life forms. Each transition manifests at great cost, the ultimate sacrifice to succumb to the seedbed of the dream.

What is our way into the journey, then, if not to suffer the internal pressures and the imploding of the familial, if not to endure the loss of the good and the holy and to relinquish what gave meaning, voice, and position within the culture? What is our path to the wild realms of grace if we do not risk the incredible intimacy that opens the "tender membrane" of mind and heart? How else can we bear witness to that ever-creating primitive discernment that life adapts by changing form?

The too-small context that has held our religious story must die to itself and so, too, the orders that identify with too narrow a worldview. In listening to the ancient voices we heard a new language that draws us into new alliances. Risking a fundamental shift in our primary relationships, we no longer see land expressed in the marvel of multivalent forms as our economic security or a onetime capital gain. Conserving our lands can appear a simple, perhaps even an insignificant action, one very small endeavor among the huge aching needs in our world. Yet this radical shift in our spiritual practice—*contemplatives return to the land*—is the "single greatest service that women religious can make to the larger destinies of the human, the Chris-

---

1. Vincent McNabb, *The Church and Land* (Norfolk, VA: IHS Press, 2003), 31.

tian, and the Earth community."[2] We hear a mayor speak of contemplation to the people of an industrial city and a land specialist invite the state representatives and all those gathered to embrace a new manner of reconciliation. New communities, new "ordering" appears on the landscape.

There on the floodplains of the Connecticut River where, over the centuries, rising waters deposited the silty soils that delight a farmer's heart, a generation of new entry-level farmers sow and reap, re-create community and restore eroding landscape. In the widening embrace of shorebirds and maple swamps, fertile soils and expansive flyways, the human heart endures the tensions of a love so great it destroys past loyalties and new alliances appear. Once again, and yet ever new, within the floodplains of the human heart, the grain of wheat must die.

Along rocky coastlines and high on wooded hillsides, along scenic riverfronts and the broad sweep of open farmlands, the ancient religious wisdom of our planet nestled on this continent again takes root.

---

2. Thomas Berry, *The Christian Future and the Fate of Earth*, ed. Mary Evelyn Tucker and John Grim (Maryknoll, NY: Orbis Books, 2009), 80.

21

Everyday Eating in Eucharistic Life

Food, Communion, and Moral Communities
in the Anthropocene

JAMES JENKINS

Five-month-old Annie sleeps on my chest during the Yale Sustainability spring brunch. As we wait to be recognized for Nourish New Haven—the *Journey of the Universe*–inspired local food justice and sustainability conference—Mary Evelyn Tucker, John Grim, Greg Sterling, and I discuss Sunday's *New York Times Magazine* feature on environmental activist and writer Paul Kingsnorth: "It's the End of the World as We Know It . . . and He Feels Fine." Kingsnorth speaks of mourning, grief, and despair in the "age of ecocide"; confesses his "longstanding faith in environmental activism draining away"; thinks to himself, "I can't do this anymore"; and asks, "So what do I do?"[1]

Climate science and ecological reports implicate all of us in the destruction of Earth's ecosystems. How do we cultivate hopeful determination to change the story? How do we encourage adaptation and collaboration, rather than separation and indifference? How do we foster human habits to live harmoniously in creation? What world will my daughter Annie inhabit when she reaches my age?

Indifferent economic and political systems deplete our spirits and destroy our world. In *Confessions of a Recovering Environmentalist*, Kingsnorth frames our condition around food:

> That we are both hollow men and stuffed men, and that we will keep stuffing ourselves until the food runs out, and if outside the dining room door we have made a wasteland and called it necessity, then at

---

1. Daniel Smith, "It's the End of the World as We Know It . . . and He Feels Fine," *New York Times Magazine*, April 17, 2014.

least we will know we were not to blame, because we are never to blame, because we are the humans.[2]

Metaphorically, food illuminates the self-centered consumption, convenience, wastefulness, and addictions of human behavior. Kingsnorth alludes to T. S. Eliot's poem "The Hollow Men"—"this valley of dying stars . . . *This is the way the world ends*"[3]—to indict us.

Agriculture and eating statistics are even worse. The global food system contributes to 30 percent of all greenhouse gas emissions. Industrial agriculture has led to a loss of 75 percent of crop diversity since the early 1900s. Thirty to 40 percent of all food is wasted. Seventy percent of fresh water in the United States goes to agriculture. One billion people go to bed hungry each night, and another 1.5 billion suffer from being overweight to obese.[4]

Journey of the Universe reveals the human disconnect from natural rhythms into industrial patterns: "With billions of humans hooked into this vast machine, material production rises but the cost is self-destructive. In addition to the chronic stress, ill-health, and alienation that humans feel inside the machine, there is the unintended consequence of ruining the foundations of every human economy."[5] This unsustainable reality cannot end the human story. To paraphrase Proverbs, with no vision, the planet perishes.[6]

How might this story change and help us understand what to do? *Journey of the Universe* and the Christian tradition provide complementary insights into the power of food, communion, and moral communities for the future. Drawn from science, religion, and living communities, this vision is shared and invites all humans to come together at the table to reestablish ecological health and integrity.

"What is needed is courage to live in the midst of the ambiguities of this moment without drawing back into fear and a compulsion to control," *Journey of the Universe* summons. "Are there guarantees? No, none. But there are reasons for confidence."[7] *Journey* offers not only evolutionary history of the universe but also creative hope for an emerging "new order of well-being," illuminated by eating together to cultivate codependence and generate

---

2. Paul Kingsnorth, "Confessions of a Recovering Environmentalist," *Orion Magazine*, January–February 2012.

3. T. S. Eliot, "The Hollow Men," in *Poems: 1909–1925* (London: Faber and Faber, 1925).

4. *Food Tank Annual Report 2013–2014*, http://foodtank.com/about.

5. Brian Thomas Swimme and Mary Evelyn Tucker, *Journey of the Universe* (New Haven, CT: Yale University Press, 2011), 108.

6. Proverbs 29:18. All biblical passages are drawn from *The New Oxford Annotated Bible* (*NRSV*), ed. Michael Coogan (New York: Oxford University Press, 2010).

7. Swimme and Tucker, *Journey*, 117.

flourishing across diverse communities.[8] *Journey of the Universe* followers and Christian communities, joining with others in shared practices, may replace destructive systems through resilient ecological communities living in relationship. Biblical narratives contain these alternative, transforming shared meals that culminate in the Last Supper—inspiring the essential Christian activity of communion and Eucharist. In *The Future of Ethics: Sustainability, Social Justice, and Religious Creativity*, Willis Jenkins offers a strategy for hope, not only for Christians, but also to Kingsnorth and others who care about our planetary future:

> Ethics can work from the tactics moral communities devise to sustain the possibility of a moral life in the face of problems that would defeat it.... The challenge to Christian ethics, then, lies in finding or inventing the practices that sustain possibility for living the faith amidst anthropocene powers. Insofar as those practices overcome features of problems that would defeat moral agency they will matter to wider publics looking for analogies of adaption.[9]

Eating in the Christian tradition and throughout evolutionary creation teaches humans the value of shared meals and communion, from the intimacy of family gatherings to the impact of the global food system.

Examining food within the biological evolutionary story reveals the importance of adaptation, collaboration, and learning. The first simple cells began in a darker place: 4 billion years ago, most likely in deep sea vents, under extreme conditions, thermophyllic bacteria used heavy metals as food.[10] Photosynthesis, the process of creating food from the sun, developed over tens of millions of years and signals "one of the deepest tendencies in the universe"—communion.[11] Life adapts to flourish and catalyze new creation.

Basic cells discern dietary choices: "Is this a risk worth taking? Is this food nourishing? Will this increase the chances of remaining alive?"[12] These cells learned survival strategies to coexist through relationship.[13] Rather than "survival of the fittest" thinking that dominates our politics, economics, and individual psyches, *Journey* shows that a cosmogenesis continues to emerge through giving for the whole and reorganizing through relationships. Darwin

---

8. Ibid.

9. Willis Jenkins, *The Future of Ethics: Sustainability, Social Justice, and Religious Creativity* (Washington, DC: Georgetown University Press, 2013), 7.

10. Swimme and Tucker, *Journey*, 48.

11. Ibid., 51.

12. Ibid.

13. Ibid., 54.

and other early evolutionary thinkers such as Alfred Russell Wallace understood the codependence needed for ecological flourishing—a truth often lost in modern individualism but basic in the food we eat[14] (keep in mind the cook, the picker, the farmer, the field). Because of our cellular ancestors' ability to adapt and remember, our food is "transformed into our skin, our muscles, and our organs."[15] Science shows that one individual did not make these discoveries, but rather "life's whole process of adaptation and memory" enabled us to eat.[16] It is life as a whole with increasing complexity that learned to digest its various foods.[17]

As we face the urgent challenge "to construct livable cities and to cultivate healthy food in ways congruent with Earth's patterns," *Journey* highlights the importance of Earth's evolutionary wisdom for discerning how to eat.[18] Humans were not created in perfection but have evolved within changing ecosystems. Our technological power—particularly in food production—currently outpaces our spiritual discipline to live within Earth's natural systems. Our hubris to see ourselves over nature may also be transformed into humility with the Earth. Science leads us deeper into the mystery and wonder of who we are—still creating and changing over billions of years since the Earth's inner warmth and the sun's light brought forth life. Learning to eat in life-enhancing ways biologically changes us while also reconnecting us to the first Earth eaters 4 billion years ago—thermophyllic bacteria.

Ecologically, *Journey* illustrates how interdependent communities arise out of suffering and death. The story encourages us to orient ourselves creatively in the midst of destructive processes. Food is essential in all life processes—from stars to seeds to five-month-old Annie. The story of Jesus and the "fundamental mystery in which the small self of the individual dies into and nourishes the whole community," told in *Journey*, unites the wisdom of ecology and Christianity. Furthermore, parenting affirms the natural

---

14. Paul White, "Darwin's Church," in *God's Bounty? The Churches and the Natural World*, ed. Peter Clarke and Tony Claydon (Rochester, NY: Boydell, 2010), 351. White's descriptions of nineteenth-century Anglican priests and thinkers, including Darwin, resonate with *Journey*'s emphasis on wonder and humility: "While Darwin's undirected and seemingly purposeless natural selection was radically different from Paley and Wilberforce's designed universe, all three men viewed the natural world as 'wonderful.' . . . Unlike some of his contemporaries and acolytes, Darwin did not intend his theories to exalt the position of humans in the universe. In fact, in a style Paley and Wilberforce would have praised, Darwin suggested that the laws of nature should make humans more humble."

15. Swimme and Tucker, *Journey*, 58.

16. Ibid., 57.

17. Ibid., 60.

18. Ibid., 116.

self-giving desire of human beings who recognize their calling. How might focusing on the wider community's nourishment transition humanity away from self-destruction? How might Jesus's message of eternal punishment and eternal life apply literally and ecologically? "Come, you that are blessed by my Father, inherit the kingdom prepared for you from the foundation of the world; for I was hungry and you gave me food. . . . I was a stranger and you welcomed me. . . . Truly, I tell you, just as you did it to one of the least of these who are members of my family, you did it to me."[19] How might we see the flesh of God throughout creation? How might theology and ecology enhance our notions of family responsibility and sacrificial love? It is not as hard as one would imagine. We feed one another to survive.

Biblical food accounts reinforce radical sharing and communal sacrifice culminating in Jesus's Last Supper, shared with followers, family, and friends gathered along the way. Exodus features the Passover meal and manna in the wilderness—food stories contrasting with the social oppression in Egypt and creating the Israelite community along the journey from slavery and scarcity to the promised land. Levitical laws outline customs for leaving a portion of the harvest for the poor[20] and providing for the redemption of the land.[21] God's commandment is not too hard or far away but very near "in your mouth and in your heart."[22] Prophets challenge empty rituals of fasts and feasts. Isaiah envisions "spending yourselves in behalf of the hungry and satis-fying the needs of the oppressed" to become like a "well-watered garden, like a spring whose waters never fail" and a restorer of dwellings.[23] In the Gospel according to John, after the feeding of the five thousand, Jesus tells the dis-ciples, "Gather up the fragments left over, so that nothing may be lost"[24] and speaks in depth about the bread of life: "For the bread of God is that which comes down from heaven and gives life to the world."[25] Food builds com-munity along the journey. Others, including the land, are considered within the social law. Prophetic voices remind communities that what and how one eats do not always demonstrate moral integrity. Not only does Jesus share the Last Supper meal, but he returns from the dead to eat with his disciples

---

19. Matthew 25:34–46.
20. Leviticus 23:22.
21. Leviticus 25:24.
22. Deuteronomy 30:11–14.
23. Isaiah 58:9–12.
24. John 6:12.
25. John 6:32.

and restore their vision after facing uncertainty.[26] Scripture presents food as a source for moral, communal action within a challenged world.

Arising from biblical tradition, sacramental practice and theology magnify eating for becoming a new creation. Andrew McGowan gives the historical account of the evolution of Eucharistic meals in early Christian communities:

> The Eucharist is a field of Christian practice characterized (like early Christian doctrine) by diversity and not just a single idea represented in bread and wine. It presents rich and varied themes of memory, presence, celebration, and sacrifice—and there is no stronger theme than thanksgiving, *eucharistia* itself.... Its messages and its purpose are in fact those of communal Christian existence: of incorporation, challenge, transformation, and hope, centered on the message and meaning of Jesus.[27]

As with early Christian practice, today's Eucharistic meals are communal and contextual within the larger issues of the world. J. G. Davies's *Worship and Mission* explains that the bread and wine shared in worship "represents the true relationship of the whole to God" as a cosmic sacrament to enable humans "to understand that all matter can be a sacrament of communion with God."[28] Eucharistic action leads to a "community of reconciliation, not of alienation," Davies wrote in the 1960s amid the civil rights movement. The same principle applies to ecological reconciliation now. "The Eucharist itself, as worship, bore witness to the impossibility of understanding worship from any other standpoint than the response of the believer in the totality of his life in the world."[29] Sacrament, from the beginning, is not an object but a living action of communal relationship spreading beyond a single community. In *Being Christian*, Rowan Williams emphasizes this totality with global consequences for seeing "what a Christian attitude to the environment might be" by experiencing "humanity and the whole material world in a fresh way, seeing things sacramentally, seeing the depth within them, where the giving of God is always at work."[30]

Sacramental activity, with symbolic consciousness, leads to the recognition of a sacred universe working collaboratively in continual processes of

---

26. Luke 24:31; Mark 16:14; John 21:13.

27. Andrew B. McGowan, *Ancient Christian Worship: Early Church Practices in Social, Historical, and Theological Perspective* (Grand Rapids: Baker Academic, 2014), 62–63.

28. J. G. Davies, *Worship and Mission* (London: SCM, 1966), 96.

29. Ibid., 101, 102.

30. Rowan Williams, *Being Christian: Baptism, Bible, Eucharist, Prayer* (Grand Rapids: Eerdmans, 2014), 50, 55.

creation. Experiencing sacramental ecology abundantly, humans will discover through science and *Journey* that the Earth's processes can teach us how to evolve as a planetary species. Our humility within creative dynamics becomes our motivating hope. As the *Journey* film emphasizes, "Not biology but symbolic consciousness is the determining factor" to learn new ways of sustaining life through biological evolution. Spiritual hunger may become nourished through ecological creativity and Eucharistic care. *Journey* asks, "Could it be that our deeper destiny is to bring forth new coherence within the planet as a whole, as the human community learns to align itself with the underlying dynamics of Earth's life?"[31] Sacramental life throughout the diversity of Christian traditions helps imagine the religious realignment that is possible with the world. E. O. Wilson hypothesizes that only the power of a sacrificial, *religious* motivation can sufficiently redirect human will to undertake the changes we need to save us from ecological disaster.[32]

The incarnation of Christ is the ultimate sacrament in which Jesus shares God's gift for future generations—individual death transitioning to resurrection for the world. Modeled through meals with Jesus, self-giving hospitality shows how humans might eat with one another and throughout the planet. In "Sacraments of the New Society," Williams states,

> The objects of the world, seen in the perspective of the Eucharist, cannot be proper defense of one ego or group-ego against another, cannot properly be tools of power, because they are signs of a creativity working by the renunciation of control, and signs of the possibility of communion, covenanted trust and the recognition of shared need and shared hope.[33]

The sacramental activity modeled by the Eucharist offers the clearest case for God's giving and hospitality—signs of kingship that would work well to reform our biblical notions of dominion, not to mention our modern use and abuse of creation, including other people. Ecology reinforces this theological vision. The Anglican mystic Evelyn Underhill also defines incarnation as the "supreme sacrament" that "discloses in visible and temporal terms the nature of the Eternal God; and thus declares the true significance of the Universe, as a means whereby the Transcendent Spirit is self-revealed to those spirits

---

31. Swimme and Tucker, *Journey*, 66.

32. Summarized by Sarah Coakley, *Sacrifice Regained: Reconsidering the Rationality of Religious Belief* (Cambridge: Cambridge University Press, 2012), 27.

33. Rowan Williams, *On Christian Theology* (Oxford: Blackwell, 2000), 218.

whom He has made for Himself."[34] While humans may be the only creatures able to comprehend sacramental presence, all creation takes part. Williams emphasizes the Eucharistic method for changing our lives: "We go from the table to the work of transfiguring the world in God's power: to seeing the world in a new light, to seeing human beings with new eyes, and to working as best we can to bring God's purpose nearer to fruition in the world. God the Giver is an invitation for our free participation with God and one another."[35] Jesus's sacramental activity creates community by not only giving but also receiving hospitality from others. Williams explains, "Every act must speak of God, but not in such a way as to suggest a satisfying of divine demands, an *adequacy* of response to God's creative act. What we do is now to be a sign, above all, of a gift given for the deepening of solidarity."[36]

Sacramental life is for activity in this world and originates in the symbolic power of food. Williams writes, "We do not encounter God in the displacement of the world we live in, the suspension of our bodily and historical nature. . . . God acts in emptiness by bringing resurrection and transforming union, not by lifting us to 'another world.'"[37] That our bodies physically transform what and how we eat reveals the biological truth of this theology. Christians need not strive for other places or other times, for here now is the moment when God is ready to enter or reveal that divine nature given to us for the whole world.

John Polkinghorne and Sarah Coakley argue that "relation" and "causation" provide the philosophical categories for the "holistic connectivity" in the universe.[38] Polkinghorne explains how the language of theology and the language of science relate through "consonance" and "conceptual congruity" that mutually illuminate "science's picture of the relational nature of the physical world and theological belief in the Trinitarian nature of God."[39] The great work of science to understand the physical dynamics of an unfolding, evolutionary universe allows Christians to recognize how community religious practice illuminates the cosmic process of divine co-mingling. This divine awareness may also be found in everyday eating.

34. Evelyn Underhill, *The Evelyn Underhill Reader*, ed. Thomas Kepler (New York: Abingdon, 1962), 177.

35. Williams, *Being Christian*, 57.

36. Williams, *On Christian Theology*, 204.

37. Ibid., 207.

38. Sarah Coakley, "Afterward: 'Relational Ontology,' Trinity, and Science," in *The Trinity and an Entangled World: Rationality in Physical Science and Theology*, ed. John Polkinghorne (Grand Rapids: Eerdmans, 2010), 195.

39. John Polkinghorne, "The Demise of Democritus," in Polkinghorne, *Trinity and the Entangled World*, 12.

Wendell Berry and Michael Pollan remind us that eating is an agricultural act.[40] "The impulse to share food is basic and ancient," Sara Miles writes in *Take This Bread*. "Food is what people have in common, and it is, precisely, common."[41] From food pantries to potlucks, fasts to feasts, coffee hour to the communion table, Eucharist to everyday eating, Christian communities share food. Eating is valued across world religions and cultures. Twenty-first-century obsessions with food relate to human hunger for true nourishment and take on religious characteristics in some contexts. When making our food choices and asking what to eat, we would be wise to follow Jesus's principles for the sacramental whole of creation: "Strive first for the kingdom of God, and his righteousness, and all these things will be given to you as well."[42]

Journey of the Universe teaches us more precisely how basic and ancient eating has been developed through billions of years of evolutionary history. These food stories connect humanity to the ecological patterns that create, sustain, and nourish life. With Christianity, then, eating might be restored as a Eucharistic act of supreme thanksgiving and symbolic power. The sacredness of eating can nourish us communally into new life together and can unite humanity within Earth's ecosystems now and beyond. We might take the central faith of Christianity—that Jesus gave his life as food for the entire world—to orient ourselves to care for one another, including the whole creation. Even hollow and stuffed men can wonder about food. The church's and the Earth community's shared future depends first not upon action but rather a new way of being conscious of relationship, symbolized in everyday eating in Eucharistic life. *Journey of the Universe* and Christianity arrive together at the right question: "So what do *we* do?" Moral communities spiritually alive and ecologically integrated will devote themselves to living into answers. May Paul Kingsnorth, Annie, and all of us eat among them, confident to continue the great work.

---

40. Michael Pollan, *The Omnivore's Dilemma: A Natural History of Four Meals* (New York: Penguin, 2006), 9, 10. See also Wendell Berry, "The Pleasures of Eating," in *The Art of the Commonplace: The Agrarian Essays of Wendell Berry*, ed. Norman Wirzba (Emeryville, CA: Shoemaker and Hoard, 2002), 321. Pollan explains the connection between food and agriculture succinctly: "At either end of any food chain you find a biological system—a patch of soil, a human body—and the health of one is connected—literally—to the health of the other. Many of the problems of health and nutrition we face today trace back to things that happen on the farm, and beyond those things stand specific government policies few of us know anything about.... The way we eat represents our most profound engagement with the natural world. Daily, our eating turns nature into culture, transforming the body of the world into our bodies and minds.... Our eating also constitutes a relationship with dozens of other species—plants, animals, and fungi—with which we have coevolved to the point where our fates are deeply entwined."

41. Sara Miles, *Take This Bread: A Radical Conversion* (New York: Ballantine, 2007), 49.

42. Matthew 6:33.

22

Living Water

NANCY G. WRIGHT

And in that place I saw the fountain of righteousness, which was inexhaustible: And around it were many fountains of wisdom: And all the thirsty drank of them. And were filled with wisdom, and their dwellings were with the righteous and holy and elect.[1]

In June 2012 I stayed at the Hotel Argylle, on the island of Iona, Scotland.[2] On this island, the famous gray-stoned Christian monastery, founded by Irish priest St. Columba during the fourth century, hovers over the sea, always drawing pilgrims from far and wide. A sign in the picturesque hotel's dining room reads, "To reach the heart of Iona is to find something eternal—fresh visions and new courage for every place where love or duty or pain may call us." *This is exactly where I need to be,* I thought. On this sabbatical, in order to be refreshed and renewed, I need to be in a natural area that is valued and protected, and I need courage to be where love and pain call me.

The historical beginnings of the monastery are shrouded in mist, literally and figuratively. It is said that Columba and his followers landed on what is now Columba's Bay in 563. He was escaping Ireland for a reason not entirely known—perhaps penance for involvement in clan battles or for making an illicit copy of the psalms. He sailed in a leather-bound boat known as a curragh, letting the sea take him where it would, as our life journeys do. In the centuries following, untold numbers of pilgrims walked a pilgrimage route on the island, to inwardly focus on their life's path and even their soul's calling. *Could I find the same in my own sabbatical journey?* I wondered and hoped. At Columba's Bay, we walkers on the weekly pilgrimage route threw a stone

---

1. Enoch 48:1.

2. I deeply thank Mary Evelyn Tucker, Rebecca Gould, Stephanie Kaza, and Mary Coelho for much help in writing this paper. I also give my thanks to the Lilly Endowment sabbatical grant and to my church, which made the sabbatical possible.

212

into the water to symbolize what to leave behind and picked up another to symbolize a new life commitment.[3]

The sabbatical pilgrimage gave three overall blessings. The first blessing inspired a cosmic awareness of water as the matrix of life, the second granted a deeper Christian reflection on living water, and the third strengthened a conviction of the need in myself and our culture for a deeper union (reunion) with water.

Creation: The First Blessing

The Universe Story

The living universe—how privileged we are to live in a century when the story of the 13.8-billion-year-old living universe is being told in new, unprecedented ways, including through *Journey of the Universe*. Discoveries in 1924 by Edwin Hubble of the expanding universe filled with billions of galaxies, Darwin's staggering explanations about the evolution of life, cell biologists' revelations about DNA, and Alfred Wegener's presentation in 1920 of the details of plate tectonics—we are heirs to great and *new* news. An immense new awareness fills many with wonder and humility at our privilege of living at this time and also fear and trembling at our unprecedented impact in the age of the Anthropocene on the gorgeous and finite Earth.

The Blessing of Water

Reading the *Journey of the Universe* book[4] and watching the film and *Conversations* series undergirded and expanded the sabbatical. For the first time I began to inquire about the cosmic origin of water. With a new sense of wonder, I realized that the Big Bang created hydrogen and stellar evolution reformulated this element into oxygen, the two combining into the H_2O molecule. In interstellar space, water and other substances over eons condensed and froze, coating planetesimals with ice. In our solar system, water is or was present on each of the sun's planets, with different fates.

---

3. This has resonance with the Yom Kippur practice of throwing bread or stones into a river, to carry the community's and individual's sins away.

4. Brian Thomas Swimme and Mary Evelyn Tucker, *Journey of the Universe* (New Haven, CT: Yale University Press. 2011).

Earth cooked and cooled over millions of years, with water vapor and dust spewed up by volcanoes: then steam and rain fell. As Earth stabilized, oceans wrapped the Earth, held by a thin layer of atmosphere, and cellular life developed, perhaps in deep vents in the ocean floor. The ancient poets who created the Genesis narrative knew something of this grand history: "In the beginning when God created the heavens and the Earth, the Earth was a formless void and darkness covered the face of the deep, while a wind from God swept over the face of the waters."[5]

That early wisdom bespeaks wonder and wisdom about Earth, summoning an attitude of gratitude to the Creator for the gifts of the heavens, Earth, and water, the matrix of all life. A storied relationship with Earth and water that acknowledges a Creator of All-That-Is, such as the biblical and other cosmic narratives, evokes wonder toward the gift of all life because creation is seen as fostered by a divine, caring presence. Thus, human behavior toward creation, the Bible suggests, should model that of God's intimate knowledge and care. As God says to Job, "Do you know when the mountain goats give birth? . . . Can you number the months that they fulfill . . . ?"[6]

The Second Blessing: Living Water and Baptism

Two years ago, at the sabbatical start, I needed bathing in the sacrament of water. As a pastor sometimes preaching on the environment, I knew the rigors of trying to convince some people about climate change and other "environmental issues." I was depleted.

I was worried and anxious, both about my parish and about Earth, especially about water and water justice issues. I, like many, knew too well the facts about water: that 71 percent of Earth's surface is water but only 1 percent is fresh water in rivers and lakes; that all organisms, including humans, consist chiefly of water, the matrix of all cellular activity; that edible ocean fish have declined by 90 percent in the past fifty years. Human/social concerns proliferate around water. Wars have been and will increasingly be fought over water (in the Bible, an ongoing theme is conflict related to wells running dry). Further, unbelievably, most water now is polluted, including sacred rivers such as the Ganges and Jordan, and Lake Champlain. All this lies heavy on our hearts.

---

5. Genesis 1:1.
6. Job 39:1–2.

The Original Blessing of Water as Living Water

As I left for sabbatical, I felt myself intrigued by the story of living water told in the Gospel of John, chapter 4. The story takes place on a hot day at Jacob's well. A Samaritan woman encounters Jesus, a Jew (as the story tells us, at that time, Jews did not deal with Samaritans, nor did men talk publicly to women). Jesus asks her for water, and she, perplexed and astonished, asks, "How is it that you, a Jew, ask a drink of me, a woman of Samaria?" Jesus answers, "If you knew who was asking you for water, you would ask him for living water." She inquires how he can draw the living water, because he has no bucket and the well is deep. Then a theological conversation ensues between Jesus and the woman about living water. I needed living water, I knew. But what is this living water? How could I find it?

My sabbatical took me from Iona to the Holy Land, where Jesus sat in boats on the Sea of Galilee, and where he was baptized in the Jordan by John the Baptist, receiving his sense of call as the Beloved Son of God.[7] As I immersed myself in the Jordan, I felt ambiguous and wary: how contested and small this river is now in comparison to Jesus's time. How difficult it is with current tensions and injustice over land and water to feel inspired by a message of spiritual transformation. Still, I felt renewed by bathing in the Jordan and was surprised by that. Is water living, spiritual, despite being diminished and degraded?

I learned that for the first several centuries, Christians said no blessing over the water prior to baptism, because it did not need a blessing. In their understanding, the condition of the water as pure and flowing (living) was itself the blessing.[8] Further, baptism in the early centuries of the church was believed to restore the baptized to awareness that this Earth is still paradise. Fourth-century hymn writer and theologian Ephraim of Syria wrote, "In

---

7. Earlier meanings of ritual immersion in the Hebrew scripture referred to cleansing, rites of passage, and purification; water immersion marked liminal times (life/death experiences, reentering the community after war, postmenstruation). John's emphasis on conversion and repentance forecast a coming Day of Judgment. The "coming one" was to "baptize with the Holy Spirit and fire" (Matthew 3:11–12 = Luke 3:16–17; see also the judgment imagery in Daniel 7 and Revelation 20:11–15). Through time, the meanings given to water change and are elaborated, which is one of the key arguments in this paper. The Christian meaning of baptism is a reentry into the life/death/resurrection of Jesus Christ and receiving of his Spirit. In contemporary times, living water imagery includes an awareness that water itself may need to be purified (cleansed from pollution in order to carry its true meaning as living water).

8. Foreword by David N. Power in Linda Gibler, *From the Beginning to Baptism: Scientific and Sacred Stories of Water, Oil, and Fire* (Collegeville, MN: Liturgical Press, 2010), ix.

baptism did Adam find that glory which had been his among the trees of paradise."[9]

The early Christians saw and marveled at water, seeing how it adapts itself and ministers to living beings. Of this "ministry" of water, the fourth-century Christian Cyril of Jerusalem wrote (referring to Jesus's statement in John 7:37–39[10]):

> And why has He [Christ] called the grace of the Spirit by the name of water? Because by water all things subsist. . . . For one fountain watered the whole of the Garden, and one and the same rain comes down upon all the world, yet it becomes white in the lily, and red in the rose, and purple in the violets and pansies, and different and varied in each several kind; so it is one in the palm tree, and another in the vine, and all in all things . . . adapting itself to the nature of each thing which receives it, it becomes to each what is suitable.[11]

The *Didache* is a mid-first-century pastoral program for training converts to prepare them for baptism. A passage famously reads, "Concerning baptism, baptize thus: Having said all these things [the Beatitudes, the Golden Rule, Ten Commandments—in sum the 'ways of life and death'] beforehand, immerse in the name of the Father and of the Son and of the holy Spirit in flowing water—if, on the other hand, you should not have flowing water, immerse in other water [that is available]; and if you are not able in cold, [immerse] in warm [water]; and if you should not have either, pour out water onto the head three times."[12] Thus, the phrase "flowing water," living water, is used very early for Christian baptism as the preferred baptismal medium, and no blessing is needed over the water, only over the person being baptized.

Knowing that living water naturally ministers, it follows that those who receive its blessing with a sincere and grateful heart then, remarkably, have the power to give it to others. So, the woman said to Jesus, "Give me this water, so I never have to draw again." (She takes her place among the millions of women and girls through the ages who have drawn water, needed for their

---

9. Ephraim of Syria, *Epiphany* 12.1, quoted in Rebecca Ann Parker and Rita Nakashima Brock, *Saving Paradise: How Christianity Traded Love of This World for Crucifixion and Empire* (Boston: Beacon, 2008).

10. "While Jesus was standing there, he cried out, 'Let anyone who is thirsty come to me, and let the one who believes in me drink. As the scripture has said, "Out of the believer's heart shall flow rivers of living water."' Now he said this about the Spirit, which believers in him were to receive."

11. Cyril of Jerusalem, *Catechetical Lectures*, quoted in Parker and Brock, *Saving Paradise*, 138.

12. Quoted in Parker and Brock, *Saving Paradise*, 19.

families, the great and terrible weight of buckets filled with water capable of disfiguring their bone structures.) And she drops her bucket and becomes a minister (my words), proclaiming to her townspeople, "Come and see a man who told me everything I have ever done! He cannot be the Messiah, can he?"[13]

The effect of a relationship to Jesus, like wonder elicited by the cosmic history of water, can open one's eyes and heart to the marvelous nature of water and to a new, creative sense of identity. Receiving living water in either of these ways might then strengthen our urgency to alleviate women's burden of fetching and physically bearing water for their families and empower their needed life-giving leadership in communities around the world.

Loss of Original Blessing

When water is not seen as ministering to us humans and all life, when it is not seen as our sister (St. Francis), or as representative of Spirit or of God, or of cosmic Presence, it dies. We lose our identities as water-beings and creatures of spirit.

Such loss began as Roman Christianity became established and deepened up to the present day. First, as Christians were no longer baptized in streams or rivers, baptismal fonts grew smaller, and the said blessings grew longer, with God's original blessing of water replaced by a priest's blessing: Augustine of Hippo wrote in his *Treatise on the Gospel of John*, "Take away the word and what is water but water?"[14] Second, strict scientific interpretations of water, and use of water solely as a commodity and waste sink, catastrophically simplified and disfigured the actual principles and movements of water. We became "de-ranged" (separated from our true home and identity).

The scientific story tells us of the essential role of water in the evolving universe. It can give the fascinating and scientific understanding of H_2O. But the tendency toward reductive science suggests no metaphor or resonant quality for water. Such scientific scrutiny can be pitiless—the arrogant gaze, all mind and no heart—especially when combined with unfettered capitalism, which treats water only for its practical usefulness. We need a retelling of the scientific story, told to evoke meaning, beauty, awe, as well as gratitude (the purpose of *Journey of the Universe*), to see water clearly and find our part in the blessing. To see water in its spiritual, living essence is to discover spiritual

---

13. John 4:28–29.
14. Gibler, *From the Beginning to Baptism*, 25.

radiance! Sages, artists, and now many scientists, as reflected in *Journey of the Universe*, have always tried to convey this truth.

Union with Water: The Third Blessing

Our conscious awareness of being water beings and spiritual beings means a deeper union, a marriage, with water. Jesus and the Samaritan woman at the well mirror the marriage between Rebekah and Isaac, who met at a well:[15] two couples at the well, then, signifying and prefiguring our marriage to water, or a marriage between the beautiful scientific story of water and the spiritual story.

We have a choice about our relationship with water. Do we see ourselves as people with little freedom and responsibility, and little connection to the natural world, and to water, or do we allow ourselves to feel full of potential and wonder, singing the glories of Earth? Do we strive to maintain the status quo, or do we create social and political situations in which all who seek a life of awe, wonder, and fulfillment of potential have the opportunity to do so?

The Lutheran word for sin is *curvitas* (curved inward; narcissism). Baptism turns natural self-interest outward toward abundant life and grace, through an inner marriage of body, mind, and soul. Narcissus looked into the pool and only saw himself (*curvitas*); Helen Keller recognized water and her whole perception of life changed (conversion). Helen Keller writes in *The Story of My Life*,

> We walked down the path to the well-house, attracted by the fragrance of the honeysuckle with which it was covered. Someone was drawing water and my teacher placed my hand under the spout. As the cool stream gushed over one hand she spelled into the other the word water, first slowly, then rapidly. *I stood still, my whole attention fixed upon the motions of her fingers. Suddenly I felt a misty consciousness as of something forgotten—a thrill of returning thought; and somehow the mystery of language was revealed to me. I knew then that "w-a-t-e-r" meant the wonderful cool something that was flowing over my hand. That living word awakened my soul, gave it light, hope, joy, set it free! There were barriers still, it is true, but barriers that could in time be swept away.* I left the well-house eager to learn. Everything had a name, and each name gave birth to a new thought. As we returned to the house every object which

---

15. Genesis 24.

I touched seemed to quiver with life. That was because I saw everything with the strange, new sight that had come to me.[16]

How are we all Helen Kellers? How are we blind about water and need to see? Is water the clue to a new understanding of ourselves and the universe? I believe so.

I finished my sabbatical with a contemplative kayaking retreat with the Reverend Kurt Hoelting (a previously ordained Christian minister and now a practicing Buddhist), who directs a program called Inside Passages. Kurt encouraged us to simply notice, hour after hour, watching for anything in any moment that filled our attention. (Ripples from raindrops radiated out in infinity of concentric circles, all connecting, overlapping, and mesmerizing me in the spell of water.)

Conclusion

My congregation also participated in the sabbatical. They cleaned up Bartlett Brook behind our church. To foster water justice, they and I supported Lutheran World Relief in water projects around the world—for example, to create simple check dams that capture rainwater and restore arid landscapes capable of three crop rotations, with egrets flying over the watered fields.

My sabbatical, focused on living water, took me from a bitterness and fatigue to a shimmering awareness of water and renewed responsibility. Such a gaze is made possible by an interior freedom. The world seems full of radiance, of the love of Christ. Through this incarnated mystery, brought to life through living water, we can celebrate the universe and give thanks for the chance to contemplate who we are as beings, water beings, who evangelize—tell the great news of—the miracle of water, life's matrix.

---

16. Helen Keller, *The Story of My Life* (1903; New York: Bantam Classics–Random House, 1990), 15–16 (emphasis added).

Earth Jurisprudence for the Earth Community

∽

23

Foundations for an Earth Jurisprudence

Law's Revolution from Order to Justice

BRIAN EDWARD BROWN

On this centennial morning of the birth of Thomas Berry and in the spirit of *Journey of the Universe*, it is appropriate to consider an Earth jurisprudence that emerges from within the corpus of his Great Work.

For orientation, one might turn to Harold J. Berman and his study *Law and Revolution: The Formation of the Western Legal Tradition*. In it he identifies six great revolutions beginning with the Papal Revolution of 1075 and including the German Reformation of 1517, the English Revolution of 1640, the American Revolution of 1776, the French Revolution of 1789, and the Russian Revolution of 1917. Though distinct, Berman notes a commonality among these six historic moments:

> Each has marked a fundamental change . . . a lasting change in the social system as a whole. Each has sought legitimacy in a fundamental law, a remote past, an apocalyptic future. Each took more than one generation to establish roots. Each eventually produced a new system of law, which embodied some of the major purposes of the revolution,

and which changed the Western legal tradition, but which ultimately remained within that tradition.[1]

Elsewhere I have attempted a more complete examination of how Thomas Berry's vision of law conforms to Berman's analysis.[2] But time limits my focus here to the tension between law as order and law as justice as a way of appreciating Berry's critique of law's failure and its simultaneous promise.

To start, law provides a sense of clarity and codification of values around which a community organizes itself and finds coherence with settled expectations about an entire range of appropriate behavior. Law's stability and order afford fundamental security for the community's functional transactions within itself and with others beyond. Varying in detail and complexity, law articulates the conditions for ordered communal existence. But if law is the guardian of continuity and tradition, conserving those values that its consensual duties and responsibilities protect, its vitality is measured by its creative response to the changed circumstances that the particular community confronts in its identity as an embodied movement into the future. According to Harold Berman, this dynamic nature of society and the failure of law to respond to critical changes in a timely fashion are common elements in all six revolutions within the Western legal tradition.[3]

Beyond that, he argues that such failures reflect

an inherent contradiction in the nature of the Western legal tradition, one of whose purposes is to preserve order and another is to do justice. Order itself is conceived as having a built-in tension between the need for change and the need for stability. Justice also is seen in dialectical terms, involving a tension between the rights of the individual and the welfare of the community. . . . In the great revolutions, the overthrow of the preexisting law as order was justified as the reestablishment of a more fundamental law as justice. It was the belief that the law was betraying its ultimate purpose and mission that brought on each of the great revolutions.[4]

---

1. Harold J. Berman, *Law and Revolution: The Formation of the Western Legal Tradition* (Cambridge, MA: Harvard University Press, 1983), 19.

2. Brian Edward Brown, "The Earth Jurisprudence of Thomas Berry and the Tradition of Revolutionary Law," in *The Intellectual Journey of Thomas Berry: Imagining the Earth Community*, ed. Heather Eaton (Lanham, MD: Lexington Books, 2014), 195–222.

3. See Berman, *Law and Revolution*, 21.

4. Ibid., 21–22.

Here, Thomas Berry's critique of law is congruent with those moments of revolutionary transformation that have preceded it. His determination of law's inadequacy is its relative failure to address the harms to the expanded Earth community and the impoverished sense of justice accorded it.

Consistently, Berry deplores the devastation that imperils the Earth. The severity of planetary demise is as stark as the human destructiveness that perpetrates it is clear. Earth presently suffers the very "disintegration"[5] of its biological integrity. Berry writes,

> Seldom does anyone speak of the deficit involved in the closing down of the basic life system of the planet through abuse of the air, the soil, the water, and the vegetation. . . . This deficit . . . is . . . the death of a living process, not simply the death of a living process, but of the living process. . . . We are determining the destinies of the earth in a comprehensive and irreversible manner. The immediate danger is not possible nuclear war, but actual industrial plundering.[6]

Against this enormity, law's failure is the measure of its negligible silence. If one of its fundamental functions is to preserve the stability and security for meaningful coexistence among members who subscribe to the values that its sanctions uphold, contemporary law betrays its inadequacy even as it exposes the impoverished sense of community it deems to protect. Varying in origin, current legal systems ground their legitimacy in the sovereignty of the respective nation-state whose values they claim to embody. Commonly asserted, however distinctive the efficacy, is law's intent on securing the defense and advancing the welfare of those living under its particular jurisdiction.[7]

But the narrowness of this national allegiance to citizenry well-being has been the very focus of a collective anthropocentric disregard for the integrity of life in its planetary fullness. Within the confines of so many discrete

---

5. Thomas Berry, "The Cosmology of Religions," in *The Sacred Universe: Earth, Spirituality, and Religion in the Twenty-First Century,* ed. Mary Evelyn Tucker (New York: Columbia University Press, 2009), 119. See also Thomas Berry, "The Petrochemical Age," in *Evening Thoughts: Reflecting on Earth as Sacred Community,* ed. Mary Evelyn Tucker (San Francisco: Sierra Club Books, 2006), 95.

6. Thomas Berry, "Economics as a Religious Issue," in *The Dream of the Earth* (Sierra Club Books, 1988), 72.

7. Since Berry often directs specific attention to the values asserted in the US Constitution, it may be instructive to consider its preamble: "We the people of the United States, in order to form a more perfect union, establish justice, insure domestic tranquility, provide for the common defense, promote the general welfare and secure the blessings of liberty for ourselves and our posterity, do ordain and establish this Constitution for the United States of America."

national boundaries, human concern for law's protectiveness has been largely domestic and self-referential. Such concern has not only marginalized care for Earth's fauna and flora but has presumptively conferred entitlement over their disposition and very existence through the conceit of property. That instrument, intrinsic to the self-definition of national sovereignties, removes human intimacy with the animating presence of Earth's water, soil, and air, acquiescing their despoliation as mere resources for exploitation. The consequence of such an orientation, condoned by the rhetoric of self-protective national interests, has led to the deadly deficit of Berry's metaphor. The self-regard of human law, long propertised and extensively commercialized, has rendered it insensible and mute to the cataclysm that terminates the Earth in its Cenozoic florescence. Of this magnitude of change Berry writes, "Something much greater is happening than is generally realized. . . . We are witnessing nothing less than the dissolution of the planet Earth and all its living systems in consequence of this strange distortion of our human role in the Earth process."[8]

Implicit in Berry's assessment is law's inadequacy to appropriately respond to the very nature of the harm that threatens the viability of planetary life. In that regard, the failure of presently constituted legal systems, anthropic in their concern and fragmented in their native allegiances, is consistent with past revolutionary moments in the Western legal tradition. All were instigated by failures within preexisting legal and governance structures to make timely response to the changed sociopolitical circumstances of the respective communities from within which the specific revolution arose. As Berman has written,

> Each revolution represents the failure of the old legal system that the revolution replaced or radically changed. These systems were failures if only in the sense that they were, in fact, replaced or radically changed. . . . The old law was also a failure in another sense: it proved incapable of responding, in time, to the changes that were taking place in society. . . . If, in short, the inevitable had been anticipated and necessary fundamental changes had been made within the pre-existing legal order— then the revolutions would presumably have been avoided. To change in time is the key to the vitality of any legal system that confronts irresistible pressure for change. Thus the great revolutions . . . represent explosions that have occurred when the legal system proved too rigid to assimilate new conditions. . . . They also constituted a great release of

---

8. Berry, "Earth as Sacred Community," in *Evening Thoughts*, 47.

energy, which, to be sure, destroyed much of the past but also created a new future. Ultimately, each of the great revolutions may be seen to have been not so much a breakdown as a transformation.[9]

The conditions for those past transformative revolutions are broadly consistent with the conditions of the present. From Berry's perspective, the old laws of the human community, yet insufficiently informed by the new story of cosmogenesis, are failed legal systems. Without that critical cosmology in which the creativity of the universe, shaping the solar system from within the dynamics of the Milky Way and bringing forth the singularity of Earth as the primary community of water, soil, and atmosphere and blooming into the profusion of life that enabled its further self-reflective emergence as human, human law has become dysfunctional. Like the preexisting legal orders of past revolutionary moments, the present laws of Berry's criticism prove inattentive to, and ineffective for, the most profound of changes yet witnessed in human and, necessarily, in Earth history. However irresistible the evidence and the biosystemic pressure for revolutionary change across the full range of cultural expression within the primacy of the integral Earth community, the rigidity of law as presently constituted order resists the urgencies of law as justice.

As noted, an ambivalence within law manifests in the tension on the one hand to establish and preserve a consistency and stability, and its vital responsiveness, on the other, to move beyond the rigidity of that conservative tendency and to adopt timely changes that the protectiveness for the community's welfare dictates. Given the expansive understanding of the human-Earth community whose integrity has been established by the coherence of the cosmic narrative, and the devastation from an unreformed exploitive anthropocentrism, law's conservatism is without justification.

Indefensible from the gravity inflicted and the risk imposed, the assaultive human further compounds its illegitimacy as self-destructive. Every diminishment to planetary waters, soil, and atmosphere that hinders Earth's self-renewal and creativity in the full complementarity of its fauna and flora erodes human well-being in its physical and psychic capacities. It is Berry's consistent contention that, as the natural world recedes under the onslaught of a technocratic dominance as progress, the human repudiates engagement with the source of its deepest meaning. Speaking of the compound jeopardy to its own sense of purpose and capacity for fulfillment, as well as its experience of

---

9. Berman, *Law and Revolution*, 20–22.

numinous reality, Berry underscores the current harm from, and necessity to move beyond, the commercial-industrial Earth devastation:

> There is a special need in this transitional phase out of the Cenozoic to awaken a consciousness of the sacred dimension of the Earth. For what is at stake . . . is the meaning of existence itself. Ultimately it is the survival of the world of the sacred. Once this is gone the world of meaning truly dissolves into ashes. . . . A desolate Earth will be reflected in the depths of the human.[10]

Thus, Berry exposes a legal order that so forfeits its legitimacy by so abandoning its authenticity to protect against harm, as does the law of the terminal Cenozoic. Remaining within their originating national identities, legal codes of the early twenty-first century are largely uniform in their neglect for the common good of the integral Earth community. To that extent, they have remained inured to the violation perpetrated by the commercial-industrial ethos of extraction and consumptive exploitation that Berry consistently renders "plunder," the coercive seizure and possessory deprivation of goods to which the belligerent is not entitled, an aggressive taking "without just right."[11]

Yet, in Berry's divestiture of law's moral stature, exposed in its hollow protectiveness for organic Earth, there remains an implicit recognition of law's capacity to actualize its potential beyond preserving the order of things, and to claim its identity as justice. That law can sustain assessment of its own fidelity to the fundamental charge of protection against harm; that it can evaluate the evidence of a violated and threatened planetary body; that it can scrutinize and indict the commercial-industrial obsession with progress toward some wishful wonderworld; that it can render its conviction that such an enterprise is wanton plunder and indefensibly criminal; and that it can issue a cease-and-desist order and demand restorative restitution; that law can fulfill these inherent charges, is to step beyond its conservative inclinations and invigorate itself as revolutionary justice, as ecological justice. If the Western legal tradition has witnessed past moments of transformative potential where

---

10. Brian Swimme and Thomas Berry, *The Universe Story: From the Primordial Flaring Forth to the Ecozoic Era—A Celebration of the Unfolding of the Cosmos* (New York: HarperOne–Harper Collins, 1994), 250.

11. "Plunder: To pillage or loot. To take property from persons or places by open force, and this may be in course of a war, or by unlawful hostility, as in the case of pirates or robbers. The term is also used to express the idea of taking property from a person or place, without just right, but not expressing the nature or quality of the wrong done" (*Black's Law Dictionary*, 5th ed. [St. Paul, MN: West Publishing, 1979], 1039).

law, instigated by and responsive to fundamental changes within its respective setting, rejected the conditions of a preexisting order only to extend reformative measures into a new one, the present crisis surpasses each and all of them in the dangers that evoke law's most urgent repudiation and expansive protectiveness. But while activated through the evidence of the terminal Cenozoic, the revolutionary dynamism of law's potential will be proven by the persuasiveness of the case it argues and the commitment it demonstrates against the denial and resistance to the accusations it alleges. For Berry, the intensity of that denial is acute and reflects the thorough self-identification of the human with its technocratic dominance and the progress it vaunts. To suggest that the pursuit of human satisfaction through the commercial-industrial consumption of the natural world is the very culprit of its destruction is too threatening to entertain.

Yet law's fidelity to its authenticity as protectiveness responds to such denial by its unwavering attention to the testimony of harm's severity and the enormity of the risk that further denial threatens. While not sufficient in itself, law's dedication to penetrate the cultural delusion by exposing its calamity is a significant measure of its identity as ecological justice. Its own vitality in the starkness of its task is conveyed in Berry's analogy with addiction. "The remedy," he writes, "would seem to emerge, as in denial situations generally, in a crash so severe that we are suddenly confronted with a choice between death and abandoning our addictive mode of functioning. . . . The crash that faces us . . . is in some manner the crash of the Earth itself."[12]

But if law's efficacy is its resolve to confront cultural diversion from the consequences of its behavior, it must further claim its revolutionary identity by recognizing its circumscription within the confines of its past. Here, Berry's critique of the American Constitution illustrates his censure of law's anthropocentric focus on exclusively human rights. However lofty its political achievement, however prized a model for democratic aspirations, its charter of liberties is the expression and vehicle for human self-aggrandizement with complete disregard for the limitations imposed by the larger body politic of organic Earth. While protecting autonomy from tyrannous regimes, the constitutional scheme enabled a culture that knew no restraint toward other than human members of the North American community, all subsumed under the homogeneity of property. Disenfranchised and subjugated under that unexamined assumption, the living rivers and forests, prairies and peaks, and the host of beings inhabiting them, became the commodified objects of possessory human disposition. Law, in its expression as constitutional freedom,

_____
12. Swimme and Berry, *Universe Story*, 255.

restricted and confined as a human prerogative, left unprotected those living communities beyond the pale of its recognition. In fact, it facilitated their ruin through its constrictive self-identification with the commercial-industrial commitments of the corporation enterprise. In making this charge Berry cites Morton J. Horwitz's two-volume study, *The Transformation of American Law*. Its findings on the collaboration of the legal profession and judiciary with the commercial entrepreneurial ambitions in the first decades after the American Revolution presaged the rise and dominance of an increasing corporate control. Horwitz writes,

As political and economic power shifted to merchants and entrepreneurial groups in the post-revolutionary period, they began to forge an alliance with the legal profession to advance their own interests through a transformation of the legal system. By around 1850 that transformation was largely complete. . . . Legal regulations . . . were increasingly subordinated to the disproportionate economic power of individuals or corporations. . . . Law, once conceived of as protective, regulative . . . and, above all, a paramount expression of the moral sense of the community, had come to be thought of as facilitative of individual desires and as simply reflective of the existing organization of economic and political power. This transformation in American law both aided and ratified a major shift in power in an increasingly market-oriented society. *By the middle of the nineteenth century the legal system had been reshaped to the advantage of men of commerce and industry at the expense of farmers, workers, consumers, and other less powerful groups within the society.* Not only had the law come to establish legal doctrines that maintained the new distribution of economic and political power, but, wherever it could, it actively promoted a legal redistribution of wealth against the weakest groups in the society.[13]

Berry, confirmed by Horwitz's research, exposes the moral failure of law as it succumbed to the narrow interests of the commercial-industrial enterprise and its ultimate expression in the multinational corporation. For Berry, the legal prerogatives favoring the interests of corporate dominance not only excluded and marginalized whole classes of citizenry, but further removed law's protective recognition for the other than human world, more effectively bound as the objects of commerce's advance. But the injustice of a legal-eco-

---

13. Morton J. Horwitz, *The Transformation of American Law, 1780–1860* (Cambridge, MA: Harvard University Press, 1977), 253–54 (emphasis added). The italicized sentence is quoted in the first of Berry's several essays.

nomic consortium as that which emerged early, however subtly, within an American context to become the model for a current global control, is implicitly betrayed not only by its elitist subversion of democratic principles of personal freedom and participatory government. It is likewise condemned by the indefensible jeopardy of its assault on the viability of planetary life, the constant witness of Berry's prosecution. Here again, law as revolutionary justice is called to reclaim its authenticity over its discredited collusion with, and at the behest of, a privileged corporate network and the dysfunctional order of its permissive Earth plunder. Law's capacity to yet register harm, weigh evidence, and sustain a claim against the wanton violence of a system that it had facilitated but now renders unsustainable is the reclamation of its own revolutionary foundations and the necessary step toward its own reinvention.

24

Earth Jurisprudence in a Cosmological Perspective

Sometimes It Takes a Joker

PATRICIA A. SIEMEN

My life's journey made a significant turn in 2003 when I read Thomas Berry's "The Origin, Differentiation, and Role of Rights."[1] His articulation of the analogous rights of every being to exist, to have natural habitat or home, and to fulfill its purpose in the course of evolutionary life deeply disturbed me. It made ultimate sense, but my legal training resisted: "This is not practical; it's foolish; how could this be implemented?" While I had studied Berry and Brian Swimme's *The Universe Story*,[2] describing the evolutionary journey of the universe, it was Berry's applying a functional cosmology for law and governance, an Earth jurisprudence, that bothered me. Ultimately, it became the foundation for inspiring me to create the Center for Earth Jurisprudence in 2006.[3] Today I use *Journey of the Universe* in all my classes to set the context for an emergent Earth jurisprudence.

The emergence of Earth jurisprudence, of governance from an ecocentric perspective, where human rights are an interdependent subset of Earth rights, is essential for a continuing viable life on Earth. Earth jurisprudence is an expression of the evolutionary creativity expressed in the story of the universe. It is because of a single coevolutionary process that we humans are kin and related to all other beings who share Earth as home. The universe needs the emergence of Earth governance included in its ongoing story of creativity and viability.

---

1. In Cormac Cullinan, *Wild Law: A Manifesto for Earth Justice* (London: Green Books / Gaia Foundation, 2003), 115.

2. Brian Swimme and Thomas Berry, *The Universe Story: From the Primordial Flaring Forth to the Ecozoic Era* (San Francisco: HarperCollins, 1992).

3. The Center for Earth Jurisprudence is a program of Barry University School of Law, Orlando, Florida. Thomas Berry was a mentor until his death in 2009.

Earth Jurisprudence: Court Jesters Need Apply

Earth jurisprudence is a new arena within the field of environmental law that many people do not yet understand. It takes both a jurisprudential and practical approach. Its philosophical approach asks what "ought" the law to be or do, and its practical approach explores how we extend legal protection to all entities who are members of the Earth community. Earth jurisprudence shifts the dominant environmental law paradigm from a property law frame to one that respects and protects the inherent rights of all beings to exist and flourish under the law. This is a huge jurisprudential shift for Western-oriented people. Those proposing that Earth and its entities have the right to moral and legal consideration are often deemed ridiculous and clearly not living in the "real" world of commerce and legal realities. Clearly another emergent energy is needed to break this tension, so let me introduce the archetypal energy of the *court jester*.

The court jester is a ubiquitous character whose role includes risking his tenure in order to tell the truth to the king or ruler.[4] In European medieval times, the jester even dared to tell the king when he was in error or blindsided. If the king listened, there was a chance that he might be saved from doing grave harm to the kingdom. Today, Earth needs court jesters, perceived as fools, of every discipline, to step out and speak basic truths to world and corporate leaders about the disruption of Earth's evolutionary processes and the gross climate and environmental devastation taking place. Today's jesters need to dramatize the inadequacy of the current legal and economic policies and inspire imaginative alternatives that are grounded in the emergent functions of the universe. Laws and economic policies are needed that are modeled on Earth's laws and self-regulating patterns to sustain long-term health for all species. Court jester archetypes need to offer a vision for redesigning our governance and legal systems in ways that support and protect the Earth community rather than legitimate its destruction. So much is at stake. We can afford to do no less.

Today's jester has a challenge in identifying the actual "ruler" or system to whom he or she should be speaking. There are so many circles of intertwined governance and economic "kingpins" reigning at local, regional, and global levels. Many are responsible for extractive decisions destroying Earth's functional integrity and health. Proposing to them a framework that is embedded in a wholistic scientific and cosmological context is essential for a viable

---

4. See chapter 1 of Beatrice K. Otto, *Fools Are Everywhere: The Court Jester around the World* (Chicago: University of Chicago Press, 2001). See also http://press.uchicago.edu/Misc/Chicago/640914.html.

future, even if labeled naive. Imagining that we can successfully adopt legal systems that respect and balance nature's intrinsic rights to exist and flourish in sufficient time to counter climate justice is revolutionary. Yet the consequences of climate change and the acceleration of extinction of major species demonstrates the urgency for such systemic shifts. As Bill McKibben stated at a talk in Vermont in 2013, "The laws of physics do not wait."[5]

Current Legal System Woefully Inadequate

Since the beginning of the industrial era most of the Western world has been living out of a legal system narrowly framed by a property law paradigm. This framework considers every entity to be a potential "resource" that is available to become a commodity to be bought and sold. The laws of most nations are premised on upholding economic activity based on an antiquated concept of the public good. There is limited consideration of the cost to the commons, or to the public good, when natural limits are reached. To convince governance and corporate rulers that today's "public good" requires recognition of the interdependent role played by all members of the Earth community can be an audacious task. Translating Berry's concept that all members of the Earth community are "a communion of subjects, not a collection of objects,"[6] into legal and economic public policies is indeed a worthy challenge, a great work. Nevertheless, this work is happening around the world. A shift is occurring as indigenous peoples, social justice activists, lawyers, artists, academics, theologians, poets, musicians, environmentalists, economists, educators, engineers, and people of faith and goodwill address the mounting violations of environmental and human rights within a framework of interconnected wholes.[7] Multiple diverse, yet related movements are gaining momentum. Some, such as the Global Alliance for the Rights of Nature[8] and the Earth Charter Initiative Council[9] are driving into law the moral duties and obli-

---

5. Bill McKibben at the Our Children, Climate, and Faith Symposium (public address, Strafford Town Hall, Strafford, VT, August 16, 2013).

6. Cullinan, *Wild Law*, 115, quoting Thomas Berry, "Origin, Differentiation, and Role of Rights."

7. See Earth Law Center for mapping of intersections of environmental and human rights violations: http://earthlawcenter.org/what-we-do/mapping-violations-of-human-and-environmental-rights/.

8. The Global Alliance for the Rights of Nature is a worldwide movement of individuals and organizations creating human communities that respect and defend the rights of nature.

9. The Earth Charter Initiative (ECI) Council oversees the Earth Charter International Secretariat and advises on the broader Initiative. See http://www.earthcharterinaction.org/content/pages/Council.html/.

gations of humanity to live in balance and right relationship with the rest of the larger Earth community. For example, in November 2014, two more communities—in Athens, Ohio, and Mendocino County, California—adopted the Community Bill of Rights banning corporate rights to conduct hydro-fracking in their communities, securing rights to a healthy environment and local self-governance, and recognizing rights of nature.[10] These visionary activists who are advancing community rights are contemporary archetypal court jester personas entering the realm of public policy and Earth jurisprudence. And they are not the ones looking foolish in light of collapsing ecosystems. Rather, it is the intractable governance and economic behemoths who continue to legitimate the exploitation and commodification of nature that are tragically and foolishly wrong.

Implications of Earth Jurisprudence

So how would an Earth jurisprudence governance system be different from current law? First, it is important to understand that prevailing US environmental laws provide very little protection of the *integrity* of the natural world, other species (unless already threatened and nearly extinct), atmospheric health, or ecosystem resiliency. For example, neither the US Clean Air Act of 1970,[11] nor the Clean Water Act[12] gives priority protection to *the air* or *the water*. Rather these laws are designed to primarily protect the interests of humans rather than the actual health of the water or air. To illustrate further, if a lawsuit alleging environmental damage to freshwater springs because of excessive nitrate pollutants is to be brought, the plaintiff has to prove that there is significant *human* injury due to the nitrate pollution. It is not sufficient to establish that the *springs or springshed* is significantly impaired. Current laws generally do not recognize the actual harm to a natural entity unless threatened or endangered species are involved.[13] As Christopher Stone raised in his groundbreaking 1972 law review article, *Should Trees Have Standing? Toward Legal Rights for Natural Objects*,[14] under US laws the environmental-

---

10. See the Community Environmental Legal Defense Fund, http://www.celdf.org.
11. 42 U.S.C. §7401 et seq. (1970).
12. 33 U.S.C. §1251 et seq. (1972).
13. Endangered Species Act, 16 U.S.C. §§1531–1544.
14. Christopher D. Stone, "Should Trees Have Standing? Toward Legal Rights for Natural Objects," *Southern California Law Review* 45 (1972): 450; Christopher D. Stone, *Should Trees Have Standing? Law, Morality, and the Environment*, 3d ed. (New York: Oxford University, 2010).

ly injured entity has no standing to get its claim into court.[15] That was true in 1972 and is still the prevailing law today. Earth jurisprudence would expand standing to other-than-human species and ecosystems in order to protect the health and viability of the natural entity in itself.

In the United States, and in nearly all countries except for Ecuador[16] and Bolivia,[17] there is no recognized, enforceable right of the natural world to be protected from substantive damages to its existence and health. As Thomas Berry noted in *The Great Work*, the US Constitution, gives all rights to humans

with no legal protection for the natural world. The jurisprudence supporting such a constitution is profoundly deficient. It provides no basis for the functioning of the planet as an integral community that would include all its human and other-than-human components.[18]

Adopting laws that protect the health of an ecosystem or watershed, for example, and providing standing for the natural entity through an appointment of a guardian to speak for the common good of the ecosystem would be a significant step forward for an Earth jurisprudential framework.

The Universe Story as Context for Framing Earth Jurisprudence

So what type of legal system could be adopted that *would* respect and protect the emergent dynamics of living systems and the "land community"[19] as described by Aldo Leopold? In Berry's Ten Principles for Jurisprudence Revision[20] he set forth foundational principles on which a new legal system

---

15. See a counterargument in Justice William O. Douglas dissent in *Morton v. Sierra Club*, 405 U.S. 727, 741 (1972).

16. Ecuador was the first country in the world to recognize constitutional legal rights of nature (Pachamama) when it ratified a new constitution in 2008. Chapter 7, Article 71 states, "Nature or Pachamama, where life is reproduced and exists, has the right to exist, persist, maintain and regenerate its vital cycles, structure, functions, and its processes in evolution." It also provided standing to nature and authorizes the public to bring enforcement actions: "Every person, people, community, or nationality will be able to demand the recognitions of rights for nature before the public organisms." Article 72 authorizes that nature has a right to restoration as well.

17. Bolivia adopted a Law of the Rights of Mother Earth (Law 071) in December 2010.

18. Thomas Berry, *The Great Work: Our Way into the Future* (New York: Bell Tower, 1999), 74.

19. Aldo Leopold, *Sand County Almanac* (1949; Oxford: Oxford University Press, 2001).

20. Thomas Berry, *Evening Thoughts: Reflecting on Earth as Sacred Community*, ed. Mary Evelyn Tucker (San Francisco: Sierra Club and the University of California Press, 2006), 149.

could be constructed that would be coherent with Earth's story of its evolutionary processes and inner dynamics. The first principle states "Rights originate where existence originates. That which determines existence determines rights."[21] He continues, "The natural world on planet Earth gets its rights from the same source that humans get their rights: from the universe that brought them into being."[22] Thus Berry sets the context for legal recognition and protection of the inherent rights of nature and rebalances human rights within the context of Earth rights. And since humanity has evolved from the dynamic complexities of the universe's journey, human rights are nested within Earth rights. That is why the Universal Declaration on the Rights of Mother Earth[23] adopted at the Peoples Convention on Climate Change and the Rights of Mother Earth[24] was intentionally modeled on the pioneering work of the UN Declaration on Human Rights.[25] As internationally renowned physicist and environmental leader Vandana Shiva says, "Earth rights are human rights. Humans have no rights if Earth is not viable."[26]

Given an inherent Earth rights *framework*, Earth jurisprudence will develop laws and governance that respect and protect the intrinsic rights of every member of the Earth community to exist, to have a natural habitat, and to fulfill its role in the evolutionary processes of the Earth community.[27] This will require a complex balancing of macro-level biogeochemical cycles of ecosystems, other species, and diverse cultural perspectives. Modes of participatory decision-making by affected parties will be essential. It is a complex and exciting challenge. Some may even think it is foolish, worthy of something only a court jester would suggest. It will require a massive shift in the perception of human relationships with the rest of the Earth community. We have much to learn from indigenous peoples who traditionally see life as an interconnected web and a part of "all my relations."

---

21. Ibid.

22. Ibid.

23. See Universal Declaration on the Rights of Mother Earth, http://therightsofnature.org/universal-declaration/.

24. The Peoples Convention on Climate Change and the Rights of Mother Earth took place in Cochabamba, Bolivia, in April 2010. It was called by President Morales after the collapse of climate talks at the UN Conference of Parties at the UFCCC in Copenhagen, Denmark, December 2009.

25. See UN Declaration on Human Rights, http://www.un.org/en/documents/udhr/.

26. Personal conversation with author after a presentation given at the Center for Earth Jurisprudence, Miami, Florida, July 14, 2010.

27. Berry, *Evening Thoughts*, 149. Paraphrasing Berry's fifth principle: "Every component of the Earth community, both living and nonliving has three rights: the right to be, the right to habitat or a place to be, and the right to fulfill its role in the ever-renewing processes of the Earth community."

At its essence, Earth jurisprudence prefers that humans voluntarily self-regulate their own behavior, relate humbly with others sharing Earth, and live in balance and harmony with the larger Earth community. The question for legal systems is, How do we create governance and economic systems that support a celebration of *vivir bien*, or well-being,[28] for diverse future generations?

It is critical that we recognize that our current legal system is woefully inadequate to the challenges and promises facing an Anthropocene Era.[29] Earth can no longer sustain humans doing "business as usual." The next section addresses the creation of laws that recognize the responsibility that humanity has to protect the interdependent relationships embedded within all members of the Earth community.

Moving toward an Earth Jurisprudence

Thomas Berry and Brian Swimme proposed that the universe is organized around three foundational patterns: differentiation, subjectivity, and communion.[30] Each pattern can be found within the emergent capacities of Earth's processes and components. The work of Earth jurisprudence is to create human governance systems grounded in these natural dynamics.[31] Law by its nature is conservative and meant to protect society's cultural values and structures from arbitrary change. However, when legal systems do not respond to shifting cultural values, a revolution in society and law can happen, as was seen with the civil rights struggle in the United States. Today an emerging ecological consciousness is driving many grassroots organizations to consider new legal initiatives respecting Earth's integrity. The UN *Report*

---

28. The concept of "living well in harmony with Nature" has emerged in the past ten years primarily within the indigenous communities of the Andes. It speaks of indigenous cosmovision, in which the concept of material prosperity and consumption does not exist or take priority. See the *UN General Assembly Harmony with Nature Report*, A/69/322, August 18, 2014, http://www.harmonywithnatureun.org/content/documents/285N1450929.pdf.

29. The term "Anthropocene Era" was coined by Nobel Laureate Paul Crutzen and Eugene F. Stoermer in a newsletter in 2000. See "Dawn of the Anthropocene Era," International Geosphere-Biosphere Programme, Newsletter 41 (2000). It indicates a new geological epoch when humanity becomes a significant and sometimes dominating environmental force.

30. Brian Thomas Swimme and Thomas Berry, *The Universe Story: From the Primordial Flaring Forth to the Ecozoic Era* (San Francisco: HarperSanFrancisco, 1992), 74. See also Thomas Berry, *Great Work*, 162, and Berry, *Dream of the Earth* (San Francisco: Sierra Club Books, 1988), 45.

31. Judith Koons, "Key Principles to Transform Law for the Health of the Planet," in *Exploring Wild Law: The Philosophy of Earth Jurisprudence*, ed. Peter Burdon (Kent Town, South Australia: Wakefield Press, 2011), 46.

of the Secretary-General on Harmony with Nature released in August 2014 provides a historical overview of global initiatives wherein different governmental entities are adopting laws and policies providing legal consideration of nature or Mother Earth.[32]

The crisis of Earth calls us to critique the legal system from a cosmological perspective, even as the capacity for economic behemoths to resist and co-opt such emergent responses seems almost unlimited. The voice of the jester is indeed needed. But the universe will have its own way, and the laws of physics cannot be stopped by economic power. Local and regional movements, seen as tiny cells and organisms, continue to create pockets of receptivity and resilience where the community organizes around the values of diversity, unity, and respect for local watersheds and ecosystems, for example. These are seedbeds for evolutionary change and adaptation. The power of local initiatives builds as networking creates a bonding of cooperation. This is the work of people who are willing to be foolish for the sake of the Earth. Indigenous peoples model this way of Earth jurisprudence by their traditional ways of being "kin" to all that exists. Today they are teaching the rest of humanity how to live in harmony with the rest of Mother Earth.

Adopting an Earth Jurisprudence from a Cosmological Context

In exploring methodologies to apply the cosmological core principles to current US laws, I draw upon the work of my friend and colleague Judith Koons, professor of law at Barry University School of Law in Orlando. In a 2009 law review article,[33] Koons, who draws directly from the work of Berry and Swimme in *The Universe Story*, sets forth a vision of Earth jurisprudence through the lens of subjectivity, communion, and differentiation.[34] She proposes that this interior self-organizing, self-regulating capacity can be applied within an Earth jurisprudence paradigm, demonstrating an intrinsic worth of nature. This intrinsic worth warrants moral consideration, which for Koons is a stepping-stone to legal consideration and legal rights.

It should be noted that legal consideration and the recognition of the intrinsic value of Nature is not the same as a governance system "bestowing" rights on nature, because the intrinsic value of nature *already* exists. Under an Earth jurisprudence system, the legal system recognizes these already existing

---

32. See http://www.harmonywithnatureun.org/content/documents/285N1450929.pdf.

33. Judith Koons, "Earth Jurisprudence: The Moral Value of Nature," *Pace Environmental Law Review* 25 (2009): 263.

34. Ibid.

rights and then provides a range of legal considerations and protections of those rights. One of the options is seeing nature as a rights-bearing entity in itself. Another is the granting of legal consideration by some legal authority, which, as Christopher Stone argued in 1972, is not the same as holding rights.[35] Koons then presents a range of legal consideration options that can be applied to a nonhuman entity once it is given a jural person status, such as corporations, churches, trusts. She writes,

> An entity may be granted rights, be given duties and responsibilities, be the recipient of immunities and privileges, or be held liable—all of which are intermediate, operative notions that flow from the broader principle of legal considerableness. Having legal status means being enabled to participate in the legal system, although not necessarily as a rights holder.[36]

Koons proposes that one way in which subjectivity of Earth entities could be granted legal protection is through the doctrine of standing. Standing is the gatekeeper as to whether one can bring a legal action into court. Numerous legal scholars are reconsidering the standing doctrine that currently prohibits other-than-human species and natural entities the right to sue in their own name. Currently, only human plaintiffs can allege injury and be awarded damages or restitution. There are many arguments for expanding the standing doctrine to include natural entities, especially when nonliving entities such as corporations and trusts are already granted legal standing. The doctrine of guardianship is well established in civil law and could serve to protect the best interests of natural entities in legal proceedings. Focusing on addressing the *actual* harm to other members of the Earth community, rather than on a contorted legal result granting some human relief, would strengthen the integrity of law as well.

The principle of communion, as expressed in the interdependent, nested relationships that humans have with the rest of the Earth community, asserts that we are one part of the whole. This unity is described by Cormac Cullinan: "Western physicists confirm that the same atoms and sub-atomic particles may be part of the soil on Monday, a plant on Tuesday and us on Wednesday."[37] Failing to see our nested relationship to other beings has resulted in our current generation bringing forth the conditions for massive extinction

35. Stone, "Should Trees Have Standing?" 78.
36. Koons, "Earth Jurisprudence," 49.
37. Cullinan, *Wild Law*, 146.

of species and major climate disruptions, as well as major environmental deterioration.

Koons proposes that the dynamic of interdependency and communion "can be translated into jurisprudence as a principle of relational responsibility."[38]

* * *

Because of humanity's capacity for reflective consciousness, we can muse on the wonders of the universe and know that Earth nurtures and provides for us. We are the ones who create legal and ethical systems to govern relationships between ourselves and other humans. Therefore, we have the ability to develop systems of law and governance that create a duty of care to protect Earth's natural systems.

As noted earlier, our current legal system allows for guardians and trusteeships for people and entities that cannot care for, or protect, themselves. There is the public trust doctrine[39] that could be expanded so as to create a trust/trustee relationship with Earth's natural systems and entities. Normally the state would be the trustee, or designate an appropriate body, held to a fiduciary duty to care for the body or corpus of the trust. The public trust has historically been applied to care for the waterways and shorelines, although under Roman law it extended to air and land that were deemed held in common by the people, or by the state for the good of the people. As Stone noted in 1972, it is conceivable that a guardianship model also could be established to represent the best interests of threatened species and natural entities. Guardians, appointed by local or state authorities, would be people who have a deep knowledge of the needs, capacities, and functions of those beings, as related to their larger natural habitat.[40]

There is a potential contradiction in trying to protect natural entities through a public trust structure, however, as the corpus or body of a trust, under law, is considered a *res*, or a "thing" to be protected. Trust law does not infer that the body of the trust (in this case, a natural entity such as a freshwater spring or a geological formation considered sacred to indigenous peoples) is a subject in its own right. Therefore, the rules of guardianship might be better applied. Neither of these approaches supports the concept of

---

38. Koons, "Earth Jurisprudence," 51.

39. See Joseph L. Sax, "The Public Trust Doctrine in Natural Resources Law: Effective Judicial Intervention," *Michigan Law Review* 68 (1970): 471; see also Joseph L. Sax, "Liberating the Public Trust Doctrine from Its Historical Shackles," *U.C. Davis Law Review* 14 (1980–1981): 185.

40. See Stone's discussion in *Should Trees Have Standing?* 103–12.

nature being a rights-bearing entity in and of itself. However, legal precedent recognizes that both a guardian and trustee must represent and advocate for the best interests of the one not able to speak for itself.

The third principle of differentiation is an expression of the incredible creativity manifested by the universe. The universe obviously loves diversity! Nothing is the same: no two molecules, no two microbes, no two minnows, no two meadows, no two mountains, nor any two men or women. Every being is a differentiated being with its own internal capacity for autopoeisis, self-organization, and relationship. Koons supports Cullinan's suggestion that the diversity of the Earth community and natural self-regulating systems form a type of Earth Democracy.[41] International environmental leader, physicist, and author Vandana Shiva describes Earth Democracy as a "living democracy," a type of self-governance that "grows like a tree from the bottom up."[42] Shiva says that "Earth Democracy connects the particular to the universal, the diverse to the common, and the local to the global."[43]

Earth Democracy is premised on the idea that the best decisions for nature's well-being are those decided by the principle of subsidiarity: decisions should be made at the lowest (closest to the parties) and most appropriate level of governance. Some decisions are best made at the local level (e.g., regulating building codes), others at a regional level (e.g., control of pollution in waterways and the extraction of water), and others at a global level, such as atmospheric pollution. And some decisions, such as sources of clean energy and energy use, need to be made at all three levels. The Community Environmental Legal Defense Fund (CELDF) under the direction of Thomas Linzey and Mari Margil provide excellent examples of Earth Democracy being implemented at local levels.[44]

They work with organized communities to help draft local ordinances banning corporate actors from bringing into the community unwanted commerce such as factory hog farms and hydraulic fracking for natural gas. They work with local communities to expand the home rule provisions of local townships, cities, and counties to grant constitutional rights to natural eco-

---

41. Cullinan, *Wild Law*, 91.

42. Vandana Shiva, *Earth Democracy: Justice, Sustainability, and Peace* (Boston: South End Press, 2005), 10.

43. Ibid.

44. For an overview of CELDF's work, see http://celdf.org/. It is the premier organization in the United States working with local communities and also with international partners in using the law to protect the rights of people and natural communities to determine health and well-being and to resist and challenge the rights of corporations to overrule the rights of community members. They were the key legal advisors to the drafting of Ecuador's constitutional protections for the rights of nature.

systems while stripping corporations of constitutional rights. Their work is essential for an applied Earth Democracy.

Earth Democracy is a facet of Earth jurisprudence. It calls upon humans to assume a duty of care for the health of Earth, not just to remedy its illnesses. Earth jurisprudence and the preservation of Earth can only be advanced and adopted when local, regional, national, and global governance systems decide to take responsibility for the current and future well-being of Earth, our home. It requires a major shift from solely anthropocentric-focused forms of governance to one where humans and nature are both honored and respected.

Conclusion

Earth jurisprudence is the work of all people called at this time to care for Earth's greatest good. Not to act with loving attention, resilience, and resistance may be the most foolish decision we have ever made. Hope for survival and celebration of Earth's bounty and beauty lies in building momentum to care for Earth as demonstrated at the Climate March in September 2014. Over 625,000 people worldwide gathered to illustrate their demands that global leaders take action to curb and control greenhouse gas emissions. The people came to say that if the global representatives do not act, the people will. This is sometimes the best and most effective way that laws and culture change: by exercising a democracy of our feet and hearts.

So much is at stake. It is time that we all become court jesters, willing to tell the truth boldly to the rulers of our day. Let us not be afraid to be seen as foolish or dreamers. So let us put on our jesters' garbs. It is only when we freely and humbly embrace loving service to all our kin and "make the path by walking"[45] that the ever-creating jester spirit can use us to bring about harmony and balance with all of creation.

---

45. Attributed to Anthony Machado, "Traveler, There Is No Path." See Eric McNulty, Richer Earth, http://www.richerearth.com/.

25

Hope for Law and Other Animals in a More-Than-Human World

PAUL WALDAU

To the children, to all the children,
to the children who swim beneath the waves of the sea,
to those who live in the soils of the earth,
to the children of the flowers in the meadows and the trees in the
    forest,
to all those children who roam over the land and the winged ones
    who fly with the
winds,
to the human children too,
that all the children may go together into the future in the full diver-
    sity of their regional communities.[1]

This foregrounding of "children" in the dedication to Thomas Berry's *The Great Work* projects, from one vantage point, a common sense that is both communal and ecological. Yet there are circles in today's frictiony world where talking about children in this way or about a multispecies communion of subjects would be taken as nothing less than a taunt that hints at, as an American Evangelical put it, "moral equivalency" that violates what *often* has been held to be humans' *truly* "great work," namely, rising above "animality."[2] Those who proclaim frankly, "I am an animal," thus are perceived today in *many* powerful circles to threaten not only human superiority and privilege over other animals but also human dignity itself.

One of these unfriendly circles is modern law, which is dominated by a stark dualism between "legal persons" (humans) and "legal things" (all other

---

1. Thomas Berry, *The Great Work: Our Way into the Future* (New York: Bell Tower, 1999), v.

2. Personal communication at the gathering of American evangelicals at the Humane Society of the United States, Washington, DC, November 4–5, 2010.

living beings and all objects).[3] There is also much evidence that would reveal, if included here, how "higher" education, public policy circles, mainline religious institutions, and any number of science domains are, like modern legal systems, decidedly unfriendly "to *all* the children" but especially those *who* "swim beneath the waves of the sea . . . live in the soils of the earth . . . and . . . fly with the winds."

For these reasons, here I hope to address those who continue to be inclined to separate humans, to hold them apart and above any and all nonhuman animals. Most particularly I want to invite, to make community, to dialogue with and companion those of us who regard the strategy of separating human from other-than-human lives as advisable. I argue, instead, that *any* such strategy defeats *all* life, both human and nonhuman animals alike, and plants, too. Although I present a number of different arguments about the many ways separation is a worse-than-bad strategy, the tenor of all these arguments can be discerned in the voice of a medical doctor whose focus is *humans alone*: "Much of the damage that we inflict on ourselves, on others, and certainly on the natural world stems from extreme adherence to the notion of human exceptionalism."[4]

The breadth and depths of our origin and embeddedness in the world—indeed, in the entire universe—are eloquently explored in not only the book and film versions of *Journey of the Universe*,[5] but also in the remarkable interviews included in the supporting educational materials. This trio elaborates the moving insights that Berry set out in second chapter of *The Great Work,* which carries an elegantly simple title of "The Meadow across the Creek." In this chapter, one comes home to the embeddedness of all living beings in our shared and ever-adapting world. Thereby, one senses why the word "children" works so encompassingly in *The Great Work*'s dedication. Similarly, *Journey of the Universe* takes us even further into understanding why the powerful image of children is both nimble and deep enough to allow us to marvel at our multispecies heritage and community. Talking as Berry does, then, is making community, and whenever we make community, we

---

3. This dualism, which is described more below, is addressed throughout Paul Waldau, *Animal Studies—An Introduction* (New York: Oxford University Press, 2013); and Paul Waldau, *Animal Rights* (New York: Oxford University Press, 2011).

4. John Ratey and Richard Manning, *Go Wild: Free Your Body and Mind from the Afflictions of Civilization* (New York: Little, Brown, 2014), 8. The "afflictions of civilization" referred to in the subtitle are listed and explained at 39ff.

5. Brian Thomas Swimme and Mary Evelyn Tucker, *Journey of the Universe* (New Haven, CT: Yale University Press, 2011); and Brian Thomas Swimme and Mary Evelyn Tucker, *Journey of the Universe*, DVD directed by David Kennard and Patsy Northcutt (2011; New York: Shelter Island, 2013).

make morals.[6] Said another way, recognizing our memberships in a number of nested communities, including those replete with other-than-human animals in our shared, more-than-human world, affirms our embeddedness in the cosmos and sparks, in our immediate and day-to-day lives, personal moral development.

Over recent millennia, humans' need to notice and take seriously the other-than-human members of our Earth community is an insight that individuals and groups have repeatedly developed and nurtured, only to have their descendants ignore this insight, eventually to forget it entirely. Thankfully, "discovery" of the importance of noticing and taking other living beings seriously has been made again and again. An early version is the first-millennium BCE insight of the Axial Age sages as summarized by the irenic scholar Karen Armstrong.

> Nearly all the Axial sages realized that you could not confine your benevolence to your own people: your concern must somehow extend to the entire world. . . . Respect for the sacred rights of all beings—not orthodox belief—was religion.[7]

Twentieth-century versions of this insight abound. Some frame the issue as one of our human possibilities, as in Viktor Frankl's recognition that "self-actualization is possible only as a side-effect of self-transcendence."[8] Going beyond the important realm of *individual* self-actualization are insights about *species-level challenges*. Said simply, our species *needs* to make community with the more-than-human world. A moving statement of this key insight is Aldo Leopold's mid-twentieth-century exhortation for humans to evolve "from conqueror of the land-community to a *plain member and citizen* of it."[9]

Here in the early twenty-first century, as many begin to discern that we are an integral part of the journey of the universe, debates about our own health

---

6. The "making morals means making community" notion is explained in Waldau, *Animal Studies*, 209n27: "This observation is taken out of context, for it comes from an analysis of how the very earliest Christians made meaning in the hostile Roman world in which they lived." See Wayne A. Meeks, *The Origins of Christian Morality: The First Two Centuries* (New Haven, CT: Yale University Press, 1993), 5. See also Waldau, *Animal Studies*, 305, for the comment, "Making community is at once an ethical, personal, and practical set of problems."

7. Karen Armstrong, *The Great Transformation: The Beginning of Our Religious Traditions* (New York: Knopf, 2006), xiii–xiv.

8. Viktor E. Frankl, *Man's Search for Meaning: An Introduction to Logotherapy*, 4th ed. (Boston: Beacon, 1992), 115.

9. Aldo Leopold, "The Land Ethic" in *A Sand County Almanac, with Essays on Conservation from Round River* (New York: Ballantine, 1991), 240 (emphasis added).

now prompt us to "go wild." Harvard Medical School's John Ratey uses this mantra to bring home the insight that "we are wild animals" in order to give us a prescription for health and flourishing that contrasts greatly with the diseases of civilization that the subtitle of his 2014 book calls out.[10] As or even more important than our adult health are the interests of our own species' children, for something beyond their "mere" medical health is at stake. As Richard Louv has pointed out for decades, at stake is the healthy and full cognitive development of our human children. This is the message of Louv's groundbreaking 2005 *Last Child in the Woods*, which carries the subtitle *Saving Our Children from Nature-Deficit Disorder*.[11]

So let me give thanks, count blessings, point out good news: the Forum on Religion and Ecology, *Journey of the Universe*, Thomas Berry, the leadership of Mary Evelyn Tucker and John Grim, and so many of the educators, scholars, activists, and religious leaders involved in the worldwide religion and ecology movement have helped our human communities arrive at the "rediscover" stage needed yet again because our species suffers *so repeatedly* from the affliction of ignoring and forgetting the key insight that we are but one life form among an astonishingly rich and decidedly more-than-human community. All of these efforts help us once again to *real-ize*, that is, to make real, inclusivist, community-building efforts that bring us home to our larger community. Such efforts nurture eudaimonic flourishing in a communion of subjects and permit us a clear awareness that humans are animals, will always be animals, and become remarkable *only* when we acknowledge such realities fully.

So I invite you to go on the journey—really, a rejourney—that helps us repudiate the now long-standing tendencies that have prompted humans to pretend that we are, in any sense, separate and apart from our Earth community and its nonhuman citizens. We are exceptional animals, surely, but we are at our most exceptional—that is, in Frankl's terms we self-actualize—*only* when we transcend our own species by humbly acknowledging our larger community. The irony, of course, is that we are exceptional only when we, in Leopold's framing, recognize our *"plain membership" in* "the land-community." In these ways, we challenge the dysfunctional exceptionalist tradition that has deceived us into claiming that we are separate from and superior to our natural home.

---

10. Ratey and Manning, *Go Wild*.

11. Richard Louv, *Last Child in the Woods: Saving Our Children from Nature-Deficit Disorder* (Chapel Hill, NC: Algonquin, 2005).

Law—Dreaded or Enchanted Future?

As hinted already, today's legal systems, their legal establishments, and their educators, despite some noteworthy trends, have yet to scale the arduous paths that lead to fundamental change in the dysfunctions that ensue when human exceptionalism and its underlying speciesism prevail.[12] Today's legal systems are bastions of the modern world's forgetfulness and denial on the question of our larger community, and thereby they prompt many to commit the fallacy of misplaced community.[13] As much or more than other key modern institutions, the legal system holds in place the modern world's dismissal of other-than-human animals and the more-than-human world, thereby ensuring our species' related slide away from the role of plain citizen and responsible community member in many of the nested communities of which we are an integral part. In this, contemporary law reminds one of Virgil's sobering admonition:

Easy is the descent to hell;
Day and night the doors stand open;
but to retrace one's steps,
to come out again into the upper air
this is the task and burden.[14]

The basic conceptuality that today dominates the legal systems and education of virtually all modern industrialized states is the dualism of "legal persons versus legal things." This is the stark and inadequate framing projected onto reality by the law we have inherited. Like so many forms of education that "equip people merely to be more effective vandals of the earth,"[15] law does more than harm us. It is complicit in our failure to self-transcend and thus our failure to self-actualize. The harms are multifaceted, but here I raise five particular themes about features of these harms.

---

12. Human exceptionalism is examined throughout Waldau, *Animal Studies*, while "speciesism" as a term and concept is analyzed in Waldau, *The Specter of Speciesism: Buddhist and Christian Views of Animals* (New York: Oxford University Press, 2001).

13. See various arguments on this theme in Waldau, *Animal Studies*, 16, 18, 222, 249, and 302.

14. *Aeneid* 6:126–129; the translation is the author's from the Latin version, which is available at Project Gutenberg (http://www.gutenberg.org): [126] . . . *facilis descensus Averno*; [127] *noctes atque dies patet atri ianua Ditis*; [128] *sed revocare gradum superasque evadere ad auras,* [129] *hoc opus, hic labor est.*

15. David Orr, *Earth in Mind: On Education, Environment, and the Human Prospect* (Washington, DC: Island Press, 1994), 5.

Theme 1: Any tendency to separate humans from other animals will inevitably harm humans, whether this is done via moral discourse, spiritual discourse, metaphysical analyses, or legal systems. This is the thrust of the challenge to "human exceptionalism." I hasten to add one possible exception—namely, a certain manner of exceptionalizing humans that starts and ends with humility. It is not at all hard to imagine that, in the manner of Socrates being held "the wisest man in Athens" because he alone was willing to admit his own ignorance, we humans might fairly deem ourselves "the most special of animals" in but one particular sense. This claim is true only when we are fundamentally humble about being one animal citizen, albeit a powerful and insightful one, in a more-than-human world. I personally welcome ways of speaking and acting that *insist* we be integrated as delightful and very interesting members of "the larger community that constitutes our greater self." I think such a way of speaking honors humans, and I am impressed by those who achieve such humility as perhaps the most special animals I know. A corollary of this is that *separation cannot be functional*. In an ecological universe, there can be "delightful and very interesting members of our larger community," but there is no separation and no functional version of dismissal of any and all nonhuman living beings.

Theme 2: Our heritage is impoverished because exceptionalism is now the foundation of many modern legal systems. Law's dualism of "legal persons (that is, humans) and *mere* legal things (everything else)" remains all too congenial to a persisting Cartesian dualism, escapist forms of spirituality, and the industrial-scale harms accepted by both religious and secular societies.

Theme 3: There is some good news. Legal education has for a decade and a half been newly hospitable to a variety of voices challenging the human-animal dualism as wrongheaded, arrogant, and ultimately a misconceived division. The best example is the emergence of courses in animal law, which now are offered by three-quarters of American law schools.[16]

There is another front outside of legal education that is providing good news: contemporary scholars in many different fields today are hard at work on identifying the constant rediscovery of humans' connections to and needs for community with other-than-human animals.[17]

Theme 4: Avow our animality, and be plain citizens. Berry's dedication puts human children in the extraordinary, nonhuman company of "*all* the [other]

---

16. The emergence of this phenomenon since 2000 is described in Waldau, *Animal Studies*, 114–15.

17. Waldau, *Animal Studies*, for example, examines the fields of education, history, law, politics, public policy, philosophy, religion various creative arts, comparative studies, critical and cultural studies, and social and natural sciences.

children" *who* "swim beneath the waves of the sea . . . live in the soils of the earth . . . and . . . fly with the winds," and thereby claims our rightful place in the more-than-human community to which we plainly belong.

As animals, our task is twofold—surely to be fully ourselves by realizing all our potential and yet also to be plain citizens in our more-than-human community. Both tasks can be accomplished, as I suggest above, only by a Socrates-like humility. As noted in Theme 5 below, whether our species and modern societies develop this character is a matter of choice—and our choices will project onto future humans and nonhumans alike an imagined future that can be better or worse than what our societies are doing today.

Before exploring that choice, note some *even more good news* here: thinking about humans as members of a more-than-human community is now a possibility. We can be more imaginative because, through decades of increased attentiveness, we now know far more about many nonhuman animals than we did only half a century ago. Complementing that increase in awareness, we also are far more open to the wisdom exhibited by many of the human cultures that imperialism, colonialism, and consumerism have harmed greatly and, sadly, often destroyed entirely.

But consider one piece of wisdom that helps us see features and limits of even our most remarkable philosophical thinking. Ludwig Wittgenstein famously said, "If a lion could talk, we could not understand him."[18] But consider a remarkably different attitude quoted by Ratey and Manning when focusing exclusively on *human* health: "Well-being comes from learning to talk to the lions."[19] Rather than explain what accounts for this difference, I urge you to explore the original context of each of these statements. The former is informed by a leading, altogether insightful philosopher's penchant for theorizing about language, which certainly represents well the intellectual and academic communities' fascination with conceptual constructs and language-based insights. The other, "talk to lions" point of view, which is found in the concluding line in the penultimate chapter of Ratey and Manning's human-focused analysis, is from a humble indigenous people.[20]

Suffice it to say that, *if we pay attention to ourselves*, to our animal bodies, to other animals' realities, and to why talking to lions and other plain and yet not-so-plain citizens of our shared world is important to us, we may see further than did the impressive Wittgenstein, so analytically gifted and willing to "question everything." We may indeed again recognize why *our*

---

18. Ludwig Wittgenstein, *Philosophical Investigations* II (New York: MacMillan, 1953), 223.
19. Ratey and Manning, *Go Wild*, 239.
20. The !Kung strategy of talking to lions is described by Ratey and Manning, *Go Wild*, 225.

own well-being is richly connected to possibilities like "learning to talk to the lions."

Ratey and Manning explain such talking as one way of "re-wilding,"[21] namely, re-wildling ourselves by getting us in touch with our evolutionary heritage. We are, simply said, a remarkably complex and interesting *animal* who has, for some centuries now, been ignoring our animality because we have misconstrued our differences from other animals as superiority rather than what it in fact is. We are one of the very developed forms of animality, and in this we are an uncontestable testament to the fact that animality can be altogether fascinating. *There are, if we but look, other fascinating forms of animality*; indeed, there are many. For these, "We also need our most careful forms of thinking and the best of our soaring imagination because at one and the same time, we are in some respects like all other animals, like only some other animals, like no other animal."[22]

Theme 5: Recover and re-project our law, thereby projecting an imagined future upon reality that is ethically attuned, scientifically literate, spiritually alive, and ecologically integrated. If we do this, perhaps there will be, to echo the title of this paper, hope for law and other animals in a more-than-human world.

---

21. Ratey and Manning, *Go Wild*, 10. See also ibid., 187, 239, and 249–50. Different meanings of the term "re-wildling" are presented in G. Monbiot, *Feral: Searching for Enchantment on the Frontiers of Rewilding* (London: Allen Lane, 2013), 9–10.

22. Waldau, *Animal Studies*, 1.

PART IV

*Evolving Christianity within
an Emergent Universe*

26

Roman Catholicism and Journey of the Universe

DENNIS O'HARA

It might be argued that Roman Catholics were early contributors to and proponents of studies on evolution. The work of Roman Catholic Jean-Baptiste
Lamarck (1744–1829) on complexifying and adaptive forces provided one
of the first truly cohesive articulations of evolution. The study by Augustin-
ian monk Gregor Johann Mendel (1822–1884) of dominant and recessive
traits in breeding fathered the field of genetics. With the publication of *Le
Phénomène Humain* (written from 1938 to 1940 but published posthumous-
ly in 1955), Jesuit paleontologist Pierre Teilhard de Chardin (1881–1955)
traced the evolution of matter through increasing complexification from
cells to complex species to humans and prospectively to union with God.

However, such a positive assessment of Roman Catholicism's embrace of
evolution would oversimplify the record. For instance, *La Civiltà Cattolica*,
the Jesuit journal that today provides us with enticing interviews with Pope
Francis and has long had a reputation for providing an unofficial report of the
views of Vatican authorities, aggressively attacked evolutionary theories in
the decades following the First Vatican Council (1869–1870). Henri Berg-
son's *Creative Evolution*, published in 1907, would be placed on the infamous
Index Librorum Prohibitorum by the Vatican and would remain there until
that list was abolished in 1966 (undoubtedly to the disappointment of pub-
lishers everywhere). During his life, the core of Teilhard's work could not be
published since church authorities, concerned about the unorthodoxy of his
notions, refused to give the required imprimatur; he was also removed from
teaching positions, most notably at the Institut Catholique de Paris.[1]

---

1. Teilhard was removed from his teaching post at the Institut Catholique de Paris by the
magisterium in the 1920s due to concerns about the orthodoxy of his views, and his books could
not be published during his lifetime because his works could not receive the required *nihil obstat*
and imprimatur from church authorities. Yet his work was eventually praised by Joseph Ratzinger

At best, one can say that the Roman Catholic embrace of evolution over the past centuries has been rather mixed. Today, for those Roman Catholics who outrightly eschew the notion of evolution, or for those who rather schizophrenically hold to a more literal reading of the Genesis story of creation while concurrently acknowledging the scientific explanation for the origin of the universe, *Journey of the Universe* might continue to prod a gradual reframing of their understanding of creation and humanity's place in it. However, for Catholics who have accepted the new cosmology as recounted in *Journey of the Universe*, the book, film, and teaching aids provide further insight and language for their understanding to this wonderfully creative and mysterious story.[2] Like Teilhard, they celebrate the sacredness of matter and echo the Jesuit practice of finding God in all things.[3] For them, Earth becomes a sacramental commons.[4] Like Passionist priest and cultural historian Thomas Berry (1914–2009), they recognize the need for a new cosmological story, a new epic of evolution that can deeply engage the minds, hearts, and imaginations of people in order to foster ecological and social change so that we reinvent the human and begin to live in ways that are mutually enhancing for us and the rest of creation.[5] For these Catholics, *Journey of the Universe* is fuel for those transformations; it aids their desire to read and reflect upon both books of revelation—namely, scripture and creation.

In this chapter I briefly trace recent papal comments on evolution. I argue that while papal proclamations maintain that no inherent conflict exists between a contemporary scientific understanding of evolution and church doctrine, those proclamations contrast with papal positions that seem to ignore the shift from a pre-Copernican universe to the epic of evolution. That is, while recent popes might declare that they (and the church) no longer hold that the Earth (and, by extension, humanity) is the center of the universe, they speak and act as if they did. More specifically I focus on the notion of human ecology, since this has become a lens through which the

in *Introduction to Christianity* (London: Burns and Oates, 1968) and in Joseph Ratzinger, *Spirit of the Liturgy* (San Francisco: Ignatius Press, 2000), 29, and subsequently in 2009 when Ratzinger had become Pope Benedict XVI (John Allen Jr., "Pope Cites Teilhardian Vision of the Cosmos as a 'Living Host,'" *National Catholic Reporter*, July 28, 2009). One might conclude that the magisterium's assessment of Teilhard's work had undergone some degree of reformation.

    2. In this chapter, the use of the word "Catholic" refers to the Roman Catholic Church rather than the Anglican Catholic Church.

    3. Pierre Teilhard de Chardin, *Hymn of the Universe* (New York: Harper and Row, 1965).

    4. John Hart, *Sacramental Commons: Christian Ecological Ethics* (Lanham, MD: Rowman and Littlefield, 2006).

    5. Thomas Berry, *The Dream of the Earth* (San Francisco: Sierra Club Books, 1988), 29–30; Brian Swimme and Thomas Berry, *The Universe Story* (San Francisco: Harper, 1992), 250–51.

Vatican has, in recent papacies, viewed evolution and the ecological crisis, and since Pope Francis, uses this same expression in his encyclical, *Laudato Si'*. I believe that the magisterial application of the notion of human ecology is problematic, and I further believe that a better appreciation of the epic of evolution can contribute to its correction.

Although this chapter pays particular attention to magisterial positions on evolution and ecology, most of the discussion on evolution, ecology, and theology by Roman Catholics occurs in academic, not magisterial circles.[6] This is particularly true for most of the innovative work in ecological theology that considers new scientific research and retrieves insights from earlier Christian sources, using either or both of these dialogue partners to fashion new theological responses to contemporary ecological challenges. Nevertheless, the magisterium is a central teaching authority in the Roman Catholic Church; therefore, academics often refer to its positions, which can be a natural starting point for theological discussions. Thus, in this chapter, the discussion starts with magisterial positions and then considers related positions held by Roman Catholic theologians.

Roman Catholic Magisterium—Evolution and Ecology

With his encyclical titled *Humani Generis*, Pope Pius XII declared in 1950 that the question of evolution was open for Roman Catholics and they were therefore free to form their own opinions on the matter. While "the Teaching Authority of the Church [did] not forbid . . . research and discussions . . . with regard to the doctrine of evolution," it advised that opinions "favourable and those unfavourable to evolution [must] be weighed and judged with necessary seriousness." However, all Catholics, including those who favored evolution, must hold that "souls are immediately created by God"; that all are descended from Adam, from whom all are marked by original sin; and that all must ultimately "submit to the judgment of the Church."[7]

In an address to the Pontifical Academy of Sciences in October 1996, Pope John Paul II recalled, "In his Encyclical *Humani Generis* (1950), my predecessor Pius XII had already stated that there was no opposition between

---

6. The magisterium is the teaching authority of the Roman Catholic Church. Not every statement made by a member of the magisterium is a statement of doctrine or dogma and, as such, does not require *obsequium religiosum* or religious assent of the faithful. Sometimes the magisterium is commenting on "open questions" that permit further exploration, such as questions pertaining to evolution and ecology.

7. Pope Pius XII, *Humani Generis*, August 12, 1950, §36–37, http://www.vatican.va/holy_father/pius_xii/encyclicals/documents/hf_p-xii_enc_12081950_humani-generis_en.html.

evolution and the doctrine of the faith about man [sic] and his vocation, on condition that one did not lose sight of several indisputable points."[8] John Paul II was continuing speculation that he had previously explored with the director of the Vatican Observatory, Fr. George Coyne.

If the cosmologies of the ancient Near Eastern world could be purified and assimilated into the first chapters of Genesis, might not contemporary cosmology have something to offer to our reflections upon creation? Does an evolutionary perspective bring any light to bear upon theological anthropology . . . and even upon the development of doctrine itself? What, if any, are the eschatological implications of contemporary cosmology, especially in light of the vast future of our universe? Can theological method fruitfully appropriate insights from scientific methodology and the philosophy of science?[9]

Even though Pope John Paul II observed that cosmology could be "assimilated into" scripture rather than that scripture might have been articulated using a cosmology contemporary to the time of its writing, he nevertheless speculated that the new evolutionary perspective that has emerged from modern science might inform our understanding of theological anthropology and have implications for the reimagining of doctrine. Unfortunately, he did not move significantly beyond this speculation to treat the Universe Story as a new context for reflecting on and rearticulating theological matters. That is, while his appreciation of science, the created order, and the seriousness of the ecological crises would prompt him to call for an ecological conversion[10] to

---

8. Pope John Paul II, "Magisterium Is Concerned with Question of Evolution for It Involves Conception of Man," Address to the Pontifical Academy of Sciences, October 22, 1996, §3, http://inters.org/John-Paul-II-Academy-Sciences-October-1996. These "points" included his position that there are several versions of the theory of evolution and not a singular certain theory, that the church's teaching on *imago Dei* could not be displaced by conclusions derived from evolutionary studies, and that while the human body might have evolved from matter, the human soul is created by God.

9. Pope John Paul II, "Letter of His Holiness John Paul II to Reverend George V. Coyne, S.J., Director of the Vatican Observatory," June 1, 1988, http://w2.vatican.va/content/john-paul-ii/en/letters/1988/documents/hf_jp-ii_let_19880601_padre-coyne.html.

10. Pope John Paul II, "General Audience: God Made Man the Steward of Creation," January 17, 2001, §4, http://w2.vatican.va/content/john-paul-ii/en/audiences/2001/documents/hf_jp-ii_aud_20010117.html.

address ecological sin,[11] he would not concurrently call for theological conversion of a comparable nature.[12]

Similarly, Pope Benedict would note "that there was no opposition between faith's understanding of creation and the evidence of the empirical sciences" on evolution,[13] yet he would admonish that "obedience to the voice of the earth is more important for our future happiness than the voices of the moment, the desires of the moment [since] . . . existence itself, our earth, speaks to us, and we have to learn to listen."[14] His theologizing on questions of evolution and ecology merely echoed earlier teachings and did not seem to be advanced by new scientific insights. This theological perdurability seemed to align with his writing on evolution, which emphasized order rather than novelty, cosmos rather than cosmogenesis.[15]

> To "evolve" literally means "to unroll a scroll," that is, to read a book.
> . . . It is a book whose history, whose evolution, whose "writing" and meaning, we "read" according to the different approaches of the sciences, while all the time presupposing the foundational presence of the author who has wished to reveal himself therein. . . . The world, far from originating out of chaos, resembles an ordered book; it is a cosmos. . . . The human mind therefore can [study this] . . . "cosmology" discerning the visible inner logic of the cosmos. . . . Experimental and philosophical inquiry gradually discovers these orders. . . . And thanks to the natural sciences we have greatly increased our understanding of the uniqueness of humanity's place in the cosmos.[16]

Although both of these popes wrote more often about environmental issues than any prior popes (undoubtedly because of the more pressing nature of environmental problems in their times) and admonished humans to be less destructive of the planet, and although an evolutionary perspective had be-

11. Pope John Paul II, "Homily, Zamosc, Poland," June 12, 1999, §3, http://w2.vatican.va/content/john-paul-ii/en/homilies/1999/documents/hf_jp-ii_hom_19990612_zamosc.html.

12. Donal Dorr, "'The Fragile World': Church Teaching on Ecology before and by Pope Francis," *Thinking Faith* (February 26, 2014), 3.

13. Pope Benedict XVI, "Address of His Holiness Benedict XVI to the Members of the Pontifical Academy of Sciences on the Occasion of their Plenary Assembly," October 31, 2008, http://press.vatican.va/content/salastampa/it/bollettino/pubblico/2008/10/31/0685/01691.html.

14. Pope Benedict XVI, "Meeting of the Holy Father Benedict XVI with the Clergy of the Dioceses of Belluno-Feltre and Treviso," July 24, 2007, http://w2.vatican.va/content/benedict-xvi/en/speeches/2007/july/documents/hf_ben-xvi_spe_20070724_clero-cadore.html.

15. Cf. "Finally, in the twentieth century, the universe revealed itself as an emergent evolutionary process, not as cosmos, but as cosmogenesis" (Berry, *Dream of the Earth*, 28).

16. Benedict XVI, "Address to the Pontifical Academy of Sciences," October 31, 2008.

come increasingly common in public discourse during their pontificates, neither pope significantly advanced theological teaching on evolution and care for creation. The topics of ecology and evolution were treated separately as content for theological commentary. However, many Christian,[17] and more specifically, many Roman Catholic[18] ecotheologians understood that the discovery of an evolutionary cosmology provided a new context (not merely additional content) for doing theology, and as such, required a fresh examination of theological notions that had been framed within a prior, now outdated, cosmological worldview. Accordingly, they would explore how evolution could "bring . . . light to bear upon theological anthropology . . . and even upon the development of doctrine itself." Indeed, they would argue that an evolutionary understanding of creation and humanity not only revises and enhances our understanding of theological anthropology and doctrine, it also informs our understanding of the causes and responses to the ecological crisis. Thus, while Popes John Paul II and Benedict XVI are recognized for contributing to a growing Roman Catholic interest in ecological crises and for their meek acceptance of evolution, they did not add to the scholarship that has linked these two disciplines as Roman Catholic ecotheologians have. Even while these ecotheologians reimagined the role of humanity within the Universe Story—that is, the role of humans as one of many players contributing to the unfolding of the Universe Story—Popes John Paul II and Benedict XVI reasserted humanity's primacy in that story and used the notion of human ecology to do so.

Magisterium and Human Ecology

When Pope John Paul II wrote about human ecology,[19] his anthropocentric understanding of the relationship between humans and the rest of creation

---

17. See, for example, Gordon Kaufman, "Re-conceiving God and Humanity in Light of Today's Evolutionary- Ecological Consciousness." *Zygon* 36, no. 2 (June 2001): 335–48; Ernst M. Conradie, *An Ecological Christian Anthropology: At Home on Earth?* (Burlington, VT: Ashgate, 2005); Sallie McFague, *A New Climate for Theology: God, the World, and Global Warming* (Minneapolis: Fortress Press, 2008).

18. See, for example, Denis Edwards, *Ecology at the Heart of Faith: The Change of Heart That Leads to a New Way of Living on Earth* (Maryknoll, NY: Orbis Books, 2008); Ilia Delio, *The Unbearable Wholeness of Being* (Maryknoll, NY: Orbis Books, 2013); John Hart, *The Sacramental Commons* (Lanham, MD: Rowman and Littlefield, 2006).

19. The notion of human ecology can be traced back to the 1920s. Amos H. Hawley notes, "Emerging abruptly in the early 1920s, human ecology quickly became, as an otherwise unkind commentator puts it, 'one of the most definite and influential schools in American sociology.'" Amos H. Hawley, "Ecology and Human Ecology," *Social Forces* 22, no. 4 (May 1944): 398, quoting

judged the value of the latter to be determined by the former. Humanity was to use its intelligence and freedom to dominate the planet and thereby fashion a "fitting home."[20] While humanity, he noted, should reject "the irrational destruction of the natural environment, we must also mention the more serious destruction of the *human environment,* something which is by no means receiving the attention it deserves. . . . Too little effort is made to *safeguard the moral conditions for an authentic 'human ecology.'*"[21] Pope Benedict affirmed this assessment in both his encyclical *Caritas in Veritate*[22] and in his World Day for Peace message for 2010. In the latter document, he warned that "a supposedly egalitarian vision of the 'dignity' of all living creatures [abolishes] the distinctiveness and superior role of human beings [opening] the way to a new pantheism tinged with neo-paganism."[23] For Pope Benedict, issues affecting the natural ecology could not be properly addressed until issues affecting the human ecology were first corrected. He argued that "*when 'human ecology' is respected within society, environmental ecology also benefits. . . . In order to protect nature . . . the decisive issue is the overall moral tenor of society.*"[24]

Those who criticize the magisterial understanding of human ecology that fashions a hierarchical placement of the human over the rest of creation are

---

M. A. Alihan, *Social Ecology: A Critical Analysis* (New York: Columbia University Press, 1938), xi. Robert Ezra Park defined human ecology as "an attempt to investigate (1) the processes by which the biotic balance and the social equilibrium are maintained once they are achieved and (2) the processes by which, when the biotic balance and the social equilibrium are disturbed, the transition is made from one relatively stable order to another." See Robert Ezra Park, "Human Ecology," *American Journal of Sociology* 42, no. 1 (July 1936): 15.

20. "The original source of all that is good is the very act of God, who created both the earth and man, and who gave the earth to man so that he might have dominion over it by his work and enjoy its fruits. . . . The earth, by reason of its fruitfulness and its capacity to satisfy human needs, is God's first gift for the sustenance of human life. . . . It is through work that man, using his intelligence and exercising his freedom, succeeds in dominating the earth and making it a fitting home" John Paul II, *Centesimus Annus,* May 1, 1991, §31, http://www.vatican.va/holy_father/ john_paul_ii/encyclicals/documents/hf_jp-ii_enc_01051991_centesimus-annus_en.html.

21. Ibid., §38; cf. §39 (emphasis in original). Cf. John Paul II, "God Made Man the Steward of Creation," §4.

22. "But it should also be stressed that it is contrary to authentic development to view nature as something more important than the human person. This position leads to attitudes of neo-paganism or a new pantheism." Pope Benedict XVI, *Caritas in Veritate,* June 29, 2009, §48, http:// www.vatican.va/holy_father/benedict_xvi/encyclicals/documents/hf_ben-xvi_enc_20090629_ caritas-in-veritate_en.html.

23. Pope Benedict XVI, "If You Want to Cultivate Peace, Protect Creation," World Day of Peace Message, January 1, 2010, §13, http://www.vatican.va/holy_father/benedict_xvi/messages/ peace/documents/hf_ben-xvi_mes_20091208_xliii-world-day-peace_en.html.

24. Benedict XVI, *Caritas in Veritate,* §51 (emphasis in original).

concerned that such an approach is too anthropocentric and favors an out-dated notion of stewardship, among other concerns. Thus, this problematic understanding of human ecology can unintentionally reinforce perspectives that have provided the context for humanity's devastation of the planet. A more evolutionary and ecocentric understanding of humanity, shaped by reflections on the implications of adopting the epic of evolution as our new cosmology, can provide a context for formulating mutually sustainable responses to the ecological crisis.

Responding to Anthropocentrism

Rather than declaring humanity to be the "master" of creation who must "subdue the earth,"[25] a more contemporary perspective acknowledges that humanity is the product of 13.8 billion years of cosmogenesis, emerging from and remaining dependent upon Earth's evolutionary dynamics. If the history of the universe were recorded in twenty-eight volumes of five hundred pages each, with each page representing a million years of history, Earth's formation would be celebrated in volume twenty-five and humanity in its most primitive form would appear on the latter half of page 500 in volume twenty-eight; human civilization would be recorded within the last word on that page. As one of the more recent additions to Earth's epic of evolution, humanity could not have emerged within that story until earlier players had created the constituent elements and the conditions that would both permit humanity's emergence and sustain its existence. As late arrivals in the story, we are dependent on the contributions of the earlier players like plankton, plants, and insects to oxygenate the planet; capture and store the energy of the sun so that others could consume it; and create dynamically flourishing ecosystems within which we might thrive. While our extinction would not significantly harm the existence of plankton, plants, or insects, their departure would mark our end. As Thomas Berry rightly notes, "The human community and the natural world will go into the future as a single sacred community or we will both perish. . . . We have been trying to go into the future as a human community in an exploitative relationship with the natural community without any sense of being integral with this natural world as a sacred community."[26] Looking

---

25. Pope John Paul II, *Redemptor Hominis*, March 4, 1979, §15, http://w2.vatican.va/content/john-paul-ii/en/encyclicals/documents/hf_jp-ii_enc_04031979_redemptor-hominis.html.

26. Thomas Berry and Thomas Clarke, *Befriending the Earth: A Theology of Reconciliation between Humans and the Earth*, ed. Stephen Dunn and Anne Lonergan (Mystic, CT: Twenty-Third Publications, 1991), 43.

back on the lengthy expanse of the epic of evolution, we can now realize that rather than being the masters of creation who are perched on the peak of a hierarchical pyramid, we are actually among the most vulnerable players in that story. With the other contributors to the story line that has created Earth as we know it, we are part of an interdependent, interrelated web of life. With all of the other players in the Universe Story, we form a communion of subjects who populate that epic. Earth is not a stage festooned with props set for a one-character play of the human story. Instead, "The universe is a communion of subjects rather than a collection of objects."[27]

A certain degree of self-centeredness is both normal and necessary for humans, as it is for all species. Every species is compelled to seek its own self-preservation and reproduction, to live its life according to its nature, and to claim the resources that it needs for doing so. When we considered Earth to be the center of the universe and ourselves to be the most important creature on Earth, it was not surprising that we concluded that we were of primary importance and everything else held the importance that we deemed fit to grant it. However, over recent centuries, science has taught us that neither Earth, nor our sun, nor our solar system, nor our galaxy is the sole center of the universe. "What we have come to realize is that there is not one center, but millions. Each supercluster of galaxies is at the very center of the expansion of the universe. We live in a multicentered universe and are only now awakening to this new discovery."[28] This new knowledge is not only decentering, it disorientates us from our previous understanding of the order of the universe. With the new telling of the Universe Story and our emergence within that story, with the revelation of the dependence of humanity on the creative dynamics and players that preceded its emergence in that story, and with our growing appreciation for the profound interconnectivity of all of creation, we discover that Earth is primary and the human is derivative. Anthropocentrism and especially androcentrism that favor the human at the expense of the rest of creation not only contradict our understanding of the new cosmology, they contribute to the conditions that have given rise to our current ecological crises.[29] Correctives that favor a more ecocentric outlook do not threaten the unique role of humanity in the Universe Story. Our dignity is preserved since all players in that story have intrinsic dignity merited

---

27. Swimme and Berry, *Universe Story*, 243.

28. Brian Thomas Swimme and Mary Evelyn Tucker, *Journey of the Universe* (New Haven, CT: Yale University Press, 2011), 21.

29. Berry and Clarke, *Befriending the Earth*, 96–103; Denis Edwards, "Human Beings within the Community of Life," in *Ecology at the Heart of Faith* (Maryknoll, NY: Orbis Books, 2006), 7–26.

by their various unique contributions to the evolution of the universe, and our identity has become more authentic rather than inflated since we better appreciate our important supporting role within cosmogenesis.

Responding to Stewardship

Similarly, a retrospective view of the Universe Story also reminds us that Earth is a self-organizing, self-regulating, self-creating planet that has repeatedly adapted to changing circumstances, overcome numerous obstacles, and flourished for billions of years prior to the arrival of humanity, growing in ever-increasing complexity, diversity, and profound interdependence.[30] We are only now beginning to grasp the details of its history, the complexity of its relationships, and the genius of its problem-solving. We stand in wonder and humility before such magnificence.[31] This is not a planet that has either relied on or required human stewardship in order to function. This is a planet that is too complex in its diverse functions for our very limited managerial skills. Magisterial claims that we are stewards of creation perpetuate an anthropocentric worldview that has permitted humanity to devastate the planet.[32] Earth does not require our stewardship; it requires our noninterference with its self-organizing abilities. Magisterial teachings would be more useful if they focused on integrating human culture within Earth's networks of relationships in ways that are mutually enhancing for humanity and the rest of creation rather than situating humanity as a steward-manager over all of the planet. A perspective that adopts a more ecocentric approach informed by Franciscan spirituality,[33] and develops an appreciation of cosmotheandric relationships,[34] will better describe humanity's role in the Universe Story and initiate a more authentic reading of creation as a book of revelation. The epic of evolution provides a context for writing not only a new ecological anthropology but also a new ecotheological anthropology.

---

30. Swimme and Tucker, *Journey of the Universe*, 56, 106.

31. Ibid., 113–14.

32. Cf. Pope John Paul II, "General Audience: God Made Man the Steward of Creation," and Benedict XVI, "If You Want to Cultivate Peace." For a critique of the magisterial position, see Christopher Hrynkow and Dennis O'Hara, "The Vatican and Ecospirituality: Tensions, Promises and Possibilities for Fostering an Emerging Green Catholic Spirituality," *Ecozon@* 2, no. 2 (2011), http://www.ecozona.eu/index.php/journal/article/view/241.

33. Ilia Delio, "Is Spirituality the Future of Theology? Insights from Bonaventure," *Spiritus* 8 (2008): 148–55.

34. Rohan Curnow, "Theos, Cosmos, and Anthropos: Trinity, Incarnation, and Creation in the Framework of Raimon Panikkar's Cosmotheandric Vision," *Australian eJournal of Theology* 6 (February 2006): 1–17.

Pope Francis—Evolution and Human Ecology

Rather than giving a tepid acceptance of evolution by declaring that there is no contradiction between some theories of evolution and Roman Catholic understandings of the origins of the universe, Pope Francis has taken a more affirmative approach.

> The Big Bang theory, which is proposed today as the origin of the world, does not contradict the intervention of a divine creator but depends on it. Evolution in nature does not conflict with the notion of Creation, because evolution presupposes the creation of beings who evolve. . . . [God] created beings and he let them develop according to the internal laws with which he endowed each one, that they might develop, and reach their fullness. He gave autonomy to the beings of the universe at the same time in which he assured them of his continual presence, giving life to every reality.[35]

Francis's position echoes Thomas Berry's assertion that God creates a universe that creates itself via "a capacity of self-articulation inherent in the universe."[36] Accordingly, humans evolve by means of an integral relationship with the evolutionary processes of cosmogenesis, and must live in accordance with the planet's terms. How well we adhere to this limitation, Berry cautions, can have serious implications for our future.[37] Similarly, Francis admonishes that not only are we capable of destroying creation, we are actually doing so.[38] Thus, he warns, "If we destroy Creation, Creation will destroy us! . . . God always forgives. . . . We men, women, we forgive sometimes. But . . . Creation never forgives!"[39]

Therefore, human "interventions on nature" need to be beneficial to all of creation, "with respect for the beauty, finality and usefulness of every living being and its place in the ecosystem."[40] Humanity, argues Francis, because

35. Pope Francis, "Address to the Pontifical Academy of Sciences," October 27, 2014, https://w2.vatican.va/content/francesco/en/speeches/2014/october/documents/papa-frances-co_20141027_plenaria-accademia-scienze.html.

36. Berry and Clarke, *Befriending the Earth*, 25.

37. Ibid., 42, 45, 46.

38. Pope Francis, "Homily, Solemnity of All Saints, 1 November 2014," http://w2.vatican.va/content/francesco/en/homilies/2014/documents/papa-francesco_20141101_omelia-ognis-santi.html.

39. Pope Francis, "Audience: If We Destroy Creation, It Will Destroy Us," May 21, 2014, http://www.news.va/en/news/pope-at-audience-if-we-destroy-creation-it-will-de.

40. Pope Francis, "Fraternity, the Foundation and Pathway to Peace," World Day of Peace message, January 1, 2014, §9, http://w2.vatican.va/content/francesco/en/messages/peace/docu-ments/papa-francesco_20131208_messaggio-xlvii-giornata-mondiale-pace-2014.html.

of its present capacity to devastate the planet, must instead choose to "safe-guard creation" and "care for it . . . for the benefit of all, always with great respect and gratitude."[41] For too long, industrialized societies have chosen to dominate, possess, manipulate, and exploit the planet. Concurrently, he notes, they have lost an "attitude of wonder, of contemplation, of listening to creation."[42]

For these reasons, Francis links discussions of human ecology with environmental ecology, concluding that climate change and ecological crises are human-made, that the same attitudes and values that foster ecological destruction are also destructive of humans. Therefore, "Human and environmental ecology go hand in hand. I would therefore like us all to make the serious commitment to respect and care for creation, to pay attention to every person, to combat the culture of waste and of throwing out so as to foster a culture of solidarity and encounter."[43]

When Francis speaks about human ecology, his approach has been somewhat more ecologically friendly than that of his predecessors. Rather than using human ecology to claim a role for the human that places it above the rest of creation, he describes a more bilateral approach that places the human "hand in hand" with the rest of creation. Alternatively, Thomas Berry includes "human ecology within nature ecology, rather than the other way around."[44] Even if Francis's understanding of human ecology is less ecocentric than Berry's, he remains sensitive to the human-Earth connection. In his apostolic exhortation *Evangelii Gaudium*, he warns that in economic systems that place a primacy on "increased profits, whatever is fragile, like the environment, is defenseless before the interests of a deified market."[45] The exploitation of the environment and the exploitation of people are linked.[46] Eschewing a rigid separation of humanity from the rest of creation (or above the rest of creation), he declares, "Thanks to our bodies, God has joined us so

---

41. Francis, "Audience: If We Destroy Creation."

42. Pope Francis, "General Audience—5 June 2013," http://w2.vatican.va/content/frances-co/en/audiences/2013/documents/papa-francesco_20130605_udienza-generale.html.

43. Ibid.

44. Berry and Clarke, *Befriending the Earth*, 97.

45. Pope Francis, *Evangelii Gaudium*, November 24, 2013, §56, http://w2.vatican.va/con-tent/francesco/en/apost_exhortations/documents/papa-francesco_esortazione-ap_20131124_evangelii-gaudium.html.

46. Francis argues that "an economic system centered on the god of money also needs to plun-der nature, plunder nature, in order to maintain the frenetic pace of consumption inherent in it" (Francis X. Rocca, "Pope Urges Activists to Struggle Against 'Structural Causes' of Poverty," *Catholic News Service* (October 28, 2014), http://www.catholicnews.com/data/stories/cns/1404449.htm.

closely to the world around us that we can feel the desertification of the soil almost as a physical ailment, and the extinction of a species as a painful disfigurement."[47] With a new solidarity with the poor of creation and with creation as the "poor," with a growing awareness and experience of our profound kinship with and dependence on the other-than-human, Francis seemingly wants us to develop an intimacy with creation to overcome traditional forms of anthropocentrism.

Francis's position on ecological issues suggests that he will continue to describe "human ecology" with different language than his predecessors. He seems to appreciate that as long as human interests are placed above ecological interests, certain economic systems will invariably be given a priority that favors some people and indeed some ecosystems at the expense of others. Respect for the flourishing and integrity of creation necessarily includes and fosters a respect for the flourishing and integrity of people. A deeper appreciation of the epic of evolution can assist in bringing "light to bear upon theological anthropology . . . and even upon the development of doctrine itself," thereby updating or reforming of parts of the Roman Catholic tradition that need to advance into the twenty-first century.

---

47. Francis, *Evangelii Gaudium*, §215.

27

Eco-Reformation, Deep Incarnation, and Lutheran Perspectives on the Universe Story

BARBARA R. ROSSING

Universe Story, Evolutionary Cosmology, and Ecology

The new "Universe Story" of evolutionary science and living cosmology is a story that Lutherans embrace. I am deeply grateful for the ways the work of the Yale Forum on Religion and Ecology has played a part in shaping Lutheran voices. At the Lutheran School of Theology at Chicago we regularly show the film *Journey of the Universe*, through the Zygon Center for Religion and Science. We teach evolution. Our students learn the story of living cosmology through a pair of science and theology courses: the Epic of Creation, with lectures by astrophysicists, geologists, evolutionary biologists, and ecologists; and the Future of Creation, an ecological course featuring climatologists and environmental scientists and activists.[1] The goal is for seminary students and future pastors to take an understanding of evolutionary cosmology into their churches, through having engaged scientists and the science, as part of their faith.

Lutherans have long been involved in interdisciplinary reflection on ecology and theology. The pioneering ecotheological work and influence of Lutheran theologian Joseph Sittler in some ways parallels that of Roman Catholic Thomas Berry. Sittler's 1961 speech to the World Council of Churches Assembly in New Delhi constituted a kind of turning point for Protestant and ecumenical theology, challenging churches to make ecological thinking central to their life.[2] Sittler called for a Christology of nature,

---

1. Lutherans are proud of the fact that the judge who ruled against public schools' teaching of intelligent design as "science" in the 2005 Dover, Pennsylvania, court case was a Lutheran, Judge John E. Jones III.

2. Joseph Sittler, "Called to Unity," *Ecumenical Review* 14, no. 2 (2010): 177–87; reprinted in *Evocations of Grace: The Writings of Joseph Sittler on Ecology Theology and Ethics*, ed. Peter Bakken and Steven Bouma-Prediger (Grand Rapids: Eerdmans, 2000), 38–50.

emphasizing the cosmic Christ of Colossians 1, the fact that God came to reconcile not only humans but *ta panta,* "all things."[3] Sittler had already begun addressing environmental concern in his 1954 essay "A Theology for Earth," a theme he expanded in the New Delhi lecture as well as two subsequent volumes of essays.[4] Sittler's love of poetry, combined with attention to the quirky details of nature, makes his writing deeply consonant with the joy and reverence of the Universe Story.

Many other Lutherans have continued the ecotheological work begun by Sittler, including historians, systematic theologians, biblical scholars, ethicists, liturgical scholars, missiologists, and others. The Evangelical Lutheran Church in America (ELCA) issued a social statement in 1993, "Caring for Creation: Vision, Hope, and Justice," that set the framework for the church's advocacy. Subsequent statements and pastoral letters have reiterated the church's concern for climate justice, a position also underscored in testimony by ELCA bishops before the US House and Senate.[5] Lutherans Restoring Creation, founded by New Testament professor David Rhoads, provides resources for creation care and advocacy in Lutheran churches.

Not all Lutherans agree about the urgency of climate action, to be sure. One of the most extreme climate change deniers in the US Congress is a Missouri Synod Lutheran—Rep. John Shimkus from Illinois—whose position is at odds even with his own very conservative Lutheran church. In the Bakken Oil Fields of western North Dakota, or the Marcellus Shale area of rural Pennsylvania, some Lutherans view the economic benefits of oil and gas fracking as outweighing the risk of climate change. Bishop Mark Narum of the Western North Dakota Synod urges Lutherans to listen more to such views.[6] Lutherans—including conservative Lutherans—care deeply about world hunger and global health, donating generously to programs such as

---

3. Colossians 1:15–20.

4. Joseph Sittler, *The Ecology of Faith* (Philadelphia: Fortress Press, 1970); and Joseph Sittler, *Essays on Nature and Grace* (Philadelphia: Fortress Press, 1972).

5. "Caring for Creation: Vision, Hope, and Justice," http://www.elca.org/Faith/Faith-and-Society/Social-Statements/Caring-for-Creation; Presiding Bishop Mark S. Hanson, "The Urgency of Climate Change Legislation: Testimony to U.S. Senate Environment and Public Works Committee, June 2007," *Currents in Theology and Mission* 37 (2010): 136–37; Bishop Callon Holloway, "Adaptation Assistance and Climate Change: Testimony to the U.S. House of Representatives, March 25, 2009," *Currents in Theology and Mission* 37 (2010): 138–40; Presiding Bishop Elizabeth Eaton, http://elca.org/News-and-Events/7752.

6. Bishop Mark Narum, "Prairie, Petroleum, Pondering: What Does this Mean?" in *Eco-Lutheranism: Lutheran Perspectives on Ecology*, ed. Karla Bohmbach and Shauna Hannan (Minneapolis: Lutheran University Press, 2013), 144–56. One of the worst floods of the Souris River in Minot occurred in 2011, destroying many homes. The Western North Dakota Synod of the ELCA has been the recipient of the largest disaster relief funding of any ELCA synod. Yet the causal con-

the ELCA Hunger Program, Bread for the World, and the ELCA Malaria Appeal. Framing climate change in terms of hunger or human health effects, rather than as an environmental issue, may have more strategic resonances with conservative Lutherans.[7]

Globally, international Lutheran voices contribute much to ecological theology. The archbishop of the largest global Lutheran Church, the Church of Sweden, Antje Jackelen, has made climate justice advocacy the signature element of her public leadership.[8] An international Lutheran World Federation (LWF) study program has pondered theological questions related to climate change.[9] The LWF has been active in climate justice advocacy work, in partnership with the World Council of Churches. The LWF plans to observe the 2017 anniversary of the Reformation with strong attention to ecological justice, under the threefold theme of "Salvation—Not for Sale," "Humans—Not for Sale," and "Creation—Not for Sale."

A New Eco-Reformation?

Lutherans love the Reformation. As we approach the five hundredth anniversary of the event in October 1517 when Martin Luther published his Ninety-Five Theses on the Wittenberg Castle Church door, one question is how we will commemorate the anniversary. Thanks to the achievements of decades of ecumenical dialogues, including the "Joint Declaration on Justification," Lutherans and Roman Catholics no longer engage in polemics, unlike earlier anniversaries of the Reformation. An amazing joint statement, "From Conflict to Communion: Lutheran Catholic Common Commemoration of the Reformation in 2017," has been issued by both the Pontifical Council for Promoting Christian Unity and the Lutheran World Federation. At the seminary where I teach, we plan to organize a joint event with the Roman Catholic seminary in our consortium, the Catholic Theological Union. We

---

nections between flooding, more extreme precipitation, climate change, and the burning of fossil fuels have not been drawn persuasively.

7. Anthony Leiserowitz, director of the Yale Project on Climate Change Communication (http://environment.yale.edu/climate-communication), suggests that communicating the human health effects will persuade more people than an environment frame.

8. See *A Bishops' Letter about the Climate* (Uppsala: Church of Sweden, 2014); available in English at http://fore.research.yale.edu/files/A_Bishops_letter_about_the_climate_2014.pdf.

9. See the essays in *God, Creation, and Climate Change: Spiritual and Ethical Perspectives*, ed. Karen Bloomquist (Geneva: Lutheran World Federation Studies, 2009), following a visit to coastal villages of India experiencing severe flooding. My own essay in that volume is "God's Lament for the Earth: Climate Change, Apocalypse, and the Urgent Kairos Moment," 129–43.

hope this will also include reflection on Pope Francis's encyclical on climate change.

What aspects of church and society still need reforming today, leading up to the quincentenary? This is a question many of us are asking, underscoring the Reforming principal of a church that is "always reforming" (*ecclesia semper reformanda*).[10] A number of theologians have signed an open letter to our presiding bishop and sixty-five bishops urging that the anniversary be an occasion for an eco-Reformation in Lutheran churches. Commemoration of the five hundredth anniversary of the Reformation comes at an urgent *kairos* moment for creation. We believe that this is a time for a new reformation that turns the church toward Earth justice and Earth healing:

> To bring ecological justice into the ongoing Reformation of the church testifies to the living nature of the Lutheran tradition and witnesses to the scope of God's redemption of the whole world. We therefore write in solidarity with Lutherans around the world and a growing number of ELCA pastors, laity, theologians, teachers, authors, and activists whose Christian faith compels us to care deeply about creation and all inhabitants of Earth.[11]

We hope the ELCA and its bishops will make the Eco-Reformation "kairos" theme central in its worship, theology, Bible study, and other commemorative activities leading up to the Reformation anniversary.

Lutheran Theology and Ecology

Joyful Panentheism and Deep Incarnation: God in All Things

Martin Luther loved creation. He believed in the presence of God in all creation. His insistence that the finite can really hold the infinite, *finitum capax infiniti*, is a principle he argued most vehemently in his writings on the Lord's Supper against the Calvinists. "Deep incarnation" is a phrase coined by Danish Lutheran Niels Gregersen to express the idea of the radical incarnation of

---

10. German Lutheran Special Envoy for the Reformation 500th Anniversary, former Lutheran bishop Margot Kassmann, uses the phrases to title her work. See Margot Kassmann, "*Ecclesia Reformata Semper Reformanda:* Challenges of the Reformation Jubilee 2017," *Currents in Theology and Mission* 40 (2013): 413.

11. "Open Letter to Bishop Elizabeth Eaton," November 7, 2014, http://www.lutheransrestoringcreation.org/open-letter-to-bishop-elizabeth-eaton.

God in all matter.[12] Incarnational theology and sacramental theology insist that God is present, as Luther says, "in every little seed, whole and entire . . . Christ is present in all creatures, and I might find [Christ] in stone, in fire, in water, or even in a rope, for [Christ] is there."[13] Larry Rasmussen and others call this Luther's "joyous panentheism." Today, in the face of creation's suffering, we can embrace Luther's joyous insistence that God is present

> in every single creature in its innermost and outermost being, on all sides, through and through, below and above, before and behind, so that nothing can be more truly present and within all creatures than God himself with his power.[14]

To be sure, Luther's radical incarnational vision is not always voiced in Lutheran pews. Though Sittler and others already pointed Lutherans toward the cosmic Christ and Earth-embracing eschatology of Colossians more than fifty years ago, an anti-Earth escapist strand that Lutheran Norm Habel labels as "heavenism" is deeply engrained in many quarters.[15] We often fall back into thinking that the goal of salvation is to take souls to heaven after death, leaving Earth behind. Older Lutheran hymnals even contained the hymn "This Earth Is Not My Home." But this is not good Lutheran theology. Our teaching about heaven and eschatology, as well as our hymnody, needs reforming in order to fully live into "joyful panentheism" that embraces Earth.[16]

---

12. Niels Henrik Gregersen, "Deep Incarnation: Why Evolutionary Continuity Matters in Christology," *Toronto Journal of Theology* 26 (2010): 173.

13. Cited by Larry Rasmussen in "Waiting for the Lutherans," *Currents in Theology and Mission* 37 (2010): 86–98 (note 8).

14. Martin Luther, "That These Words of Christ, 'This Is My Body,' etc., Still Stand Firm Against the Fanatics, 1527," in *Luther's Works,* vol. 37, ed. Helmut T. Lehmann (Philadelphia: Muhlenberg Press, 1961), 58.

15. The term "heavenist" was coined by Norman Habel: "'Heavenism' is the belief that heaven, as God's home, is also the true home of Christians. . . . Earth, by contrast, is only a temporary 'stopping place' for Christians en route to heaven." See "Ecojustice Hermeneutics: Reflections and Challenges," in *The Earth Story in the New Testament,* The Earth Bible, vol. 5, ed. Norman Habel and Vicky Balabanski (Sheffield, UK: Sheffield Academic Press, 2002), 3–4.

16. On reforming eschatology to embrace the healing of the world, see Barbara Rossing, "River of Life in God's New Jerusalem: An Eschatological Vision for Earth's Future," in *Christianity and Ecology,* ed. Dieter Hessel and Rosemary Radford Ruether (Cambridge, MA: Harvard Center for World Religions, 2000), 205–24.

Indigenous Voices

Indigenous communities' spiritual perspectives can help us recover a Lutheran emphasis on the goodness of creation. Challenging the up-down polarity of much theology, George Tinker underscores the "deeply embedded Indian cultural notion of relatedness"—a relatedness that extends to all of life. In critiquing much of Western science and technology, Tinker's perspective is resonant with the emphases of the Universe Story.[17] Similarly, Lutheran theologian Tore Johnson, a member of the Sami community living in the Arctic region of Nordic countries, underscores the communal nature of creation, in which all living beings are seen as interrelated in a circle of life. "Sami tradition reflects the idea that creation has a voice that should be listened to.[18] Johnson calls for an eco-theological starting point that begins with creation, doing "theology from the circle of life."[19]

Theology of the Cross

Lutheran theology of the cross—the insistence that God is present also and even most of all in brokenness and pain—can help us face the structural sin of ecological devastation and formulate an analysis of both sin and redemption capable of addressing this crisis. Brazilians Wanda Deifelt and Vitor Westhelle have done important work connecting the cross to ecojustice.[20] Biblical scholar Monica Melanchthon brings Dalit perspectives from India. African American Lutheran Richard Perry and Larry Rasmussen, Cynthia Moe-Lobeda, and James Martin-Schramm, have all done important work in reflections on ecojustice.

Living Cosmology's Challenges to the Economic Claims of Extractive Capitalism

Extending the Reformation ecologically can also mean challenging the very logic of neoliberal economic globalization. An international group of Lu-

---

17. George Tinker, "American Indians and Ecotheology: Alterity and Worldview," in Bohmbach and Hannan, *Eco-Lutheranism*, 69–86.

18. Tore Johnson, "Listen to the Voice of Nature: Indigenous Perspectives" in Bloomquist, *God, Creation, and Climate Change*, 101.

19. Ibid., 106.

20. See Wanda Deifelt, "From Cross to Tree of Life: Creation as God's Mask," in Bohmbach and Hannan, *Eco-Lutheranism*, 169–76.

therans and Reformed scholars have developed a series of papers under the theme of "Radicalizing Reformation." Their ninety-four theses underscore a biblical and Lutheran priority for those who have been left out and marginalized by current economic structures.[21]

Such a radical economic and social critique is consonant with the Universe Story. The "old story" over against which we need to lift up a new Universe Story today was not only bad science and bad theology in terms of "heavenism." It was also an old story of slavery and oppression—as told from the perspective of hegemonic empires. Biblical scholars underscore that the Hebrew Bible and New Testament were written over against the cosmologies of empires—whether the Babylonian Empire, the Seleucid Empire, or the Roman Empire. Empires promoted their own cosmologies and eschatologies, telling their empire story about who we are as humans and how the world works. Over against those imperial stories, biblical authors declared a bold, prophetic no. They told a new story of how the world works, a new cosmology.

Five hundred years ago, when the capitalist economic system was still nascent, Luther's bold economic critique called for reform not only of the church but also of the debt structure of society that impoverished many. Luther's "On Usury" writings and others are more radical than we have realized, as Argentinean Guillermo Hansen points out.[22] He was clear that the worst idolatry was worship of Mammon. So too today, at a time when extractive capitalism is functioning as a sort of gospel or religion, Lutheran theology must offer a critique.

We live at a time when the "good" and "goods" are commodified, but nature and the atmosphere are still treated as a sewer or as a resource with no value, as the 1999 ELCA Social Statement "Economic Life: Sufficient, Sustainable Livelihood for All," declares: "Too often the earth has been treated as a waste receptacle and a limitless storehouse of raw materials to be used up for the sake of economic growth, rather than as a finite, fragile ecological system upon which human and all other life depends."[23]

An economic critique has not always been at the heart of Lutheran social teaching, but it is grounded in Luther's own work, as well as in the Bible. Ecology and economy are closely connected, both deriving from the Greek

---

21. See http://reformation-radical.com/index.php/en/about-radicalizing-reformation.html.

22. Guillermo Hansen, "Money, Religion, and Tyranny: God and the Demonic in Luther's Antifragile Theology," in *Market and Margins: Lutheran Perspectives*, ed. Wanda Deifelt (Minneapolis: Lutheran University Press, 2014), 31–68.

23. "Economic Life: Sufficient, Sustainable Livelihood for All," http://www.elca.org/Faith/ Faith-and-Society/Social-Statements/Economic-Life.

word *oikos* (house). We live together in the "world house," as Dr. Martin Luther King Jr. called this planet.[24]

Embodied Hope: Living Cosmology in Liturgy, Sacraments, and Local Communities

The eco-Reformation message of living cosmology takes form in faith practices and communities. Hope for the future is Christianity's greatest gift—one that is urgently needed today. A local pastor tells me that climate change is a pastoral issue: young couples who come to her say they don't want to have children. When Lutheran World Federation surveyed people about how they are experiencing the effects of climate change, in the poorest countries the pastoral question was, "Why is God punishing us?" Global warming raises profound biblical and theological questions, speaking to the heart of cosmology as well as pastoral care. Despair is a great temptation.

Churches, camps, retreat centers, church yards, and church gardens all need to be local cells of hope, community, and solidarity. Worship and the arts shape Christian imagination. The way we tell our story matters. Where the body of Christ meets to share the body of Christ, we learn to see the world differently. We learn to give our testimony. We learn to tell the story of the universe as a story of living cosmology, of a tree of life with leaves "for the healing of the world," as scripture says.[25]

One way we tell the Universe Story is liturgically. Our songs and our liturgies are profoundly ecological, as Ben Stewart reminds us about the Easter Vigil, and Lisa Dahill reminds us by taking worship outdoors. The lectionary also shapes Lutheran worship and faith. Two bold Lutheran biblical scholars, David Rhoads and Norman Habel, and a theologian/pastor, Paul Santmire, propose changing the lectionary from September 1 to October 4, St. Francis Day, to a set of scripture texts for the "Season of Creation." They have chosen a three-year cycle of different scripture texts for "Season of Creation." Sundays include River Sunday, Tree Sunday, Storm Sunday, Ocean Sunday, and more.

Symbolic action matters, as those who practice civil disobedience remind us. Luther probably never said the apocryphal statement, "If I knew the world were going to end I would plant a tree." But Lutherans do plant trees! The Northern Diocese of the Evangelical Lutheran Church of Tanzania, over-

---

24. Martin Luther King Jr., "The World House," in *Where Do We Go from Here: Chaos or Community?* (New York: Harper and Row, 1967), 177.

25. Revelation 22:2.

seen by Bishop Frederick Shoo, requires tree planting as part of wedding ceremonies, confirmations, and other liturgies. Described as a kind of Lutheran Johnny Appleseed, Bishop Shoo oversees five hundred thousand members and 164 parishes, encouraging tree planting in order to address global warming. Tree planting is an important ministry in urban churches such as Bethel New Life in inner-city Chicago. Lutherans are working to embody their ecological commitments on the ground, through wind turbines at Lutheran colleges, solar panels on church roofs, and energy conservation loan funds and initiatives.

Students preparing for ordination are expected to take creation care seriously. For some years, ELCA Lutheran seminary students have been required to subscribe to "Vision and Expectations for Ordained Ministers in the Evangelical Lutheran Church in America (ELCA)." But until 2009, and the decision to ordain gay and lesbian persons in committed relationships, scrutiny had focused on sexuality. Now we are seeking to turn the attention of candidacy committees and judicatory bodies also to the section of the document on creation care, which stipulates that "this church expects that its ordained ministers will be exemplary stewards of the earth's resources, and that they will lead this church in the stewardship of God's creation." The "Visions and Expectations" statement makes clear that candidates are expected to "speak on behalf of this earth, its environment and natural resources and its inhabitants."

To take seriously the Reformation today means to be on a reforming journey—a journey that reforms the way we tell and live our story, bringing the Universe Story and the urgency of ecological science into conversation with Lutheran confessional writings and practices. Creation is the dwelling place of God, Luther taught. The universe is Christ's home. These are radical truths, proclaiming God's radical yes—a yes of love, for the universe and for all of us.

28

Journey of the Universe *and Methodism*

While such stories will no doubt be told far into the future, a new integrating story has emerged. Even though it is only a few centuries old, it
has already begun to change humanity in crucial ways. This is the story
of the universe's development through time, the narrative of the evolutionary processes of our observable universe.[1]

In many ways, Methodism emerged as the new story of the universe was
emerging. In the eighteenth century, as scientific and technological revolutions were radically challenging ancient understandings of religion and of
God, the basic tenets of Methodism were being formed in large measure by
John Wesley's understanding of what it meant to be a Christian in the midst
of a cultural reimagining of the divine. First as a sect within the Church of
England, and then as a formal denomination after Wesley's death, Methodism is wrapped up in the Universe Story.

Methodism and Science

In considering Methodism's relationship with the Universe Story, a journey
back to its relationship with science might be an appropriate first step. Specifically, in reviewing John Wesley's relationship with science, it seems that
he might have been unfairly maligned. In 1876 Leslie Stephen wrote in the
History of English Thought in the Eighteenth Century that "we already find in
Wesley the aversion to scientific reasoning which has become characteristic

---

1. Brian Thomas Swimme and Mary Evelyn Tucker, *Journey of the Universe* (New Haven, CT:
Yale University Press, 2011), 2 (Kindle version).

of orthodox theologians."[2] Others followed who blamed Wesley for leading the religious charge against scientific thought.[3]

Yet a closer look at his publications, sermons, and theology reveal a man who was terribly curious and interested in the science of his times—absorbing many of its central tenets and pushing back against it at key intervals. He wrote extensively about medicine, electricity, and "natural philosophy," having what Robert Schofield, a professor of the history of science at Harvard who did the first comprehensive study of Wesley and science, calls a much more expansive interest and knowledge in science than most of his intellectual colleagues.[4]

In 1893, during a talk at the Chit Chat Club in San Francisco, William Harrison declared, erroneously, that Wesley anticipated Darwin's evolutionary theories nearly a century before the publication of the *Origin of Species.*[5] In 1926, during the Scopes trial, his works were even used to justify teaching evolution in schools.[6]

More recently, in the face of well-publicized debates about science and religion, Methodism has recognized the importance of the role of science in understanding our universe and our relationship to it. Methodist discipline states,

> We recognize science as a legitimate interpretation of God's natural world. We affirm the validity of the claims of science in describing the natural world and in determining what is scientific. . . . We find that science's descriptions of cosmological, geological, and biological evolution are not in conflict with theology.[7]

And in 2008 the General Conference of the Methodist Church passed a resolution "opposing the introduction of any faith-based theories such as Creationism or Intelligent Design into the science curriculum of our public schools."[8]

---

2. Randy L. Maddox, "John Wesley's Precedent for Theological Engagement with the Natural Sciences," *Wesleyan Theological Journal* 44, no. 1 (Spring 2009): 24.

3. Ibid., 24–25.

4. Ibid., 28.

5. Ibid., 25–26.

6. Ibid., 27.

7. *The Book of Discipline of the United Methodist Church* (Nashville: United Methodist Publishing House, 2012), paragraph 160(F).

8. *United Methodist Church Resolution 1027* (2008).

John Wesley and Natural Philosophy

Perhaps the most revelatory of John Wesley's many works on the theology of the Universe Story of his day was his publication in 1763 of a massive multivolume collection titled *A Survey of the Wisdom of God in Creation: or, A Compendium of Natural Philosophy.* Like *Journey of the Universe*, the survey was Wesley's attempt to be in dialogue with an expansive collection of major scientific and philosophical thinkers of his day. Wesley writes in the preface,

> But though the connecting links of nature are innumerable, and her chain universal, and infinitely extended beyond our mortal grasp, it is not therefore a reason, that we should fold our hands, and hide our talent in the earth: but it is equally our duty and our privilege, to prosecute every laudable inquiry, and attain to that various knowledge which Providence has placed within our reach.[9]

Early on, however, Wesley makes clear his starting point for this inquiry; he declares that his intention is not "to entertain an idle, barren curiosity," but rather to "display the invisible things of God, his power, wisdom and goodness."[10] Furthermore, he is selective in the works included in the survey, and, for the most part, excludes those with whom he does not agree.

Consistent with his view that faith is to be acted out in the world, John Wesley is working out, in the survey, his views on the appropriate relationship between humans and the nonhuman creation. Interestingly, however, he begins not with stewardship but rather with the importance of getting to know all aspects of creation and being in relationship with it:

> By acquainting ourselves with subjects in natural philosophy, we enter into a kind of association with nature's works, and unite in the general concert of her extensive choir. By thus acquainting and familiarizing ourselves with the works of nature, we become as it were a member of her family, a participant in her felicities.[11]

Wesley's views seem consistent with *Journey of the Universe*'s suggestion that "what we long for is profound intimacy of relationship. . . . Our human des-

---

9. John Wesley, *A Survey of the Wisdom of God in the Creation: or, A Compendium of Natural Philosophy,* vol. 1, 2nd American ed. (Philadelphia: Jonathan Pounder, 1816), viii.

10. Ibid., iv.

11. Ibid., ix.

tiny is to become the heart of the universe that embraces the whole of the Earth community."[12]

One of the most striking perspectives in the survey was its clear intention to push back against the mechanistic view of nature held by many of Wesley's contemporaries. For example, he chose not to include clergyman and natural philosopher William Derham's popular message:

> We can, if need be, ransack the whole globe . . . penetrate into the bowels of the earth, descend to the bottom of the deep, travel to the farthest regions of this world, to acquire wealth, to increase our knowledge, or even only to please our eye or fancy.[13]

While Wesley was a believer in the hierarchical view of creation expressed in the Great Chain of Being, he also believed in the inherent goodness and purpose of all aspects of creation, beyond its value to humanity. Moreover, as created, nature was in ecological balance, a paradise, "an exactly connected series of beings, from the highest to the lowest; from dead earth, through fossils, vegetables, animals, to man created in the *image of God*, and designed to know, to love, and enjoy his Creator to all eternity."[14] Wesley suggests every aspect of creation has value; mountains "are not, as some have supposed, mere encumbrances of the creation: rude and useless excrescences of the globe; but answer many excellent purposes. They are contrived and ordered by the wise Creator for this grade use in particular, to dispense the most necessary provision of water to all parts of the earth."[15]

John Wesley had a particular soft spot for animals, nurtured from childhood. He believed that an animal "deserves a more attentive consideration than has been usually given it."[16] Jane Nickell suggests that Wesley believed that in the New Creation, "God might grant to brute creatures the one thing that separates them from humanity, the capacity for relationship with God."[17] In fact, early Methodists were actively working against animal cruelty, and the Methodist discipline clearly endorses the humane treatment of animals.[18]

---

12. Swimme and Tucker, *Journey of the Universe*, 114–15.

13. Maddox, "John Wesley's Precedent," 52.

14. John Wesley, cited in Theodore Runyan, *The New Creation: John Wesley's Theology Today* (Nashville: Abingdon Press, 1998), 10.

15. Ibid., 201.

16. Ibid., 17.

17. Jane Nickell, "A Wesleyan Oikos Ethic" (unpublished manuscript), 4.

18. *The Book of Discipline*, paragraph 160(C), "Animal Life."

The Role of the Human in the Universe Story

Journey of the Universe reflects that the "great opportunity" is to "tell this new universe story in a way that will serve to orient humans with respect to our pressing questions: Where did we come from? Why are we here? How should we live together? How can the Earth community flourish?"[19] In many respects, John Wesley and Methodism attempt to respond to those same questions.

Of the three types of grace Wesley articulated and that form the heart of Methodism—*prevenient* (or preexisting) *grace, justifying grace* (God's act of restoring right relationship with humanity through the love of Christ), and *sanctifying grace*—it is through sanctifying grace that humanity responds to God's generous act of creation through a renewed heart aligned with God's purposes in the world.[20] While personal spiritual experience was important in John Wesley's theology, he believed that Christianity was to be practiced in and through community. At the heart of Methodism is a call to be the *"political image"* of God on Earth, an active force for representing God's purposes on Earth.[21] As a result, Methodists have historically been actively engaged in a wide range of social justice issues around the globe, including poverty, violence, war and peace, women's rights, and so on.

Specifically related to the relationship between the human and non-human creation, Wesley betrayed his hierarchical beliefs, calling on humanity to be "God's vice-regent upon earth" with dominion "over the works of thy hands," including "all sheep and oxen, yea, and the beasts of the field; the fowl of the air, and the fish of the sea."[22] While this position might easily lead to human dominance over nature, Wesley was always *theocentric* in his orientation. In contrast to many scientists and philosophers of his day who had fallen under the spell of Francis Bacon's view of Earth primarily as a resource for human use, Wesley viewed this vice-regency in the context of his broader belief in the goodness and purposefulness of creation *outside of its utility to humanity*:

We are now God's stewards. We are indebted to him for all we have. . . . A steward is not at liberty to use what is lodged in his hands as he

---

19. Swimme and Tucker, *Journey of the Universe*, 5.
20. Runyan, *New Creation*, 82–83.
21. Ibid., 16.
22. John Wesley in ibid., 16–17.

pleases, but as his master pleases. . . . He is not the owner of any of these things but barely entrusted with them by another.[23]

Consistent with Wesleyan teaching, the United Methodist *Book of Discipline* demonstrates a similar orientation:

> All creation is the Lord's, and we are responsible for the ways in which we use and abuse it. Water, air, soil, minerals, energy resources, plants, animal life, and space are to be valued and conserved because they are God's creation and not solely because they are useful to human beings. God has granted us stewardship of creation. We should meet these stewardship duties through acts of loving care and respect.[24]

Specific sections on water, air, soil, minerals and plants, energy resources, and global climate stewardship follow:

> We acknowledge the global impact of humanity's disregard for God's creation. Rampant industrialization and the corresponding increase in the use of fossil fuels have led to a buildup of pollutants in the earth's atmosphere. These "greenhouse gas" emissions threaten to alter dramatically the earth's climate for generations to come with severe environmental, economic, and social implications. The adverse impacts of global climate change disproportionately affect individuals and nations least responsible for the emissions.[25]

Consistent with Wesley's view that we are to be the political image of God, the *Book of Discipline* goes on to support efforts by governments "to require mandatory reductions in greenhouse gas emissions" and for individuals to voluntarily reduce their own.[26]

In the early 1970s the United Methodist Board of Church and Society and the United Methodist Women began to actively engage in energy and environmental issues following the 1971 United Nations Conference in Stockholm. As early as 1976 the General Conference of the United Methodist Church passed a resolution on environmental stewardship, and in 1980 the church hired staff to promote the care of creation. The Methodists started

---

23. John Wesley in ibid., 205.
24. *Book of Discipline*, paragraph 160.
25. Ibid., paragraph 160(D).
26. Ibid.

working in 1988 to combat climate change, the same year James Hanson first testified on "global warming" in Congress.[27]

The church's Council of Bishops released a bold statement in 2009 titled "God's Renewed Creation," which challenged Methodists to take seriously threats to the Earth and our complicity in those threats. Since then a new organization, Caretakers of God's Creation, has been formed to encourage Methodist congregations around the world to safeguard the planet, while paying particular attention to the suffering of those who are least likely to have contributed to the problem. Congregations around the globe are forming green teams, making adjustments in their churches and their theologies. At the same time, Methodist seminaries, such as Wesley, Drew, Boston University, Duke, Candler, and Methodist Theological School in Ohio are leading the effort to integrate into the fabric of theological education environmental awareness, engagement, and care for creation.

Transformation

While understanding the importance of each atom, each star, each cell, *Journey of the Universe* reminds us of the unique role of the human in how the Universe Story will play out. In exploring that role, *Journey* asks whether there is indeed "a reason for our existence? Is there a great work required of us?"[28]

For all of John Wesley's deep, rich theological ruminations, he had little tolerance for theology without practice. "For Wesley religion is not humanity's means of escape to a more tolerable heavenly realm but participation in God's own redemptive enterprise, God's new creation, 'faith working by love,' bringing holiness and happiness to all the earth."[29] As a result, Methodism—for both good and ill—is a roll-up-your-sleeves kind of religion.

Yet, one has to ask, Will the actions of good-hearted people be enough to face the seemingly insurmountable obstacles to planetary health, or is something more radical necessary? *Journey of the Universe* suggests a path forward:

Amidst such bonding and dissolution, the universe moves toward increasingly complex communities. . . . A neutron does not simply

---

27. Conversations with Jaydee Hanson, former assistant general secretary for the United Methodist Church General Board of Church and Society, September 2014.

28. Swimme and Tucker, *Journey of the Universe*, 111–12.

29. Runyan, *New Creation*, 169.

adhere to a proton. Rather, both the neutron and the proton have to undergo a transformation for the bonding to occur.[30]

Even at the most basic level, atoms must change, sacrifice, be altered for life to continue. *Journey* explains that the heart of the cosmos's ability to grow, expand, and support the marvelous complexity of life now and in the future is transformation.[31] The heart of Methodism is indeed in the same place.

John Wesley believed that through God's transforming grace, "God inaugurates a new creation, restoring the relation to which we are called."[32] We are asked to believe *into* this possibility and participate in it. Evolution now allows the human an expansive range of choice not available to other species. We can choose to allow ourselves to be transformed into citizens of the planet, who know our place, who exercise the compassion necessary to ensure that our planet continues to grow and flourish. *Journey* puts it this way: "Our role is to provide the hands and hearts that will enable the universe's energies to come forth in a new order of well-being."[33] John Wesley would agree.

---

30. Swimme and Tucker, *Journey of the Universe*, 7.
31. Ibid., 7–8.
32. Runyan, *New Creation*, 71.
33. Swimme and Tucker, *Journey of the Universe*, 116.

29

"Wonder Will Guide Us"

Reformed Theology and *Journey of the Universe*

RUSSELL C. POWELL

"Wonder will guide us." Brian Swimme utters these words at the conclusion of *Journey of the Universe* just before his boat departs and whisks him back to the Aegean Sea, where the journey began. With each successive viewing I find this idea more impactful. It comes just after Swimme has enumerated some ecological problems we face today and is suggestive of how our awe at the universe might influence our practical agency (i.e., what we might *do* in the face of rampant environmental destruction).

"Wonder will guide us." It's a hopeful, powerful statement—that the visceral, instinctive response we have at the sublimity of our world has the capacity to somehow orient our lives. It's also an idea to which the Reformed theological tradition, including the Presbyterian denominational heritage, has long given voice. As a member and inheritor of this tradition, when I was asked to envision a Reformed and Presbyterian response to *Journey of the Universe*, it seemed quite natural to take the opportunity to explore just what this idea—wonder guiding us—might mean within my particular theological context.[1]

Wonder, Religion, and Religious Experience

In the compendium volume to *Journey of the Universe*, written by Swimme and Mary Evelyn Tucker, wonder serves as "a gateway through which the universe floods in and takes residence within us." Through wonder, humans

---

1. This probably goes without saying, but what follows is certainly not the *only* way to interpret *Journey of the Universe* within a Reformed theological frame. Like other theological traditions, Reformed theology is varied, kaleidoscopic even. Hence any number of Reformed construals of *Journey* are readily procurable. Mine might just as well be considered one of many.

"participate" in the "primal energies" of the "majesty of existence"; in it, we realize that our very being is pervaded by cosmic mystery.[2] Perhaps not surprisingly, the idea of wonder evoked here by Swimme and Tucker also has clear religious connotations (Tucker is a scholar of religion, after all). In *Journey*, wonder is not only what we feel at the vastness of knowledge about Earth available to us through modern science, but additionally what we experience in the intimacy we might cultivate to Earth through more traditional modes of religious awareness. Symbolic consciousness or sacred myths and histories come immediately to mind.

I use the term "experience" in the above paragraph quite deliberately, for the concept of religion being drawn upon in Swimme and Tucker's idea of wonder is far from the rationalist spirit we often associate with religion today. Indeed, religion for *Journey* is not a question of belief or analytical consistency, but rather of the lived, affective side of human experience. It's the feeling that wells up in us when we encounter the world's resplendence. So instead of focusing upon metaphysics, which comes with the attendant epistemological issues that underpin so much contemporary discourse of religion, Swimme and Tucker rather fix their gaze upon the enchantment that the universe instills in us. This enchantment compels our *feeling* the sources of our existence, *wondering* at their greatness, and *sensing* the overwhelming dependence we share with the rest of life.

On the evening of the *Journey* film's debut in March 2011 at the Yale University School of Forestry, a short and delicious quip made by Tucker reinforced this idea that *Journey*'s religious overtones were more redolent of prereflective religious experience than any sort of dogmatic line toeing. Asked during a Q&A session following the film screening why, with all its apparent religious implications, *Journey of the Universe* never once mentioned the name of God, Tucker, with characteristic charm, answered, "The film touches on divinity, *just not by name.*"

Journey's picture of religion, then, with its unnamed and enigmatic sense of divinity, as well as its emphasis upon humans' affective responses to the cosmic powers that set the limits and possibilities of our existence, seems a far cry from the notion most of us have of the cold and calculated tradition associated with John Calvin's sixteenth-century brand of religiosity. From the Swiss Reformation centered in Geneva to the fastidious strain of Protestantism that the Puritans brought to the New World, the austerity associated with the sermons of Jonathan Edwards to contemporary Barthian neo-orthodoxy,

---

2. Brian Thomas Swimme and Mary Evelyn Tucker, *Journey of the Universe* (New Haven, CT: Yale University Press, 2011), 113, 114.

popular conceptions of the Reformed tradition, at least at first blush, appear starkly opposed to the religious qualities exhibited by *Journey's* cosmic story.

Of course, the distance that *Journey* places between its own religious attitude and the reason-based piety commonly identified with Calvinism is not coincidental. As students of Thomas Berry, Swimme and Tucker both know well their teacher's derogation of the Reformed tradition and its medieval forebears. With its "salvation mystique," Calvinism, Berry thought, diminished concern for earthly reality in favor of otherworldly redemption. This development in Western religious consciousness, which grew partially from the need to take flight from such worldly dangers as the bubonic plague during the Middle Ages, led to disastrous ecological consequences when it was married to the methods and discoveries of Baconian and Newtonian science.[3]

Still, while the rationalistic side of the Reformed tradition has historically been accentuated when discussing its contributions to Western religious thought, certain other aspects of the tradition can be retrieved and highlighted to demonstrate alternative ways that Calvinist theology might interpret dimensions of human religious experience, including those to which *Journey of the Universe* is most sympathetic. These often extenuated aspects, some of which scholars such as Belden Lane have recently worked to recover,[4] can at least serve to gesture, I think, toward a vision for an evolving Christianity from the Reformed perspective, something Thomas Berry may very well have endorsed.

Nature, Reformed Thought, and Religious Experience

I mentioned Calvin earlier as an example of the austere and overly rationalistic theological tradition he was largely responsible for creating, and for which that tradition has come to be known. Yet one dimension of Calvin's work that is not so well known is his keen attention to the natural world as a *theater of God's glory*. Recognition of this dimension, in which Calvin understood nature to serve as a kind of proscenium arch where God's orderly and all-encompassing providence could be observed and discerned, is one way of stepping beyond the prominent interpretations of Calvin's puta-

---

3. See Thomas Berry, "The New Story: Comments on the Origin, Identification, and Transmission of Values," *Teilhard Studies*, no. 1 (Winter 1978): 1–13.

4. See Belden Lane, *Ravished by Beauty: The Surprising Legacy of Reformed Spirituality* (New York: Oxford University Press, 2011).

tive world-renouncing theology.[5] Calvin's comprehension of nature, in fact, which relied heavily upon Stoic cosmology with its pantheistic understanding of God's immanent presence in the natural order, led him to go so far as to suggest that nature could be rightly analogized with *God himself.* That is, of course, if it weren't for humans' sinful predilection to conflate the Creator with his creation.[6]

In highlighting this dimension of Calvin's thought, I don't mean to somehow suggest that his name and legacy might be fully cleared of the ecologically problematic intellectual lineage underlying the Reformed theological heritage. Evidence to support arguments that Calvinism contributed to the scientific and philosophical revolution that took place in premodern Europe, during which the world became not merely a place to live in but a place to master and control,[7] is available in spades throughout Calvin's wider corpus, as well as in the writings of his theological adherents up to the present day. What I do mean to suggest, though, is that inherent to Calvin's thought are the seeds for reimagining the conventional wisdom of both the Reformed tradition's notion of religious experience and its attitude toward humans' normative relationship with the Earth. This seed, in fact, flowered in the nineteenth and twentieth centuries in the work of thinkers like Friedrich Schleiermacher, H. Richard Niebuhr, and James Gustafson, and continues in Sallie McFague's writings today. I want to suggest that these thinkers provide us a way of finding remarkable resonance with *Journey's* religious proclivities, all the while working within a broadly defined Calvinistic theological framework.

If Calvin's belief in nature's sacral standing can help us to start a conversation between *Journey of the Universe* and the Reformed tradition, Schleiermacher's *On Religion*, a book he wrote in Berlin in 1799 with the encouragement of such German Romantics as Friedrich Schlegel, has the potential to make *Journey* and Reformed theology into good friends. Indeed, *On Religion*

---

5. On nature as a "proscenium arch" in Calvin's theology, see Susan Schreiner, *The Theater of His Glory: Nature and the Natural Order in the Thought of John Calvin* (Durham, NC: Labyrinth Press, 1991).

6. Calvin writes, "I confess, of course, that it can be said reverently, provided that it proceeds from a reverent mind, *that nature is God*; but because it is a harsh and improper saying, since nature is rather the order prescribed by God, it is harmful in such weighty matters, in which special devotion is due, to involve God confusedly in the inferior course of his works" (*Institutes of the Christian Religion*, ed. John T. McNeil [Philadelphia: Westminster Press, 1960], 1.5.5 [emphasis added]).

7. Despite over twenty years passing since its publication, I know of no better study to refer readers interested in this idea than Carolyn Merchant's *The Death of Nature* (New York: HarperOne, 1990).

contains no systematic discussion of Christian doctrine, barely mentions the name of God, and is replete with reflections on the intuition and feeling of the immensity of the universe. For instance, early in the work Schleiermacher suggests that there is "something more essential" to the cosmos than mere matter—this, given the way humans are apt to wonder at its majesty. "Are we not overcome," he asks, "with reverence at the thought and sight of the world?"[8] And just as a sense of wonder leads to an understanding "that everything in the universe . . . forms a huge interconnected family that we can call 'all my relations,'" as Swimme and Tucker, drawing on Lakota peoples, put it in *Journey of the Universe*, Schleiermacher analogously says that the religious person receives life "in the infinite nature of the Whole . . . having and possessing all things in God, and God in all."[9] In essence, the authors of *Journey of the Universe* and Schleiermacher both suggest that delighting in the infinite and interconnected world necessarily results in a powerful impulse toward religious feeling.

The affinity between *Journey* and Schleiermacher is further exhibited in Schleiermacher's insistence upon religion's being not only an intense feeling, but also something that's in no way reliant upon forms of intellectual construction or moral proofs for validation. We simply respond to the world's splendor in ways that can only be described as religious. Importantly, this mode of response, like what we saw earlier in *Journey's* protoreligious sense of wonder, happens at the level of experience that is *prior to* both thought and action. Schleiermacher in fact goes to great lengths in the second of the five speeches that constitute *On Religion* to distinguish religion from metaphysics and morality. In so doing, he contributed an important piece to the larger Romantic rebellion against Kantian epistemology. Rather than uphold Kant's differentiation between the realms of necessity and freedom, or the unknowable world of things-in-themselves (what Kant called the "noumenal" realm) and the world of categories intelligible to the human mind (the "phenomenal" realm), Schleiermacher, together with his contemporaries in Berlin like Schlegel, Johann Gottfried Herder, and Novalis, sought to transcend the duality of subject and object, locating an awareness of human being and all other forms of existence as persisting *in* the Infinite. Such awareness, as we have just seen, derives from humans' a priori, innate sense of awe at the cosmos. Unlike his Romantic peers, however, Schleiermacher was the only

---

8. Friedrich Schleiermacher, *On Religion: Speeches to Its Cultured Despisers*, ed. and trans. Richard Crouter, Cambridge Texts in the History of Philosophy (New York: Cambridge University Press, 1988), 35.

9. Swimme and Tucker, *Journey*, 114; Schleiermacher, *On Religion*, 47.

one to correlate this concept of transcendence with a distinctive concept of religion and, in his case, religious experience.

Wonder's Bearing on Practical Agency

But does religious feeling or experience—wonder, in Swimme and Tucker's idiom—*guide* us? Or, to put it more in the language of our discussion here, does our awe at the proportions of the universe *arrange* and *determine* our lives, meaning that our inborn religious responses to the world somehow calibrate something as important as our *practical agency*? I believe this *is* in fact what Schleiermacher thinks, especially when he writes that religion—feeling the "infinite nature of totality, the one and all"—serves as the necessary ingredient for ethical behavior. "To want to have . . . praxis without religion is rash arrogance," Schleiermacher says.[10] The philosophical implications of this idea are great and cannot be fully elaborated upon in the space provided here. Yet it will suffice to say that, at least as it concerns *Journey's* intriguing notion of wonder guiding us, Schleiermacher suggests that our religious response to the universe entails our *practical engagement* with the world.

No longer abstracted from the world as a thing-in-itself that we can never fully grasp, by transcending the subject-object duality of Kant's metaphysics through an intuited religious response we recognize both our enmeshment in the world and the way the world, including nature, contains properties and values that make normative demands on us to engage with it in more practical and not merely detached terms. What this suggests, I think, is that our moral sentiments are not self-standing, as if our choices and actions somehow stand apart from the environment in which they are made. Rather, when grasped by the religious intuition of the feeling of wonder, we must necessarily perceive and respond to the value properties that make normative claims upon us. This means that we—when responding to the properties of the external world, which are constitutive of, prompt, and give rise to our religious experience—are *under obligation* to take the world and its innate value into account in all of our moral deliberations.[11]

---

10. Schleiermacher, *On Religion*, 23.

11. I've been thoroughly influenced on the idea of nature's inherent value and its role in determining human practical agency by the contemporary philosopher Akeel Bilgrami. Both the notion and language of being obligated to account for nature's inherent value properties in our practical agency are his. For a representative sample of his work on this topic, see his most recent publication, *Secularism, Identity, and Enchantment* (Cambridge, MA: Harvard University Press, 2014).

This is what I take the notion of wonder's guiding us essentially to mean, interpreted within the broad Calvinistic theological framework that sees God's providential being as immanently involved in the world. In this vein, numerous Reformed thinkers have followed Schleiermacher in identifying the necessity of responding to the normative demands that nature makes upon human subjects. A line can be drawn, for instance, from Schleiermacher to H. Richard Niebuhr's ethics of responsibility, wherein Niebuhr imagined a theological ethics based on the principle that "God is acting in all actions upon you. So respond to all actions upon you as to respond to his action."[12] Nature, in other words, totals the arena in which God's will in and for the world is executed, and thus comprises the entire field of human moral possibilities. Similarly, James Gustafson's *Ethics from a Theocentric Perspective*, the two-volume culmination of his career as one of the foremost Christian ethicists of his generation, demonstrates perhaps better than any other thinker the human subject's far more modest profile when we take seriously the normative demands made by the properties of value and meaning in the world.[13] And finally, it's not difficult to connect the dots from Calvin to Sallie McFague today, whose work especially lifts up the value that the Schleiermacherean branch of the Reformed tradition finds in its apprehension of the material world in a theology that conceptualizes Earth as God's body. McFague moreover investigates closely the radical political possibilities that inhere in our discernment of value in the world.[14]

In sum, the Reformed tradition gives us a helpful way of interpreting what it means for wonder to guide us, especially given the religious and theological significance that *Journey of the Universe* attributes to affective religious experience. More, the Reformed tradition also contributes a helpful strategy for interpreting another of Swimme's profound statements that comes at the conclusion of the *Journey* film: "We belong here," Swimme declares, referring to humans' place in the cosmos's 14-billion-year-old drama. Indeed, "We've always belonged here," he says. Such a statement quietly attests to the necessity of works like *Journey*, which teach us new ways to recognize our

---

12. H. Richard Niebuhr, *The Responsible Self: An Essay in Christian Moral Philosophy* (Louisville, KY: Westminster John Knox Press, 1999), 25.

13. James Gustafson, *Ethics from a Theocentric Perspective*, 2 vols. (Chicago: University of Chicago Press, 1981/1984).

14. Sallie McFague, *The Body of God: An Ecological Theology* (Minneapolis: Fortress Press, 1993). On the politics, and particularly the potential economic implications, of the Reformed tradition's brand of immanentism, see Sallie McFague, *Life Abundant: Rethinking Theology and Economy for a Planet in Peril* (Minneapolis: Fortress Press, 2000).

being at home in the universe, our *dwelling* here by dint of who we are as humans. Christian viewpoints like that of the Reformed tradition, I hope to have shown, offer important perspectives for a more complete appreciation of what this idea truly means.

30

LDS Theology and the New Story of the Universe

George B. Handley

I am grateful to be asked to give a Christian response to *Journey of the Universe*. I suppose a "Christian response" means at least two things: that I am identified, as a member of the Church of Jesus Christ of Latter-day Saints, as a Christian and that my response should also be guided by Christian principles and values—that is, that it should on some level be charitable. With regard to the former, I am grateful to be included in the Christian community as a Mormon, especially since despite LDS self-definition as a decidedly Christian sect, not all Christians have agreed to include Mormons as Christians due to doctrinal differences. Curiously, some of those doctrinal differences have a lot to do with why an LDS response to the New Story told by contemporary science in *Journey of the Universe* would be particularly positive, as I hope to explain. With regard to the latter, I am moved, challenged, and inspired not only by the film and book and the way they provoke wonder in a believer such as myself, but I also find courage and hope in the community of respondents that this symposium has gathered. As such, I can only hope that my response is as Christian as the invitation.

One thing that climate change has made clear, and that a symposium such as this also makes evident, is that we can no longer afford to engage in theological, doctrinal, or ideological arguments while the well-being of the planet is under attack. In his book *The Natural Contract*, Michel Serres describes Francisco de Goya's painting *Fight with Cudgels*, where two men both sink into the muddy soil beneath them because they are too busy fighting one another to work together. His point is simple: Climate change requires what he calls a "diligent religion of the world" in which we learn to be bound together in a common concern for the planet.[1] This is a task that must override the need to identify and fight over differences. In the realm of religion

---

1. Michel Serres, *The Natural Contract*, trans. Elizabeth MacArthur and William Paulson (Ann Arbor: University of Michigan Press, 1995), 48.

and ecology, climate change represents the great opportunity for interfaith collaboration and fellowship, something that seems far more Christian than the negative protectiveness that has sometimes characterized the internecine battles within the Christian tradition.

Perhaps for this reason, as context for my remarks that follow, I would argue that as important as the story of the universe is as told by science, our top priority should not be to persuade the human family to adopt the same story or worldview but to facilitate an exploration of the roots of environmental care in each community's traditions and then to use such interpretations as a basis for finding common ground with others. So it isn't so much a new story that we need but more attentive, creative, and environmentally motivated *rereadings*. The risk of insisting on an adoption of the same new story for everyone is the same risk of the two men wrestling one another: we may expend more of our energy trying to be right rather than trying to be good. In other words, we might expend our energy on trying to convince others of their erroneous worldviews when we should be focused on doing what is right by the planet.

In a spirit of ecumenical collaboration on behalf of Earth, then, perhaps the doctrines of creation within the LDS tradition will not be dismissed as heretical outliers but can be appreciated for their suitability for understanding and finding a Christian place in a universe shaped by evolutionary change and deep time and expanding into unimaginable dimensions. By way of summary, *Journey of the Universe* offers a portrait of a cosmos that is multi-centered, self-generating and self-organizing, rhythmic, creative, interdependent, and shaped by disequilibrium, complexification, and self-awareness. While these capacities help us to understand the emergence of life, it takes some work to reconcile such a universe with the idea of a loving, self-revealing, and intervening Creator that is so important to the Judeo-Christian tradition. I wish to do some of that preliminary work within the context of LDS doctrines of creation in order to suggest reasons why Mormons can be receptive to the story of contemporary science. I conclude with some observations about why, despite room within LDS belief for contemporary science, we continue to see tensions between science and religion in Mormon culture.

Mormons are generally proud of what they consider to be their science-friendly theology, and they have good reason to be, but it is also true that evolution, the Big Bang, and other scientific theories that have created challenges for other Christian theologies also remain somewhat misunderstood or are treated with an unwarranted amount of suspicion and even with willed ignorance. The good news is that when my students at the LDS church–owned Brigham Young University, who generally don't have previously hardened nor necessarily well-informed views of the science, have watched the film or read the book of *Journey of the Universe*, their enthusiasm for the science is almost unbounded. The church has remained officially welcoming of science but also officially neutral regarding science's particular claims, even though throughout LDS church history various leaders of the church have held strong and differing opinions about such issues as the age of the Earth and evolution. Evolution has been taught at BYU for many years, and a number of remarkable and successful scientists have come from the Mormon tradition, including Henry Eyring, a mid-twentieth-century chemist at Princeton University; Philo T. Farnsworth, the inventor of the television; Harvey Fletcher, the inventor of the hearing aid; and Paul Cox, a 1997 awardee of the Goldman Environmental Prize, to name a few. However, it wasn't until 1992 that an official packet on evolution was produced by two science deans and the dean of religious education at Brigham Young University with the official approval of LDS church leadership; the packet is now regularly distributed to students.[2] It clarified Mormon doctrine on the *whys* of the creation while allowing more neutrality on the *hows*. This has allowed the teaching of evolution to continue undeterred, even though it has resulted in an uneasy accord of deferential respect but occasional distrust between the evolutionists and the religion faculty on campus. It is probably safe to say that scientific literacy among the general membership is very uneven; despite a strong emphasis on education in Mormonism, the LDS church is a constantly growing and worldwide religion of differing levels of access to education and economic opportunity. As a result, common parlance in the church not infrequently implies that science is at odds with belief. Church leaders do not argue with contemporary science, but they do argue with the notion advocated by some atheist scientists who have insisted that the origins of life would require believers to accept that human life on Earth has emerged entirely by chance, without purpose, or beyond the control of a loving God. The church's argument, in other words, seems to be with extrascientific claims made by scientists regarding the irreconcilability of science and the idea of God, not with the science per se. However, since no official statement from church leaders has ever attempted to provide the theological grounds for accepting any of these scientific theories, these critiques of scientism linger in believers' minds as sufficient grounds for distrusting or at least remaining disinterested in the science.

---

2. You can read it here: "Evolution and the Origin of Man," http://biology.byu.edu/DepartmentInfo/EvolutionandtheOriginofMan.aspx.

This is despite the fact that LDS doctrines are not hard to reconcile with contemporary science. Mormons share with other Christians a belief in a divine and intentional creation and a creation of human beings in *imago Dei*, but they are also considered heretics within mainstream Christianity for the role that Joseph Smith played in translating, rereading, and revealing new doctrines that he claimed were restorations of original Christianity. There are at least six reasons for the perceived heresy. According to Mormon belief,

- The creation occurred out of unorganized matter that has never been created nor destroyed, only reorganized or repurposed matter into our present Earth and solar system—a creation *ex materia*, in other words, but not *ex nihilo*, or out of nothing, as most Christian belief has it.[3]
- Human agency, like all matter, can neither be created nor destroyed; we were created as spirit offspring of heavenly parents out of preexisting and coeternal "intelligences."[4]
- The world was created spiritually before it was created physically, making plants, animals and human beings alike "living souls."[5]
- Jesus Christ, as a separate being from God, was directly involved in the creation.[6]
- God and Jesus Christ inhabit physical bodies in a specific location in the universe.[7]
- Our planet and solar system are just among of the countless creations of God (allowing for a belief in the possibility of human creation on other planets).[8]

While Mormons are generally aware of these doctrinal differences, little attention has been given to their implications, especially as they apply to our understanding of contemporary science and environmental stewardship, as

---

3. In Abraham 3:24, for example, we read, "And there stood one among them that was like unto God, and he said unto those who were with him: We will go down, for there is space there, and we will take of these materials, and we will make an earth whereon these may dwell." For all LDS scriptures, including the Bible, see "The Church of Jesus Christ of Latter-day Saints" under the link "Scriptures" (https://www.lds.org/?lang=eng).

4. See Abraham 3:22, where the Lord describes the human family as consisting of "intelligences that were organized before the world was."

5. See Moses 3:19.

6. See Mosiah 3:8.

7. See Abraham 4:1.

8. See Moses 1:33. Here also the distinct roles of God the Father and God the Son are further clarified.

represented by *Journey of the Universe*. Science today implies that we live in a universe that is far more expansive, less Earth- and human-centered, and more tragic and contingent than traditional Christianity has imagined. The irony, I wish to suggest, is that even if Mormon believers choose to reject contemporary science because it implies a universe that is more chancy, less predictable, and more full of suffering than they had imagined, they would still have to confront similar implications in their own doctrines. How well-equipped Mormons and other Christians are to adapt to the scientific narrative, in other words, has as much to do with their willingness to hear the New Story of science as it does with their willingness to accept the darker implications that already exist within Christian theology. My point is that Mormons and Christians more generally aren't responding well to the Anthropocene not merely because we don't know the science; we aren't responding well because the Anthropocene asks us to shift our understanding of the cosmos in fundamental ways. The good news is that LDS doctrine was challenging Christians to do this all along.

Journey of the Universe makes a compelling case for the scientific story and gently creates entry points for Christians and other believers. There is room, for example, to imagine a divine purpose and creativity involved in the emergence of life and to find cause for and spiritual nourishment in experiences of gratitude and wonder and in our human experience of self-conscious awareness of the universe. But in its emphasis on the interdependencies and contingencies of life, on a multicentered universe, on the universe's self-generating and self-organizing capacities, and the importance of disequilibrium, the story also challenges us to rethink our assumptions about divine sovereignty and our own chances for direct appeal to intervention, healing, and revelation, all central tenets to what it means to be a Christian and certainly to be a Mormon Christian. I think Christians have been slow to respond to environmental crises because the broad and unforeseen consequences of degradation seem to suggest that nature is not the simple and manageable system we thought it was, and that individual acceptance of the responsibilities of stewardship, trust in divine purpose, or direct appeal to divine intervention are not enough. The problems of the Anthropocene require greater humility in the face of nature's unpredictability and in the face of what we can't know, a more conscientious and collective ethics than we are accustomed to in our atomized society, and an acceptance of the more tragic implications of life in a universe as chancy and contingent as ours.

The LDS Account of the Creation

It must be admitted up front that both a creation *ex materia* and a spiritual creation are heretical concepts in traditional Christian thought, which has largely to do with the ways in which these concepts condition or delimit the power of God. Without absolute beginnings, we have doubts about absolute control, and we then face the uncertainty of a future that is not absolute or definitively known. In the context of a creation *ex materia*, language—especially God's language—is not prior to or procreative of the objectivity of the world. The world is not created so much as it is "organized."[9] The implication is that our human readings of divine language are no longer as securely anchored to divine intent and become more contingent and more interdependent and that revelation is then by necessity ongoing. The future is consequently more in our collective hands than we might have hoped, but also—and perhaps this is a cause for the anxiety—less easily mastered or definitively directed or controlled. I don't mean to suggest that *ex nihilo* theologians haven't been able to make arguments for environmental stewardship nor that they all lead to dogmatic resistance to continuing revelation, but in the popular mind the *ex nihilo* dogma has often resulted in such tendencies.

Many theologians today argue that contemporary science's narrative of an expanding universe in deep time that hovers just close enough to a series of chance events to make life possible on this planet requires us to rethink the *ex nihilo* dogma and consider the possibility of an *ex materia* creation.[10] As the Mormon ethnobotanist Paul Cox argues, "Perhaps there is less incentive to protect the planet for those who assume that the world was miraculously produced in a few days and which will ultimately become the abode of the wicked. After all, if we trash this planet God could always pull another one out of a hat. Furthermore, why protect a place that is soon destined to be completely destroyed?"[11] A world subject to indeterminacy that is dynamically changing and morphing into new possibilities and realities requires a flexible theology, an adaptive ethics, and a commitment to the perpetual pursuit of scientific knowledge so that we are willing and able to work with ecological laws and processes. Our ethics, in other words, must emerge from

---

9. See Abraham 4:1.

10. For examples, see Catherine Keller, *The Face of the Deep: A Theology of Becoming* (New York: Routledge, 2003); and William Brown, *The Seven Pillars of Creation: The Bible, Science, and the Ecology of Wonder* (Oxford: Oxford University Press, 2010).

11. See Paul Cox, "Paley's Stone" in *Stewardship and the Creation: LDS Perspectives on the Environment*, ed. George B. Handley, Terry B. Ball, and Steven L. Peck (Provo, UT: Religious Studies Center, 2005), 39. Also available online at http://rsc.byu.edu/%5Bfield_status-raw%5D/stewardship-and-creation-lds-perspectives-environment.

trust in God, compassion for our fellow human beings, and hard-earned understanding of the conditions that foster life.

Although *ex nihilo* theologians have sought to argue against the deterministic implications of the doctrine, they haven't always succeeded in accepting the indeterminacy that *Journey of the Universe* makes clear lies at the heart of an ongoing creation or in articulating the grounds for the very idea of human freedom. If we don't like the idea of a world subject to chance and indeterminacy, however, we might also consider that such a world allows for the possibility of human freedom, an idea much harder to conceive of in a deterministic universe. So a universe in which we are free—and therefore to which we are answerable—is also one that appears to evolve. To say that God took preexisting matter in order to make the world is tantamount to saying that nature contains multiple and competitive energies and that God, humans, and nature are interwoven together. In this way, God becomes less a creator or inventor and instead more an artist or a recycler, and our stewardship must model his in this same way.

This is not necessarily how Mormons think about this doctrine, however. It seems as if centuries of Christian dogma and debate about the nature of God have overdetermined the terms of the discussion for Mormons, placing inordinate emphasis in Mormon culture on the idea that God acted alone and with perfect omnipotence and omniscience. This, of course, glosses over the fact that LDS theology suggests that God did not act alone but that he commissioned Jesus Christ as a separate being and, perhaps most interestingly, *us* in our premortal state to make the world and that he could neither create nor destroy our agency nor the matter out of which the world is made. As LDS theologian Sterling McMurrin wrote, "The Mormon concept of matter is that it is essentially dynamic rather than static, if indeed it is not a kind of living energy, and that it is subject at least to the rule of intelligence."[12] The world's energies stand somewhat independent of a God who orchestrates and directs but who does not mechanistically control the world. What is especially valuable about this theology is how it safeguards the nature of human freedom and moral responsibility, even though it also implies interdependency and contingency. If I am free and others are as well, we all can act but we will also all be acted upon—by God but also by others and by nature itself.

These doctrines help to ease the burden of the problem of evil and suffering by taking some of the blame off of God, a significant concern of Chris-

---

12. Sterling McMurrin, *The Theological Foundations of the Mormon Religion* (Salt Lake City, UT: Signature Books, 1965), 7.

tian theologians for centuries. However, the doctrines also introduce us into a universe that is perhaps more tragic and more chancy than we had imagined. God is harder to discern in the workings of the world. Natural theology will not suffice to trace all that God is or all that he does. That doesn't mean that he is absent wherever we fail to find him. It only means that we can just never know definitively. It means that faith, while capable of being rewarded by divine manifestations, also requires a trust in the sometimes unseen and incomprehensible God. While some believers feel uncomfortable with this ambiguity, I would submit that the ambiguity is morally and ethically valuable. In this version of Christian theology, God is sovereign, nature follows her laws, and we are free agents in a world that we can create or destroy. We are creatures whose health and well-being are not only dependent on God but also on biology and on our own agency. We are agents within the systems of life rather than as mere bystanders or observers of God's pageant of nature. If, as LDS theology has it, our intelligences, or our wills, are coeval with God, we are also cocreators of a world that is never entirely independent of us but also never entirely subordinate to our will. While this doctrine seems like hubris to some Christians, this emergence of humanity's centrality in shaping a creation that was once believed to be beyond and above human agency is, of course, what the Anthropocene now implies and is the reason why some Christians are offended by the very idea of climate change. We simply cannot escape our now central role in shaping the direction of the planet's climate and health. We must learn the same kind of stewardship by which God enacted and enacts his creative powers on the world.

LDS Resistance to These Doctrines

So why haven't these doctrines taken root and inspired a more pronounced ethic of creation care within Mormonism? There are many answers to this question, but one that interests me here is the idea that Mormons turn away from the doctrines, not simply out of apathy, but more deeply out of fear of our own freedom. To be free is to be responsible, but it is also to be responsible in the context of contingency, unpredictability, and uncertainty. But we often prefer tighter theologies and ideologies that help to reassure us that freedom means something simpler: that we can choose and act in a way, directed by God, that makes us captains of our destiny and that inoculates us against unforeseen or unwanted consequences. We should only get, in other words, what we deserve: nothing more and nothing less. But since we get far more and sometimes far less as a species, we have developed a host of theologies and ideologies to justify inequalities and sufferings that are otherwise inexplicable. We believe in and relish our freedom but we want a freedom that frees us from contingency, which is another way of saying that we don't want freedom at all. We want instead easy and predictable outcomes that, despite the temporary illusion that our consequences stem directly from our choices, are given as God's blessings, almost in advance or by guarantee.

Instead of doing the simple work of making ourselves more ecologically literate, we prefer to believe in an economy that exempts the very nature of the world we live in. Such an economy pines for a world brought entirely into being out of nothing by a single and all-powerful God, a God that is responsible for all natural events, all natural disasters, all suffering and all fortune, all transformation of energy. It is an economy that has little patience with what we have learned since Hutton and Darwin about the deep and complex story of millennia of violent deaths and extinction for unnumbered species. It is not by coincidence that many of the most vehement voices of denial regarding climate change and evolution in Christianity today speak from this kind of theological position.[13]

While loving the *idea* of nature is easy, loving the complexity, deep temporality, unpredictability, and violence inherent in nature is no easy proposition.[14] Nature can be beautiful, it can be reassuring, and it can provide a sense of communal belonging to our more-than-human kin, but it can also be ugly, destructive, and indifferent. These darker and indifferent aspects were often associated historically with Satanic or adversarial forces in many cultures. While we tend to think of ourselves as more enlightened today by scientific and secular rationality, the fact that we are no longer as frightened by nature might be a symptom of our own failure to truly understand it. While its beauty and provisions often suggest to us the caring nature of deity, nature's indifferent and perpetual regeneration of life ought to be a reminder of our death and of the persistence of tragedy. The most important point here is that, for Mormons at least and perhaps for all Christians, to believe in Christ means belief in a suffering God who weeps for sins but also for our pains and sorrows and for the pains and sorrows of all living things that attend life

---

13. Pat Robertson is a case in point. Well known for his denial of climate change, he was surprisingly quick to blame the Haitians for their suffering in the wake of the 2010 earthquake. Ken Ham's creationist theology is informed by a similar logic that precludes the possibility of suffering that is not tied to God's will. See his website: http://answersingenesis.org.

14. On this point, see, for example, William Jordan's brilliant book, *The Sunflower Forest: Ecological Restoration and the New Communion with Nature* (Berkeley: University of California Press, 2003).

on this planet.[15] When we accept the biblical declaration that God declared the world "good," we assume, falsely, that this means that nature will behave itself, that it will always help us to feel love and reassurance. Eventually experience tells us that nature is often an independent cause for human suffering and creates the conditions by which human significance is disrupted or fundamentally challenged by temporal and spatial dimensions beyond our capacity to fully comprehend. We might react with repulsion or deep resentment and thereby degrade the world that has let us down. The alternative is a truly Christian response to the story of the universe, and that is to muster the charity to accept the continued presence of the chaotic "watery deep," the *tehom*, which creates the indeterminacy at the heart of an ongoing creation. What if, instead of imagining that the world is always kind or beautiful or harmonious, we understood that even its dark side; its long and violent history of competition, death, and regeneration of life, and of catastrophe and change; and its bizarre and myriad life forms—even its seeming unpredictability—are all of intrinsic value worthy of our profound respect? These conditions are, after all, as necessary to life, if not more so, as any beautiful sunset, as *Journey of the Universe* makes so beautifully clear.

Journey of the Universe teaches that human beings find themselves part of a world that they have the responsibility and opportunity to affect, positively or negatively. Every attempt to point fingers of blame, in either a human, natural, or a divine direction, involves a shielding of the fact of mutually and shared accountability. I wish to suggest that the economy of a creation *ex materia* means we are caught in an imperfect but indispensable partnership between human agency, divine will, and natural process. And the promise of a spiritual creation that precedes the physical creation means that all of life is kin, albeit uncannily and perhaps unknowably so in any precise sense. Instead of learning the requisite Christ-like charity, forbearance, patient longsuffering, love of God, love of the Earth, and love of our human siblings that such doctrines would require, we opt for the pretense of radical autonomy, the pleasures of amassed and "private" property, and the continued abuse of the Earth's treasures to shield ourselves from the world's violence and pain.

This manner of sinning is certainly one of the most explicitly and harshly condemned in LDS scriptures. In a revelation to Joseph Smith, the Lord explains that he has left us as "agents unto [ourselves]" and that "if any man take of the abundance which I have made, and impart not his portion, according to the law of my gospel, to the poor and the needy, he shall, with the

---

15. See Alma 7:11, which describes Christ's suffering thusly: "And he shall go forth, suffering pains and afflictions and temptations of every kind; and this that the word might be fulfilled which saith he will take upon him the pains and the sicknesses of his people."

wicked, lift up his eyes in hell, being in torment."[16] Indeed, a Mormon environmentalism is certainly anthropocentric; more equitable distribution and sustainable use of natural resources matter, LDS theology suggests, because they enable us to care more properly for the poor. But this isn't to suggest that the rest of creation gets left behind. The mandate to feed the world appears not only to apply to human mouths but to the mouths of all animals. We read in another revelation to Joseph Smith, famously known as the Word of Wisdom where alcohol and drugs are prohibited, that it is God's intention and our responsibility to ensure that all life—"all wild animals that run or creep on the earth"—are sustained by the Earth, just as we are.[17] Elsewhere in the same book of revelations the Lord declares, "It is not given that one man should possess that which is above another, wherefore the world lieth in sin. And wo be unto man that sheddeth blood or that wasteth flesh and hath no need."[18] We are to use resources "with judgment, not to excess, neither by extortion."[19]

Stewardship is not easy work, especially in an ongoing creation *ex materia* in which we are perpetually and intimately bound to indeterminacy and yet always held accountable. Such a world asks us to accept limits to our managerial control and to learn to be artists, not inventors, and maybe even collaborating artists, members of a theater or dance company perhaps, more than the solitary painter or poet. We are to take of the materials of the Earth and repurpose them, to make the world anew out of unorganized matter, building in this way a sacred economy of shared nourishment. In this way we can begin the work of making a sacrament not only of all our family meals but of everything we have done with the flesh of the Earth.

---

16. Doctrine and Covenants 104:18.
17. Doctrine and Covenants 89:14.
18. Doctrine and Covenants 49:20–21.
19. Doctrine and Covenants 59:20.

31

Quakers and Journey of the Universe

LAUREL KEARNS

Both science and religion rest ultimately on our contemplation of the natural world. . . . To survey any beautiful scene without distraction is to become aware of an incredible creative process that has raised all things up from the formless dust that infuses everything with vitality and energy, that maintains balance and lawfulness, and that illuminates each order of living things with a degree of wisdom suitable to its estate. We become aware that human existence is a part of this great web, we are humbled, and we ask what response is called for from us so that we might play our role properly in this great unfolding drama. Some religionists disparage what they call nature mysticism as a counterfeit spirituality. In truth, it is not a counterfeit spirituality but the foundation, the essence, and the core of the religious sensibility.[1]

As Daniel Seeger so beautifully intimates in a recent *Friends Journal* article, Quakers should, and do, respond enthusiastically to the story told by, and the theological implications of, *Journey of the Universe*. He is not the only Quaker to learn from *Journey of the Universe*. A special edition of the *Quaker Eco-Bulletin* in 2006 was dedicated to showing how the current worldview that devalues people and creation "is now being challenged by new scientific stories about the emergence of the Universe, the planet Earth, and its life forms, and the processes that have evolved to sustain life. In many ways these scientific stories reflect and support ancient Earth-centered wisdom about the human-Earth relationship now being reclaimed and incorporated into some contemporary spiritual practice."[2] Brian Thomas Swimme, in the film

---

1. Daniel A. Seeger, "Why Do the Unbelievers Rage? The New Atheists and the Universality of the Light," *Friends Journal*, January 2011, 6–11. Pendell Hill is a Quaker study center outside of Philadelphia, Pennsylvania. Seeger is past director of Pendell Hill.

2. Keith Helmuth et al., "Changing World View and Friends Testimonies," *Quaker Eco-Bulletin*, July–August 2006, 1.

Journey of the Universe, explores the nature of this relationship: "We aren't living on the Earth, we are participating in it." This insight, and its resonances with Quaker practice, is the topic of this chapter.

Quaker Practices and Testimonies in Light of Journey of the Universe

An awareness of the life force of all things that connects us is related to the central Quaker tenet, as much as there are tenets in the Society of Friends (our more formal name): "there is that of God in everyone." For many, this statement becomes, "There is that of God in everything," or "There is that of God in all creation." This insight is the basis of Quaker ethical practice. If we become aware of that of God, then we must act in accordance with the recognition of relationship; we must seek right relations. The *Quaker Eco-Bulletin* devoted to telling the story of the universe puts it this way: "Since Earth itself, and everything on it, is an expression of this essentially unnamable, yet pervasive, fecundity of the universe, we can understand 'that of God' as moving in all forms and creatures. The motion of Creation is in every animal, every plant, every rock, every form and process of Earth. In worship, we seek to find the inner light of the divine, to be open to that of God."[3] Thomas Berry names this the "mystical dimension of the universe . . . We can see that every being in the universe is cousin to every other being in the universe."[4] In other places, Berry's language that the "universe is a communion of subjects" eloquently expresses the Quaker desire to recognize, and respond to, that of God in every thing.[5] Or as the Friends General Conference's (the organization that represents "unprogrammed" or silent Quakers) Quaker Finder webpage states, "We hold ourselves open to the Light and reach for the divine center of our being. We know the center to be a place of peace, love,

---

3. Ibid.

4. Thomas Berry, "The Universe Story: Its Religious Significance," in *The Greening of Faith: God, Environment, and the Good Life*, ed. John E. Carroll and Paul Brockelman (Hanover, NH: University Press of New England, 1997), 214.

5. Coelho et al. in *Quakers and the New Story* express this sense of immanence in relation to the New Story: "Although definitively immanent, there are also qualities of that immanent reality that are transcendent, not in a spatial sense . . . but in the sense of being the source of the manifest world; and in the sense that its expression in the manifest world does not exhaust its great fecundity. There is no longer a need to appeal to a spatially transcendent Other that intermittently intervenes in earthly affairs. . . . This is an empowering realization that invites us to explore with confidence the fullness of our personhood since, as an immanent reality, it is the depth of our being" (Phillip Clayton and Mary C. Coelho, *Quakers and the New Story: Essays on Science and Spirituality* [Cambridge, MA: New Friends Study Group, Friends Meeting at Cambridge, 2007]), 19).

and balance, where we are at one with the universe and with each other."[6] It is worth noting that this understanding of our faith and practice is centrally communicated on the webpage that is for seekers or those inquiring about the Society of Friends. It is therefore central to our identity. Thus, there is a deep openness to the revelation in understanding the story of the universe that Thomas Berry has so eloquently pronounced: "Once we realize that the human story is inseparable from the universe story, then we can see that this story of the universe is in a special manner our sacred story, the story that reveals the divine, the story that illumines every aspect of our religious and spiritual lives as well as our economic and imaginative lives."[7] The story told in *Journey of the Universe* then becomes a basis for a contemporary expression of Quaker praxis and witness, and in doing so, refutes the notion that nature mysticism, or the story of the universe, fails to evoke an ethical response. Indeed, Quakers find resonance with the *Journey of the Universe* story precisely because their own practice of seeing God in everything enables them to focus on the local, the particular, as worthy of care and action. Rather than lending to an abstraction that diverts attention from the local and immediate, as some critics worry, the Universe Story, for Quakers, leads to a deeper, immanental understanding of particularity and calls forth the need to respond.[8]

What *Journey* does is help us see the need for Earth Literacy, for understanding the intricate relationality and balance woven into the story of life's existence on our planet Earth and that planet's participation in a larger universe. Indeed, the movement for Earth Literacy, closely associated with the work of Thomas Berry and Brian Swimme, has been very influential among Friends. One popular Quaker presenter, Brad Stocker, describes his work in ways that are directly related to Berry's telling of the New Story: "My simplest definition for Earth Literacy, the elevator one I use is this: Earth Literacy begins with knowing and understanding the implications of the science story of the creation and the evolution of our Universe and Earth told with an infusion of spirituality."[9]

The special 2006 edition of the *Quaker Eco-Bulletin*, after laying out the story of the evolution of the universe, goes on to discuss "cultural evolution" and "economics and the new consciousness" before explicitly connecting it

---

6. Quaker Finder, Friends General Conference, October 2014, http://www.fgcquaker.org/connect/quaker-finder.

7. Berry, "Universe Story," 214.

8. See, for example, Lisa Sideris, "Science as Sacred Myth? Ecospirituality in the Anthropocene Age," in *Linking Ecology and Ethics for a Changing World: Values, Philosophy, and Action*, ed. R. Rozzi et al. (Dordrecht, the Netherlands: Springer, 2013), 147.

9. Brad Stoker, "Earth Literacy," in *Befriending Creation* (May–June 2014), 4–5.

with the guiding "testimonies" of Quaker life: simplicity, peace, integrity, equality, community. Since this special edition explicitly makes the connection, I will quote in its entirety the outline and interpretation of each testimony:

Simplicity—Functional approach to the arrangements of life and work; non-acquisitive; frugal; unadorned; spiritually centered; attentive to direct experiences and relationships.

Subsidiarity—Direct decision making at the most immediate level of participation on matters of local and regional concern; anchoring life and livelihood in local and regional communities; production, use, and recycling of goods and services within local and regional economies.

Peace—Nonviolent living; conflict prevention; conflict resolution; relationship building; reduction and elimination of the causes of conflict, violence, and war.

Human-Earth Relationship—Ways of life and means of livelihood that do not violate ecosystem resilience and integrity, or depend on violent and exploitative control of resources; mutually enhancing human-Earth relationship within a context of right sharing of resources.

Equality—Recognition and practice of dignity and respect; human solidarity; equitable access to the means of life and life development resources.

Ecological Footprint—Shared life space and life development resources; habitat preservation; biodiversity preservation; cultural preservation.

Integrity—Truthfulness; ethical consistency; devotion to right relationship; valuing direct experience and accurate information.

Ecological Adaptation—Ways of life and means of livelihood that are congruent with the resilience and functional integrity of the biotic environment; active enhancement of ecosystem resilience and integrity.

Community—Mutual support relationships; cooperative reciprocity; sharing of spiritual and physical commons; ceremonial representation of social life.

Social Ecology—Mutually enhancing human-Earth relationship; fully responsive to environmental processes; mindful participation in the dynamics of interdependence and ecosystem reciprocity.

Service—Life and work orientation around contribution to human betterment; e.g., human service work, education, provision of useful goods and services, public policy and civic engagement, social justice, economic security.

Stewardship—Life and work orientation around contribution to mutually enhancing human-Earth relationship; e.g., ecosystem restoration; energy use conservation; transition from nonrenewable to renewable energy and materials; local production for local use; green building; environmental education; ecological footprint reduction; overall ecologically sound economic adaptation.[10]

As you can see, this Quaker vision of the fleshing out of the Universe Story mirrors Thomas Berry's *The Great Work.* Indeed, *The Great Work, The Dream of the Earth,* and Berry and Swimme's *The Universe Story* are all recommended sources.[11]

Quakers trace their commitment to sustainability to the commitment to simplicity and integrity as "testimonies" that have been central to their faith from their beginnings in the mid-1600s in England. The testimony to simplicity meant plainness in dress, speech, buildings, and lifestyle. Plainness, as one aspect of simplicity, was seen as a tool of personal discipline and spiritual practice to cultivate an inner connection to the light of Christ in one's heart, and as a path of personal virtue to avoid the distractions of worldly things and the accumulation of wealth. Thus, there has been an anticonsumption, anticonsumerism bias built in from the beginning; George Fox, perhaps the most influential founder of the group, argued that the accumulation of wealth contributed to war and was a form of violence and advocated the "right sharing" of economic resources in the interests of social justice. The commitment to simplicity not only allows the individual to be less distracted and therefore closer to God, but also to sense the presence of God in the universe. For more contemporary Friends, simplicity is valued as an approach to a sustainable lifestyle that is more connected to nature and economic justice and less focused on consumption. This environmental side to the commitment to the simplicity testimony reaches across the Quaker world, connecting theologically liberal and evangelical organizations of Friends.

---

10. Helmuth et al. "Changing World View and Friends Testimonies," 4.

11. Thomas Berry, *The Dream of the Earth* (San Francisco: Sierra Club Books, 1988); Thomas Berry and Brian Swimme, *The Universe Story: From the Primordial Flaring Forth to the Ecozoic Era—A Celebration of the Unfolding of the Cosmos* (San Francisco: HarperCollins, 1992); Thomas Berry, *The Great Work: Our Way into the Future* (New York: Bell Tower, 1999).

Quakers and Science

In addition to this clear linking of the core Quaker testimonies, and the resonance with the Quaker theology that there is that of the divine in everything, examining the roots of the Society of Friends also illustrates the attraction to the Universe Story. From their earliest beginnings, Quakers saw science as compatible, complementary to their faith. Indeed, George Fox, a key founder, sought "unity with the creation," and a contemporary, Jacob Bauthumley, pronounced that "I see God is in all Creatures . . . and every green thing" (Bauthumley later became a Quaker, having visited Fox during one of his many imprisonments).[12] William Penn, the noted seventeenth-century Quaker behind the founding of Pennsylvania, expressed a similar interpretation in his admonition for studying nature: "For how could Man find the Confidence to abuse it, while they should see the Great Creator stare them in the Face, in all and every part thereof?"[13] This was not an uncommon sentiment among dissenting groups. John Wesley, writing roughly a century later (1748), declared "that God is in all things, and that we are to see the Creator in the face of every creature; that we should use and look upon nothing as separate from God, which indeed is a kind of practical atheism."[14]

As John Brooke and Geoffrey Cantor point out in their chapter "A Taste for Philosophical Pursuits: Quakers in the Royal Society in London," Quakers were disproportionately present in the Royal Society (for Improving Natural Knowledge) from the early decades of Quakerism, as the two were founded around the same time in the 1660s.[15] Their numbers rose steadily so that, by 1900, they made up 35 percent of the Royal Society, a preeminent association of scientists and mathematicians with a royal charter and an exclusive membership. Quakers saw no conflict in studying science, seeing it as one avenue to the truth they sought. Their distinctive approach was apparent in their embrace of Darwin's writings, a pivotal moment in science and religion, as exemplified in the writing and illustrious career of Silvanus Thompson. Brooke and Cantor describe an essay he wrote (1871) roughly a decade after the publication of Darwin's *On the Origin of Species*: "He found nothing upsetting in Darwin's theory since it did not conflict with religion.

---

12. As quoted in the article, "QEW: A Nature Walk for All Friends by Os Cresson," on the Quaker Earthcare Witness website, http://www.quakerearthEarthcare.org/article/qew-nature-walk-all-friends, where it is referenced from Nigel Smith, *A Collection of Ranter Writings from the 17th Century* (London: Junction Books, 1983), 232.

13. William Penn, *Some Fruits of Solitude in Reflections and Maxims* (1682).

14. John Wesley, Sermon 23, "Upon Our Lord's Sermon on the Mount III," I.11.

15. John Brooke and Geoffrey Cantor, *Reconstructing Nature: The Engagement of Science and Religion* (Edinburgh: T & T Clark, 1998), 218–313.

... He welcomed evolution because, like other scientific theories, it displayed God's design and purpose in the physical world."[16] Perhaps Quakers were particularly drawn to Darwin's work because of the preponderance of Quaker naturalists and botanists, observers of the natural world around them. George Fox and William Penn, key early Quakers, both emphasized that Quakers should be "competent botanists," with Fox admonishing that children be taught the "nature of herbs, roots, plants and trees."[17] William Penn admonished, "It were Happy if we studied Nature more in natural Things; and acted according to Nature; whose rules are few, plain and Reasonable."[18] Interestingly, Brooke and Cantor point out that there were very few Quaker physicists and mathematicians in the first hundred-plus years of the society, although many did research on meteorology and astronomy.[19]

The Quaker embrace of science was furthered, like that of many Protestant groups, by their approach to the Bible as a source of inspiration, but not a strict foundation and rule book for Christianity for which scientific knowledge might prove to be a challenge. Because of their commitment to the individual's responsibility to listen to their own inner light, form their own views, and from that, seek to live a life of moral responsibility, any doctrinal emphasis on biblical creationism frequently had no appeal.

Each Quaker seeks to center down, to find their own inner light, which is one manifestation of the light of Christ, of the divine, of God or however it is named. Because Quakers claim that each person finds the inner light in their own path and perception, it has long meant that Quakers are open to different spiritual paths. Although their origins are Christian, and a majority worldwide would still see themselves as Christian, Quakers are open to the spiritual insights of all religious traditions. There is no conflict with seeing the story of the universe, as so eloquently told in *Journey of the Universe*, as a sacred story capable of illuminating all aspects of our lives.

Further, the Quaker practice of discernment based on respectful listening, and seeking to hear all perspectives, meant that scientific truths were not perceived as threatening, but rather as illuminating. The film and book *Journey of the Universe* bring home this point of illumination; renowned scholar of

---

16. Ibid., 297.

17. As quoted in ibid., 302.

18. William Penn, *Some Fruits of Solitude in Reflections and Maxims*. This is the same passage that includes the statement, "For how could Man find the Confidence to abuse it, while they should see the Great Creator stare them in the Face, in all and every part thereof?"

19. The lack of physicists and mathematicians, Brooke and Cantor point out, may be due to the fact that for most of the eighteenth century, Cambridge, the primary educational institution for physicists and mathematicians, refused to grant degrees to dissenters, *Reconstructing Nature*, 302.

religion and science Philip Clayton, in the introduction to the book *Quakers and the New Story*, comments that "a variety of interpreters of science" are "suggesting that the 'new story' that science is telling actually supports something very much like traditional Quaker ways of conceiving reality."[20]

Right Relationship

A key organization of Friends' work on sustainability, Quaker Earthcare Witness (QEW), points out that the contemporary concept of sustainability is already present in the Quaker understanding of "right relationship" that can be traced back to John Woolman (a key eighteenth-century American Quaker). If humans are not in "right relationship" with the more than human natural world, then the Quaker dedication to a world without war and with just social relations cannot be achieved.[21] Perhaps this larger Quaker vision is best summed up in a QEW statement:

> We are called to live in right relationship with all Creation, recognizing that the entire world is interconnected and is a manifestation of God . . . the Truth that God's Creation is to be respected, protected, and held in reverence in its own right, and the Truth that human aspirations for peace and justice depend upon restoring the Earth's ecological integrity.[22]

This notion of right relationship can also be seen in the QEW motto: "Seeking emerging insights into right relationship and unity with nature."[23] Both the vision of that of God in every thing, in unity with nature, and the Quaker testimonies such as simplicity and right relations provide a religious cosmology, as Grim and Tucker state more generally, that locates "the human in the larger contexts of the universe and Earth processes that provide a deep framework for valuing nature."[24] *Journey of the Universe* is an exploration of humans, but not just humans, in right relationship.

---

20. Philip Clayton, "Quakerism and the New Story," in *Quakers and the New Story*, 1.

21. QEW is an outgrowth of the Friends Committee on Unity with Nature that began in 1987 in response to the work of Quaker and environmental activist Marshall Massey. The name Unity in Nature stems from George Fox's concern to be in unity with the Creation.

22. See http://www.quakerearthcare.org/QEWPastandFuture/QEW_Future/QEW_Future.htm.

23. Ibid.

24. John Grim and Mary Evelyn Tucker, *Ecology and Religion* (Washington, DC: Island Press, 2014), 154.

In the Universe Story, Quakers find their mystical insights, their testimonies as principles of living, and their centuries-long commitment to understanding the world through science united in a vision for the twenty-first century. It is a new understanding of being in right relations. *Journey of the Universe* in its rich portrayal in film is a new vision, a new understanding of what it means to be human, to be a creature of the universe. As the book version aptly states in the first chapter, "Every time we are drawn to look up into the night sky and reflect on the awesome beauty of the universe, we are actually the universe reflecting on itself. And this changes everything."[25]

The ability to see ourselves as a planet, as a result of the exploration of space, inspired Berry's work, as it certainly inspired astronauts.[26] I want to end with a poem featured on the QEW website that reflects on what it means for us, through the astronauts, to be able to see ourselves, to not only "look up into the night sky" but to be able to look back at the place in the universe that we call home. *Journey of the Universe* tells the story of that wondrous planet and our place on and in it with the same sense of awe and vision that inspired so many astronauts. Their images, and the imagery of the film, can inspire us. And that vision of the Earth as a fragile oasis, as astronauts have commented and as *Journey* shows, is part of what changes everything.[27]

Epiphany
By Angela Manno

Out of the depths I cry unto you,
Oh Presence, Maker of this perfect world
Out of this silence but for my own breath
Out of the darkness and weightlessness
The black infinite night
Encrusted with countless stars, cool, distant, white.
I turn my gaze, not ready fully to take in
The Spectacle below:
A slow procession of green and earth

---

25. Brian Thomas Swimme and Mary Evelyn Tucker, *Journey of the Universe* (New Haven, CT: Yale University Press, 2014), 2.

26. Edgar Mitchell of the Apollo 14 crew commented, "I was gazing out of the window, at the Earth, moon, sun, and star-studded blackness of space in turn as our capsule slowly rotated. . . . Gradually, I was flooded with the ecstatic awareness that I was a part of what I was observing. Every molecule in my body was birthed in a star hanging in space. I became aware that everything that exists is part of one intricately interconnected whole" (Richard Schiffman, "Bigger Than Science, Bigger Than Religion," *Yes! Magazine*, Spring 2015).

27. Space.com, "Planet Earth Is a Fragile Oasis," March 2015, http://www.space.com/6603-planet-earth-fragile-oasis-astronauts.html.

Earthen landmasses
Mountains undulating
Grazed by gentle, drifting clouds
A miraculous harmony of soft and brilliant hues
Vast oceans
Shining with sunlight
From a well of tears
Come flooding . . . a sense of Being,
Connection, and yes Love
It's all alive, life within Life.
Garden of Eden
Suspended in space,
Nothing holding it up
No fulcrum upon which it spins
But wait . . .
Two new continents have come into view
And now, suddenly
The other side of the world is before me!
I must think:
Have I some special right to gaze upon such Beauty,
The living host of all we know: all of history and music and poetry
And art and birth and death and love and tears?
What have I done to merit this moment?
This glimpse of Divinity,
Devastating Beauty,
Mother of us all?
Though I float miles above,
I am a part of that Life,
Tied to her through her breath
Which I take with me
In a tank on my back
I am afloat in the infinite sea
My heart races
There is no up or down . . .
But there is worship
There is the bursting of my heart
There is the cry from the most profound depths:
See where you live, Humanity!
See your own Self!
This tiny, miraculous island of life

Adrift in the vast cosmos
We are so alone, so fragile.
There is nothing more glorious
So said the Saint:
"Because the divine could not image itself forth in any one being, it cre-
    ated the great diversity of things so that what was lacking in one
    would be supplied by the others and the whole universe together
    would participate in and manifest the divine more than any single
    being."
And the writer of Hindu texts: "I am Beauty among beautiful things."
For all eternity there is but one Earth.
I will tell them, I will make them understand . . .
Plunging back into you in a ball of fire,
I will not forget your face,
I will remember you, Jewel of the Universe,
Most Holy Ground, Home.[28]

---

28. "Epiphany," in *Quaker Earth Care*, March 2015, © Angela Manno 2009.

Appendix I
History of Journey of the Universe

Journey of the Universe is a project that is more than three decades in the making. It is in the lineage of Thomas Berry's call for a "New Story" that appeared in his article for *Teilhard Studies* in 1978. Berry felt that we needed to bring science and humanities together in an integrated cosmology that would guide humans into the next period of human-Earth relations. Ten years later, "The New Story" was included in Berry's book *Dream of the Earth*. This inspired a ten-year collaboration between the evolutionary cosmologist Brian Swimme and Thomas Berry, resulting in *The Universe Story*, which was published in 1992. This is the first book that narrates evolution as a story with a comprehensive vision of the role of humans in the narrative. *Journey of the Universe* is the first time this story is told in film. It also was the result of a decade-long collaboration.

Working with scientists and scholars from the history of religions, Brian Swimme and Mary Evelyn Tucker wrote the film script and book. They organized several weeklong summer workshops with scientists and humanists at the Whidbey Institute in Washington State to discuss these ideas. After completing the script, they made three trips to the Greek island of Samos to film with the director, David Kennard, who was part of the *Cosmos* series with Carl Sagan. John Grim, a coproducer, was an advisor to the film, along with Thomas Berry.

Patsy Northcutt edited the film in post-production and was assisted by Catherine Butler. In addition to the film, Mary Evelyn produced the twenty-part Conversation series of interviews. These included ten scientists and historians relating the story of evolution as well as ten environmentalists illustrating how the story inspires the Great Work of change in the areas of economics, energy education, food systems, cities, and race.

The film was completed in 2011 and premiered at a conference in March at the Yale University School of Forestry and Environmental Studies. This conference brought together scientists and humanists to reflect on the cosmological implications of *Journey*. Since that time, the film has been shown in film festivals, museums, universities, as well as religious and community organizations. *Journey* premiered on KQED television in San Francisco in June 2011 and has since been broadcast on 77 percent of the PBS stations in the United States. In June 2012 *Journey* won a regional Emmy Award in Northern California for best documentary film. It has been shown on every continent, with further screenings being planned in India, China, Europe, and Latin America.

A substantial website with articles and a curriculum for teachers brings the *Journey* trilogy together: http://www.journeyoftheuniverse.org.

Appendix II
Thomas Berry (1914–2009)

Thomas Berry was born in Greensboro, North Carolina, where he spent his early childhood and where he returned when he was eighty years old. He died peacefully there on June 1, 2009. Named William Nathan after his father, he was the third child of thirteen, of whom four siblings remain. He entered the Passionist Order in high school, and upon ordination he took the name Thomas, after Thomas Aquinas, whose *Summa Theologica* he admired.

He received his PhD from the Catholic University of America in European intellectual history with a thesis on Giambattista Vico. Widely read in Western history and theology, he also spent many years studying and teaching the cultures and religions of Asia. He lived in China in 1948 where he met the Asian scholar and Confucian specialist Wm. Theodore de Bary. Their collaboration led to the founding of the Asian Thought and Religion Seminar at Columbia. Thomas authored two books on Asian religions, *Buddhism* and *Religions of India*, both of which are distributed by Columbia University Press.

For more than twenty years, Thomas directed the Riverdale Center of Religious Research along the Hudson River. During this period he taught at Fordham University, where he chaired the history of religions program. He directed some twenty doctoral theses, including those of John Grim and Brian Brown, as well as many master's theses, including those of Mary Evelyn Tucker and Kathleen Deignan. From 1975 to 1987 he was president of the American Teilhard Association, and it was from Teilhard de Chardin that he was inspired to develop his idea of a Universe Story. With Brian Swimme he wrote *The Universe Story* (HarperSanFrancisco, 1992), which arose from a decade of collaborative research.

His major contributions to the discussions on the environment are in his books *The Dream of the Earth* (Counterpoint, 2015), *The Great Work: Our Way into the Future* (Bell Tower, 1999), and *Evening Thoughts: Reflecting on Earth as Sacred Community* (Counterpoint, 2015). His final two books focusing on world religions and on Christianity were published in September 2009: *The Sacred Universe: Earth, Spirituality, and Religion in the Twenty-First Century* by Columbia University Press and *The Christian Future and the Fate of Earth* by Orbis Books. *Thomas Berry: Selected Writings on the Earth Community* was published by Orbis Books in 2014.

Berry is buried at Green Mountain Monastery in Greensboro, Vermont.

Contributors

Carl Anthony, architect, author, and urban/suburban/regional design strategist, is cofounder of the Breakthrough Communities Project. He has served as acting director of the Community and Resource Development Unit at the Ford Foundation, responsible for the foundation's worldwide programs in the fields of environment and development, and community development. He was founder and, for twelve years, executive director of the Urban Habitat Program in the San Francisco Bay Area.

Whitney Bauman is associate professor in the Department of Religious Studies and in the Honors College at Florida International University in Miami. He teaches courses in religion and science; religion, gender, and nature; earth ethics; and technology and human values. He is coeditor of *Grounding Religion: A Fieldguide to the Study of Religion and Ecology* (2010); *Inherited Land: The Changing Grounds of Religion and Ecology* (2011); and *Science and Religion: One Planet, Many Possibilities* (2014); and the author of *Theology, Creation, and Environmental Ethics* (2009), and *Religion and Ecology: Developing a Planetary Ethic* (2014). Bauman's research interests in religion, nature, and globalization have most recently taken him to Indonesia, Malaysia, the Philippines, and India. During the 2015–2016 academic year he will be in Germany on a Humboldt Fellowship, working on a book that will examine the religious influences on Ernst Haeckel's understanding of ecology.

Bede Benjamin Bidlack is assistant professor of theology at Saint Anselm College, where he teaches courses on sacramental theology, comparative theology, and interreligious dialogue. He is the author of *In Good Company: Body and Divinization in Pierre Teilhard de Chardin and Daoist Xiao Yingsou* (2015). His other publications include works in theology, philosophy, and sinology.

Rev. Stephen Blackmer is chaplain of Church of the Woods in Canterbury, New Hampshire, and executive director of Kairos Earth, a nonprofit organization dedicated to renewing the Earth by connecting conservation with spiritual practice. He also serves as priest associate at St. Stephen's Episcopal Church in Pittsfield, New Hampshire. Prior to being ordained as an Episcopal priest in 2013, Steve worked for twenty-five years to conserve the Northern Forest region of New Hampshire, Maine, Vermont, and New York, including as president and founder of the Northern Forest Center. He is an Environmental Fellow with the Robert and Patricia Switzer Foundation, was a Bullard Fellow at the Harvard Forest, and was awarded the National Conservation Partnership Award in 2003. He served also as director of conservation programs for the Appalachian Mountain Club and director of policy for the Society for the Protection of NH Forests. Steve holds a bachelor's degree in anthropology from Dartmouth College and master's degrees in both forestry and religion from Yale University.

Brian Edward Brown was an undergraduate and graduate student of Thomas Berry at Fordham University, where he earned his doctorate in the history of religions, specializing in Buddhist thought. He subsequently earned his doctorate in law from New York University. Currently he is Full Professor of Religious Studies at Iona College, New Rochelle, New York. He is the cofounder of the Thomas Berry Forum for Ecological Dialogue at Iona, as well as being one of the founding faculty of the integral environmental studies major at Iona, a joint venture of the Departments of Biology, Political Science, and Religious Studies. He teaches classes in Buddhism, the Religions of China, Religion and the Constitution, Religion and Cosmology, and Religion and the Natural World. He is the author of two principal texts: *The Buddha Nature: A Study of the Tathagatagarbha and Alayavijnana* (1991), and *Religion, Law, and the Land: Native Americans and the Judicial Determination of Sacred Land* (1999). He is coeditor of *Augustine and World Religions* (2008). Among his other publications are articles that have addressed the ecological implications of the Buddhist and Native American tribal traditions, as well as the Earth jurisprudence of Thomas Berry.

Rev. Dr. John Chryssavgis, archdeacon of the Ecumenical Patriarchate, was born in Australia, studied theology in Athens, and completed his doctorate in Oxford. After several months in silent retreat on Mount Athos, he cofounded St Andrew's Theological College in Sydney. He taught theology in Sydney and Boston. As special theological advisor to the Office of Ecumenical and Inter-Faith Affairs of the Greek Orthodox Archdiocese of America, he also coordinates the Social and Moral Issues Commission of the Orthodox Churches in America and serves as ecological advisor to Ecumenical Patriarch Bartholomew. His publications include *Beyond the Shattered Image: Insights into an Orthodox Christian Ecological Worldview* (1999); *Light through*

Darkness: The Orthodox Tradition (2004); and *In the Heart of the Desert: The Spirituality of the Desert Fathers and Mothers* (2008). He has edited three volumes containing selected writings of Ecumenical Patriarch Bartholomew, the third of which, titled *On Earth as in Heaven* (2011), deals with the patriarch's ecological vision and activities. John lives in Maine.

Anne Marie Dalton received her PhD from the Catholic University of America, completing her dissertation on the work of Thomas Berry. Her MA is from Fordham University, where she studied with Thomas Berry. She is emeritus professor in the Department of Religious Studies, with a cross appointment to International Development Studies at Saint Mary's University, Halifax, Canada. She has worked on Canadian International Development Agency (CIDA) projects dealing with the environment in China, Vietnam, and Mongolia. Her teaching and published works involve the areas of religion and ecology, women and religion, and religion and international development. She has two published books in the area of religion and ecology, *A Theology for the Earth: The Contributions of Thomas Berry and Bernard Lonergan* (1999) and *Ecotheology and the Practice of Hope* (coauthored with Henry Simmons, 2010). Anne Marie is also a member of the Interfaith Coalition on Climate Change.

Kathleen Deignan is associate professor of theology at Iona College. Her current interests are in the areas of classical and contemporary spirituality with a focus on creation theology and spirituality. Her work is devoted to transmitting the spiritual and intellectual legacies of Thomas Merton and her mentor, Thomas Berry, in a variety of academic, artistic, and pastoral formats. A GreenFaith Fellow, Dr. Deignan completed a two-year postdoctoral training for religious environmental leadership (www.GreenFaith.org). Dr. Deignan was the founder of the Iona Peace and Justice Studies Program and the Iona Peace Institute in Ireland (1988–1995). She is the founder and director of the Iona Spirituality Institute, which sponsors a variety of programs for spiritual enrichment. A sacred song writer and psalmist, Sister Deignan has composed over two hundred songs for liturgy and prayer, which have been published in a dozen CDs by Schola Ministries, a project in service to the liturgical and contemplative arts.

Ilia Delio, OSF, is Josephine C. Connelly Chair in Theology at Villanova University. She was a senior research fellow at Woodstock Theological Center in the area of science and religion, focusing on transhumanism and evolution and religion, especially the thought of Pierre Teilhard de Chardin.

She holds a doctorate in pharmacology from Rutgers University Graduate School of Biomedical Sciences and a doctorate in historical theology from Fordham University. She is the author of fifteen books and numerous articles. Her recent books include *Christ in Evolution* (2008); *The Emergent Christ: Exploring the Meaning of Catholic in an Evolutionary Universe* (2011); *The Unbearable Wholeness of Being: God, Evolution, and the Power of Love* (2013) (winner of the 2014 Silver Nautilus Book Award and a 2014 Catholic Press Association Book Award in Faith and Science); and *Making All Things New: Catholicity, Cosmology, Consciousness* (2015). She is editor of *From Teilhard to Omega: Co-creating an Unfinished Universe* (2014) and general editor of the Catholocity in an Evolving Universe series.

Heather Eaton holds an interdisciplinary doctorate in theology, feminism, and ecology from Saint Michael's College at the University of Toronto and is a professor in conflict studies at Saint Paul University in Ottawa, Canada. She works in engaging religions on ecological, social, and ethical issues. She has published extensively on ecofeminism, ecospirituality, cosmology, and ecojustice, as well as the intersection of science, evolution, and religion. Her main publications are *The Intellectual Journey of Thomas Berry: Imagining the Earth Community*, ed. (2014); *Ecological Awareness: Exploring Religion, Ethics, and Aesthetics* (with Sigurd Bergmann, 2011); *Introducing Ecofeminist Theologies* (2005); *Ecofeminism and Globalization: Exploring Religion, Culture, Context* (with Lois Ann Lorentzen, 2003); editor of *Worldviews: Environment, Culture, Religion*, special issue: "Evolution" (2007); *Ecotheology*, "Gender, Religion, and Ecology" (2006); *Worldviews: Environment, Culture, Religion*, special issue: "Thomas Berry" (2001), plus dozens of book chapters and articles. Heather works as a socially engaged academic with various national and international groups on religion, ecology, social issues, nonviolence, and peace.

Peter Ellard is the associate vice president for academic affairs at Siena College, where he teaches classes in the first-year seminar, environmental studies, and religious studies. He is currently teaching a first-year seminar on science and religion where *Journey of the Universe* and the ideas of Thomas Berry are a central component. At Siena he was named Lecturer of the Year by the Student Senate in 2011 and 2013 and Administrator of the Year in 2004 and 2012. His initial scholarship was on the ideas of the twelfth-century School of Chartres. Now he writes on the ideas of Thomas Berry and the effects of climate change. He has spent many hours among the Thomas Berry archives at Harvard, and he looks forward to the completion of his first published work on its content.

John Grim is a senior lecturer and research scholar at Yale University, where he has appointments in the School of Forestry and Environmental Studies as well as the Divinity School and the Department of Religious Studies. He teaches courses in Native American and indigenous religions and world religions and ecology. He has undertaken fieldwork with the Crow/Apsaalooke people of Montana and the Salish people of Washington State. His published works include *The Shaman: Patterns of Religious Healing among the Ojibway Indians* (1983), and, with Mary Evelyn Tucker, a coedited volume titled *Worldviews and Ecology* (1994). With Mary Evelyn Tucker, he directed a ten-conference series and book project at Harvard on "World Religions and Ecology." He edited *Indigenous Traditions and Ecology: The InterBeing of Cosmology and Community* (2001) and coedited the Daedalus volume titled *Religion and Ecology: Can the Climate Change?* (2001). He is cofounder and codirector of the Forum on Religion and Ecology at Yale with Mary Evelyn Tucker.

George Handley teaches interdisciplinary humanities at Brigham Young University, where he serves as chair of the Department of Humanities, Classics, and Comparative Literature. He has published widely in the areas of comparative literature of the Americas and environmental humanities, and has published a number of seminal essays on Mormonism and the environment; coedited *Stewardship and the Creation: LDS Perspectives on the Environment*; and authored an environmental memoir, *Home Waters: A Year of Recompenses on the Provo River*, which explores LDS theology, Mormon pioneer history, and contemporary environmental awareness in the American West. He serves on the boards of Utah Interfaith Power and Light and LDS Earth Stewardship, and is involved in various other environmental fronts. He, his wife, and four children are active in their church community.

John F. Haught (PhD, Catholic University, 1970) is Distinguished Research Professor, Georgetown University, Washington, DC. He was formerly professor in the Department of Theology at Georgetown University (1970–2005) and chair (1990–1995). His area of specialization is systematic theology, with a particular interest in issues pertaining to science, cosmology, evolution, ecology, and religion. Among his numerous books, the most recent are *Resting on the Future: Catholic Theology for an Unfinished Universe* (2015), and *Science and Faith: A New Introduction* (2012), translated into Chinese and Lithuanian. Haught has also authored numerous articles and reviews. He lectures internationally on many issues related to science and religion. In 2002 he was the winner of the Owen Garrigan Award in Science and Religion, in 2004 the Sophia Award for Theological Excellence, and in 2008 a "Friend of Darwin Award" from the National Center for Science Education. He testified for the plaintiffs in the Harrisburg, Pennsylvania, "intelligent design trial" (*Kitzmiller et al. vs. Dover Board of Education*). In April 2009 he received an honorary doctorate from Louvain University in Belgium. In Fall 2008 he held the D'Angelo Chair in the Humanities at St. John's University in New York City. He and his wife, Evelyn, have two sons and live in Falls Church, Virginia.

Mary E. Hunt is a feminist theologian who is cofounder and codirector of the Women's Alliance for Theology, Ethics, and Ritual (WATER) in Silver Spring, Maryland. A Catholic active in the women-church movement and on LGBTIQ matters, she lectures and writes on theology and ethics with particular attention to liberation issues. She is an editor of *A Guide for Women in Religion: Making Your Way from A to Z* (2004, 2014) and coeditor with Diann L. Neu of *New Feminist Christianity: Many Voices, Many Views* (2010).

James Jenkins received a masters from Yale Divinity School and Berkeley Divinity School at Yale. As sustainability coordinator at YDS, he created the Nourish New Haven food justice and sustainability conference, coordinated the drafting of the school's three-year sustainability action plan, organized events for this year's YDS theme of building sustainable communities, and helped start Ministry for the Earth Community, a leadership development program combining theological, practical, and experiential education for divinity school students. Before YDS, James taught English literature for nine years in secondary schools and completed a graduate degree from the Bread Loaf School of English at Middlebury College. He seeks to support seminaries, schools, churches, and wider communities responding to the world's environmental, social, and spiritual needs.

Willis Jenkins is associate professor of religion, ethics, and environment at the University of Virginia and is author of two award-winning books: *Ecologies of Grace: Christian Theology and Environmental Ethics* (2008) and *The Future of Ethics: Sustainability, Social Justice, and Religious Creativity* (2013).

Laurel Kearns is associate professor of sociology and religion and environmental studies at Drew Theological School and the Graduate Division of

Religion. She received her MA and PhD in sociology of religion from Emory University, and has researched, published, and given talks around the globe on religion and environmentalism for over twenty years. In addition to helping found the Green Seminary Initiative, now hosted at Drew, she has been a board member of GreenFaith and is now serving on the sustainability committees of Drew University and the Religion and Ecology group of the American Academy of Religion. A list of her publications, in addition to *EcoSpirit*, coedited with Catherine Keller, can be found on her website. She has contributed chapters to volumes such as *The New Evangelical Social Engagement*; *The Oxford Handbook on Climate Change and Society*; *The Blackwell Companion to Modern Theology* and their *Companion to Religion and Social Justice*; *Religion in Environmental and Climate Change*; *God's Earth Is Sacred*; *Love God, Heal Earth*; *Earth and Word*; the *Encyclopedia of Religion and Nature*; and *The Spirit of Sustainability*, as well as many others.

Catherine Keller is professor of constructive theology in the Theological School and Graduate Division of Religion of Drew University. Books she has authored include *From a Broken Web: Separation, Sexism and Self*; *Apocalypse Now & Then*; *God & Power*; *Face of the Deep: A Theology of Becoming*; *On the Mystery: Discerning God in Process*; and most recently, *Cloud of the Impossible: Negative Theology and Planetary Entanglement*. She has coedited several volumes of the Drew Transdisciplinary Theological Colloquium, including *Postcolonial Theologies*, *Ecospirit*, *Apophatic Bodies*, *Polydoxy: Theology of Multiplicity and Relation*, and the forthcoming *Common Good/s: Ecology, Economy, and Political Theology*. She persists in interlacing process, Continental, and ecosocial, indelibly feminist thinking with theological questions.

Chris Loughlin makes her home at Crystal Spring Center in Plainville, Massachusetts. Crystal Spring collaborates with small-scale agricultural programs, land conservation, and local efforts in becoming a living embodiment of a bio-regionally appropriate culture. Her particular work is assisting religious orders in entering a new relationship with their lands. The MA Land Trust Coalition and the Religious Lands Conservancy Project partner to preserve common values that protect and conserve land.

Beth Norcross is the founder and director of the Center for Spirituality in Nature, which offers experiences in nature designed to deepen spirituality and encourage loving relationships with all Earth's creatures (www.centerforspiritualityinnature.org). She is also the cofounder of the national Green Seminary Initiative (www.greenseminaries.org), which encourages seminaries to integrate care and engagement with creation into all aspects of theological education. Beth serves as chair of the steering committee for the Greater Washington Interfaith Power and Light (www.gwipl.org). As adjunct faculty at Wesley Theological Seminary, Beth teaches two regular classes and guest-lectures in ecotheology and ecospirituality. She has also developed a number of educational resources, including a five-session religious study guide to the Ken Burns film *The National Parks: America's Best Idea*; "Sacred Waters," an adult education curriculum; and "Building a Firm Foundation," an ecofriendly building guide for churches. She also authored an article titled "Eye on the Sparrow," which discusses Howard Thurman's close relationship with the natural world, in *Sojourners* magazine (August 2012). After receiving a BA in mathematics from Duke University as well as a master of forestry degree, Beth enjoyed a career in the environmental field. She worked as professional staff for the US Senate National Parks and Forests subcommittee, as well as vice president for conservation for American Rivers. She holds a master of theological studies degree from Wesley Seminary in Washington, DC, with a concentration in ecotheology, and a doctor of ministry degree from Wesley. Her dissertation topic was "The National Parks as Sacred Ground." Beth is the author of *Use Your Fingers, Use Your Toes* (2004), a step-by-step guide to everyday math. She has three grown children and lives with her husband in Arlington, Virginia.

Dennis O'Hara is an associate professor and the director of the Elliott Allen Institute for Theology and Ecology at the Faculty of Theology in the University of St. Michael's College, Toronto, where he is also a core faculty member in the Corporate Social Responsibility certificate program. He is also an associate member of the graduate faculty at the School for Environment at the University of Toronto. In addition to teaching courses in ecological theology, ethics, and sustainability, he has worked for the World Health Organization and Health Canada preparing policy positions and research papers. He regularly delivers popular and academic lectures in Canada and the United States, but he has also lectured in Europe and South Korea. Prior to becoming a theologian, he practiced as a chiropractor and naturopathic doctor.

M. Paloma Pavel is an author and ecopsychologist, and a frequent lecturer and keynote presenter nationally and internationally on the theory of living systems and urban sustainability. Her academic background includes graduate study at the London School of Economics and Harvard University, and she is currently visiting faculty at the UC Davis Center for Regional Change and Pacifica Graduate Institute. Recent publications include *Breakthrough Communities, Sustainability and Justice in the Next American Metropolis*

(2009) and *Climate Justice: Frontline Stories from Groundbreaking Coalitions in California* (forthcoming).

Russell Powell is a doctoral student at Princeton Theological Seminary. He has served as a research fellow at Yale University's Beinecke Library and as a research assistant with the Forum on Religion and Ecology, also at Yale. He is chaplain at Blair Academy in Blairstown, New Jersey.

Larry Rasmussen is Reinhold Niebuhr Professor Emeritus of Social Ethics, Union Theological Seminary, New York City. His most recent book, *Earth-Honoring Faith: Religious Ethics in a New Key* (2013), received Nautilus 2014 Book Awards as the Gold Prize winner for ecology/environment and the Grand Prize winner for best book overall (twenty-seven categories).

Matthew T. Riley created the Educational Curriculum for the *Journey of the Universe* project and he is active in hosting film screenings of the Journey of the Universe film. Matt is a Lecturer in Christianity and Ecology at Yale Divinity School and the Yale School of Forestry and Environmental Studies. He also teaches summer courses in Environmental Ethics and Religion and Ecology at the Yale Interdisciplinary Center for Bioethics. He is engaged as a Research Associate at the Forum on Religion and Ecology at Yale, and he serves as the Review Editor for *Worldviews: Global Religions, Culture, and Ecology*. Matt's research interests include the relationship between religious ideas and environmental values as seen through the lens of social theory, the intersection of religion and animals broadly considered, and environmental ethics. Matt's publications and dissertation focus on the legacy of Lynn T. White, Jr.

Barbara Rossing is professor of New Testament at the Lutheran School of Theology at Chicago, where she also directs the seminary's Environmental Ministry Emphasis and teaches in the Zygon Center for Religion and Science. She received the bachelor of arts degree in geology from Carleton College, the master of divinity degree from Yale University Divinity School, and the doctor of theology degree from Harvard University. She is the author of *The Rapture Exposed: The Message of Hope in the Book of Revelation* and *Journeys through Revelation: Apocalyptic Hope for Today*, as well as articles and book chapters on the Bible and ecology. She is the on-screen host of the new *Painting the Stars* DVD from Living the Questions. An ordained pastor in the Evangelical Lutheran Church in America, she has participated in international consultations on theology and climate change, including Lutheran

World Federation delegations to the United Nations Climate Change summits in Copenhagen (2009) and Cancun (2010). She chairs the Ecological Hermeneutics section of the Society of Biblical Literature.

Patricia A. Siemen works to bring forth transformational changes in law, culture, and evolutionary consciousness through serving as the director of the Center for Earth Jurisprudence, Barry University School of Law, Orlando, Florida. She is a civil attorney and Dominican Sister. Since 2006 the center has taken an interdisciplinary approach to changing the Western legal system by engaging law as well as science, ethics, economics, and the humanities through a cosmological context. CEJ believes that all of Nature has inherent rights to exist and flourish, and humanity has obligations and duties to protect and reverence those rights by living in balance and harmony with natural systems. Pat teaches Earth jurisprudence, and the center offers workshops, nature immersions, and publications that invite people to deepen their relationship with Nature. Pat holds a juris doctorate from Northeastern University, Boston; a master's in public affairs from the University of Texas, Austin; and a master's in culture and spirituality from Holy Names University in Oakland, California.

Frederick Simmons's research and teaching examine the moral implications of Christian theological commitments and the relationships between philosophical and theological ethics. He is completing a book on the ethical and potential soteriological significance of ecology for contemporary Christians, and is coediting a volume on love and Christian ethics. He is a fellow at the Center for Theological Inquiry at Princeton. He has previously taught at Amherst College, La Universidad Politécnica Salesiana, and La Pontificia Universidad Católica del Ecuador.

Brian Thomas Swimme did his doctoral work in gravitational dynamics at the University of Oregon in the Department of Mathematics. He is currently professor of cosmology at the California Institute of Integral Studies, San Francisco, California. He is creator of the DVD series "Canticle to the Cosmos," "Earth's Imagination," "The Powers of the Universe"; and is author of *The Universe is a Green Dragon* (1984), *The Universe Story* (1992) written with Thomas Berry, *The Hidden Heart of the Cosmos* (1996), and *Journey of the Universe* (2011) written with Mary Evelyn Tucker.

Mary Evelyn Tucker is a senior lecturer and senior research scholar at Yale University, where she teaches in a joint master's program between the School of Forestry and Environmental Studies and the Divinity School. She is a co-founder and codirector with John Grim of the Forum on Religion and Ecology at Yale. Together they organized a series of ten conferences on World Religions and Ecology at the Center for the Study of World Religions at Harvard. They are series editors for the ten volumes from the conferences distributed by Harvard University Press. She coedited the volumes on *Confucianism and Ecology*, *Buddhism and Ecology*, and *Hinduism and Ecology*. She is also the author of *Worldly Wonder: Religions Enter Their Ecological Phase* (2003), and with John Grim of *Ecology and Religion* (2014). She served on the International Earth Charter Drafting Committee from 1997 to 2000 and was a member of the Earth Charter International Council. In 2011 Tucker completed *Journey of the Universe* with Brian Swimme, which includes a book from Yale University Press, an Emmy Award–winning film on PBS and Netflix, and an educational series of twenty interviews.

Cristina Vanin is associate professor of theology in the Department of Religious Studies, as well as associate dean and director of the Master of Catholic Thought program, at St. Jerome's University in Waterloo, Ontario. Her areas of research include ecotheology, feminist theologies, and Christian ethics. Dr. Vanin encountered the thought of Thomas Berry when she was an associate of Holy Cross Centre, Port Burwell, Ontario, and completing a master of divinity degree at St. Michael's College, Toronto. Berry, as well as the work of Bernard Lonergan, continue to be the central focus of Dr. Vanin's teaching, research, and service. As an engaged scholar, Dr. Vanin is active on the Ignatius Jesuit Centre Board of Directors, is a past chair of the Habitat for Humanity Waterloo Region Board of Directors, the SJU representative to the Catholic Education Partnership Committee for the Diocese of Hamilton, and president-elect of the Canadian Theological Society. Her most recent publication is "Understanding the Universe as Sacred," in *The Intellectual Journey of Thomas Berry: Imagining the Earth Community*, edited by Heather Eaton (2014).

Paul Waldau is an educator, scholar, and activist working at the intersection of animal studies, law, ethics, religion, and cultural studies. A professor at Canisius College in Buffalo, New York, Paul is the lead faculty member for the master of science graduate program in anthrozoology. Paul has also taught animal law at Harvard Law School since 2002. He also teaches Harvard's summer-term course Animals: Religion and Ethics. The former director of the Center for Animals and Public Policy, Paul taught veterinary ethics and public policy at Tufts University School of Veterinary Medicine for more than a decade. Paul has completed five books, the most recent of which are *Animal Studies—An Introduction* (2013) and *Animal Rights* (2011). He is also coeditor of *A Communion of Subjects: Animals in Religion, Science, and Ethics* (2006).

Rev. Nancy Wright grew up in Colorado and received an MDiv from Union Theological Seminary, New York, in psychiatry and religion and an MA in environmental conservation education from NYU. Ordained in the United Church of Christ, from 1973 to 1984 she served as founder and executive director of the Westside Ecumenical Ministry to the Elderly in New York City. Nancy worked for nine years at two ecumenical environmental agencies, including Earth Ministry, in Seattle. From 1984 to 1993 she served St. Michael's Episcopal Church as a spiritual director. As a Lutheran pastor she has served Ascension Lutheran Church, South Burlington, Vermont, since 2006. She coauthored *Ecological Healing: A Christian Vision* (1993) and authored "Christianity and Environmental Justice" (*Crosscurrents* [June 2011]). In 2012 Nancy and Ascension received a Lilly Endowment grant for a sabbatical focused on "living water." She is environmental liaison to the New England synod for the ELCA.

Selected Bibliography

Journey of the Universe

Swimme, Brian Thomas, and Mary Evelyn Tucker. *Journey of the Universe*. New Haven, CT: Yale University Press, 2014.

———. *Journey of the Universe*. Directed by David Kennard and Patsy Northcutt. New York: Shelter Island, 2013. DVD. Originally released in 2011 by Northcutt Productions.

Tucker, Mary Evelyn. *Journey of the Universe: Conversations*. Directed by Patsy Northcutt and Adam Loften. New York: Shelter Island, 2013. DVD. Originally released in 2011 by Northcutt Productions.

Selected Works by Thomas Berry

"Affectivity in Classical Confucian Tradition." In *Confucian Spirituality*, Vol. 1 edited by Tu Weiming and Mary Evelyn Tucker. New York: Crossroad, 2003.

"Alienation in a Universe of Presence." *Teilhard Studies* 48, Spring 2004.

Befriending the Earth: A Theology of Reconciliation between Humans and the Earth (with Thomas Clark). Mystic, CT: Twenty-Third Publications, 1991.

Buddhism. New York: Columbia University Press, 1989 (originally published, 1975).

The Christian Future and the Fate of Earth. Edited by Mary Evelyn Tucker and John Grim. Maryknoll, NY: Orbis Books, 2009.

The Dream of the Earth. Berkeley, CA: Counterpoint, 2015 (original edition: San Francisco: Sierra Club Books, 1988).

Evening Thoughts: Reflecting on Earth as Sacred Community. Edited by Mary Evelyn Tucker. Berkeley, CA: Counterpoint, 2015 (original edition: San Francisco: Sierra Club and the University of California Press, 2006).

The Great Work: Our Way into the Future. New York: Bell Tower, 1999.

"Individualism and Holism in Chinese Tradition: The Religious Cultural Context." In *Confucian Spirituality,* edited by Tu Weiming and Mary Evelyn Tucker. New York: Crossroad, 2003.

"Management: The Managerial Ethos and the Future of Planet Earth." *Teilhard Studies* 3 (Spring 1980).

"The New Story." *Teilhard Studies* 1 (Winter 1978). Republished in *The Dream of the Earth,* 1988, and in *Teilhard in the 21st Century*, 2003.

Religions of India. New York: Columbia University Press, 1992 (original edition: Bruce Publishing, 1971).

The Sacred Universe: Earth, Spirituality, and Religion in the 21st Century. Edited and with a foreword by Mary Evelyn Tucker. New York: Columbia University Press, 2009.

"Technology and the Healing of the Earth." *Teilhard Studies* 14 (Fall 1985). Republished in *The Dream of the Earth,* 1988.

"Teilhard in the Ecological Age." *Teilhard Studies* 7 (Fall 1982). Republished in *Teilhard in the 21st Century,* 2003.

Thomas Berry: Selected Writings from the Earth Community. Edited and with an introduction by Mary Evelyn Tucker and John Grim. Maryknoll, NY: Orbis Books, 2014.

The Universe Story: From the Primordial Flaring Forth to the Ecozoic Era—A Celebration of the Unfolding of the Cosmos. With Brian Swimme. San Francisco: HarperSanFrancisco, 1992.

Additional Resources on Thomas Berry

Dalton, Anne Marie. *A Theology for the Earth: Contributions of Thomas Berry and Bernard Lonergan.* Ottawa: Ottawa University Press, 1999.

Eaton, Heather, ed. *The Intellectual Journey of Thomas Berry: Imagining the Earth Community.* Lanham, MD: Lexington Books, 2014.

Grim, John, and Mary Evelyn Tucker. "Thomas Berry: Reflections on His Life and Thought." *Teilhard Studies* 61 (Fall 2010).

Selected Works by Pierre Teilhard de Chardin

(Earliest English translations listed for most works)

Activation of Energy. London: Collins, 1970.

Appearance of Man, The. London: Collins, 1965.

Christianity and Evolution. London: Collins, 1971.

The Divine Milieu: An Essay on the Interior Life. London: Collins, 1960.

Future of Man. New York: Harper and Brothers, 1964.

The Heart of the Matter. New York: Harcourt Brace Jovanovich, 1974.

Human Energy. New York: Harcourt Brace Jovanovich, 1969.

The Human Phenomenon. Trans. Sarah Appleton-Weber. East Sussex, UK: Sussex Academic Press, 1999. (Originally published in English as *The Phenomenon of Man.* London: Collins, 1959.)

Hymn of the Universe. New York: Harper and Row, 1965.

"Joint Geological and Prehistoric Studies of the Late Cenozoic in India." With H. de Terra and T. T. Paterson. *Science* 83, no. 2149 (March 6, 1936): 233–36.

Let Me Explain. New York: Harper and Row, 1970.

The Making of a Mind: Letters from a Soldier-Priest, 1914–1919. London: Collins, 1965.

Man's Place in Nature. New York: Harper and Brothers, 1966.

"On the Zoological Position and the Evolutionary Significance of Australopithecines." *Transactions of the New York Academy of Sciences* 14, no. 5 (March 1952): 208–10.

Science and Christ. London: Collins, 1968.
Toward the Future. London: Collins, 1974.
The Vision of the Past. New York: Harper and Row, 1966.
Writings in Time of War. London: Collins, 1968.

Additional Resources on Teilhard de Chardin

Delio, Ilia. *From Teilhard to Omega: Cocreating an Unfinished Universe.* Maryknoll, NY: Orbis Books, 2014.

Duffy, Kathleen. *Teilhard's Mysticism: Seeing the Inner Face of Evolution.* Maryknoll, NY: Orbis Books, 2014.

Fabel, Arthur, and Donald St. John. *Teilhard in the 21st Century: The Emerging Spirit of Earth.* Maryknoll, NY: Orbis Books, 2003.

Grim, John, and Mary Evelyn Tucker. "Teilhard de Chardin: A Short Biography." *Teilhard Studies* 11 (Spring 1984). Reprinted in *Teilhard in the 21st Century,* 2003.

———. "Teilhard's Vision of Evolution." *Teilhard Studies* 50 (Spring 2005).

King, Thomas. *Teilhard's Mass: Approaches to the "Mass on the World."* Mahwah, NJ: Paulist Press, 2005.

———. *Teilhard's Mysticism of Knowing.* New York: Seabury Press, 1981.

King, Ursula. *Spirit of Fire: The Life and Vision of Teilhard de Chardin.* Maryknoll, NY: Orbis Books, 1998. Rev. ed., 2015.

King, Ursula, ed. *Pierre Teilhard de Chardin: Writings.* Maryknoll, NY: Orbis Books, 1999.

Tucker, Mary Evelyn. "The Ecological Spirituality of Teilhard." *Teilhard Studies* 13 (Spring 1985). Reissued as *Teilhard Studies* 51 (Fall 2005).

Additional Resources

Grim, John, and Mary Evelyn Tucker. *Ecology and Religion.* Washington, DC: Island Press, 2014.

Hessel, Dieter T., and Rosemary Radford Ruether, eds. *Christianity and Ecology.* Cambridge, MA: Harvard University Press, 2000.

Swimme, Brian Thomas. *Canticle to the Cosmos.* Directed by Catherine Busch. Berkeley, CA: Center for the Study of the Universe, 1990. DVD.

———. *The Earth's Imagination.* Directed by Dan Anderson. Berkeley, CA: Center for the Study of the Universe, 1997. DVD.

———. *Hidden Heart of the Cosmos.* Maryknoll, NY: Orbis Books, 1999.

———. *The Powers of the Universe.* Directed by Dan Anderson. Berkeley, CA: Center for the Study of the Universe, 2004. DVD.

———. *The Universe Is a Green Dragon.* Santa Fe, NM: Bear & Co., 1984.

Tucker, Mary Evelyn. *Worldly Wonder: Religions Enter Their Ecological Phase.* Chicago: Open Court, 2003.

Tucker, Mary Evelyn, and John Grim, eds. *Worldviews and Ecology: Religion, Philosophy, and the Environment.* Maryknoll, NY: Orbis Books, 1994.

Index